Engineering Geomorphology

Engineering Geomorphology
Theory and Practice

P. G. Fookes
Consulting Engineering Geologist and Distinguished Research Associate,
Oxford University, UK

E. M. Lee
Consulting Engineering Geomorphologist, York, UK, and

J. S. Griffiths
Professor of Engineering Geology and Geomorphology and Head of School of Earth, Ocean and
Environmental Sciences, University of Plymouth, UK

Whittles Publishing

CRC Press
Taylor & Francis Group

Published by
Whittles Publishing,
Dunbeath,
Caithness KW6 6EG,
Scotland, UK
www.whittlespublishing.com

Distributed in North America by
CRC Press LLC,
Taylor and Francis Group,
6000 Broken Sound Parkway NW, Suite 300,
Boca Raton, FL 33487, USA

© 2007 P. G. Fookes, E. M. Lee and J. S. Griffiths

Reprinted 2012, 2014

ISBN 10 1-904445-38-1
ISBN 13 978-1-904445-38-8
USA ISBN 9781-4200-5089-9

The publishers gratefully acknowledge support from Atkins
ΛTKINS

Typeset by Compuscript Ltd., Shannon, Ireland

Printed and bound in Great Britain by 4edge Ltd, Hockley.

Contents

Part 5: Investigation techniques

Foreword

Engineers under-rate geomorphology at their peril; geomorphologists ignore their potential value to engineers at their cost. Both need engineering geomorphology. That is the crucial and enduring message of this book.

Many of us learned the message the hard way. A few decades ago, engineers and geomorphologists rarely worked together, and if they met it was probably through the medium of site investigations and, perhaps, soil mechanics. Their convergence was often initiated by project failures, in which engineering solutions inadequately recognised terrain conditions. There was a promising beginning in the 1930s when *soil erosion* was seen as the impending environmental disaster facing mankind. Scientists and engineers came together to create a brilliant understanding of the problems, and huge progress was made in controlling them, especially in the United States. That experience revealed a context crucial to engineering geomorphology that is often ignored: many geomorphological hazards are deeply embedded in distinctive communities and cultures that should strongly influence the engineering responses. Thus, the solutions to soil erosion problems in Kansas may well not succeed in Kenya, although the underlying physical principles may be the same. For me, another revealing example of the convergence of different practitioners on engineering geomorphology was in the management of slope failure in Los Angeles during the twentieth century (R. U. Cooke, 1984, Geomorphological Hazards in Los Angeles, OUP), which not only reflected a progressive understanding of the geomorphological problems, but also the slow evolution of slope control regulations, in different ways, in the eighty or so affected cities. Here engineering geomorphology was incorporated slowly into the local management and their political and cultural environments.

These examples of major problems, and many subsequent studies, especially those related to land potential for agriculture, the search for aggregates, town planning and route selection, showed the importance of understanding a broader context, in which situation is as important as site, and in which dynamics of the environment are as important as its physical properties. At the same time, the recognition of potential hazards and risks on land surfaces, and their management, showed the vital need for systematic mapping, based both on remote sensing imagery and on foot-slogging field observations. This was an approach well-suited to the experience of geomorphologists. The coming together of engineers and geomorphologists revealed another important message: geomorphological evaluation of site, situation, environmental dynamics and terrain resources is a relatively cost-effective and time-saving precursor to engineering ground works.

If soil erosion was the perceived environmental threat of the 1930s, today's is climatic change. in this context, the dynamics of changing surface conditions in response to changing climate mean that engineering geomorphology is now increasingly centre stage, especially when it comes to evaluating such massive challenges as permafrost degradation, flooding hazards, avalanches, and sea-level rise, Other human-induced changes, including deforestation and desertification also create their own geomorphological problems. The problems of engineering geomorphology are certainly everywhere and there is a growing need for engineering geomorphologists to help solve them.

From the 1970s, those promoting engineering geomorphology began to codify it through case studies all over the world and through such volumes as *Engineering Geomorphology* (edited by P.G.Fookes and P.R.Vaughan, 1985, SUP) and *Geomorphology in Environmental Management* (R.U. Cooke and J.C. Doornkamp, 1974 and 1990, OUP).

Peter Fookes led the campaign and encouraged a small group of geomorphologists to reorientate their science towards application. Peter, Jim Griffiths and Mark Lee have built firmly on those foundations and their subsequent experience. This volume, together with the companion volume, *Geomorphology for Engineers* (edited by P.G. Fookes, E.M. Lee and G. Milligan, 2005, Whittles Publishing), provide an excellent state-of-the-science review, and essential reading for all those who aspire to work in engineering geomorphology.

Professor Sir Ron Cooke, York

Prologue

By the standards of the 1950s, I believe I had a 'proper' geological undergraduate education, in which I leaned towards mineralogy and petrology. Some of the staff believed that geology really ended at the top of the Palaeozoic, or at a pinch perhaps at the end of the Cretaceous—the rest was drift. This was a common view, and is still quite strongly held by some geologists. We had lectures on the Tertiary but hardly anything on the Quaternary, let alone geomorphology (then taught as physiography).

When I began to follow my chosen career – initially geology-for-engineers and culminating in engineering geology cum geomorphology – I realised that my training in geomorphology was inadequate. I quickly found that knowledge of landforms and the surface processes could be most useful in the design of sub-surface investigations, in the evaluation of findings and in the anticipation of borrow materials. I therefore sought more about this subject but found little succour from geological textbooks. On broadening my horizons with geomorphological books of the time (especially those from USA and New Zealand) and particularly by discussions with university-based geomorphologists, I found there was a new world awaiting. New techniques helped me understand the top few metres of the earth's surface where my civil engineering masters practised.

As a result, while still in the realm of engineering geology I have spent much of my life endeavouring to improve my geomorphology. I have tried to bring geomorphologists towards a closer understanding of the demands of identifying ground conditions for geotechnical engineering.

By the end of the seventies I had made friends with some of the up-and-coming young practitioners in the geomorphological world: Denys Brunsden, Ron Cooke, David Jones and John Doornkamp. They taught me some of the finer and, indeed, the coarser points of geomorphology. However, the gap between myself and a fully trained geomorphologist remains exceedingly large.

It was during this period that one of the co-authors of this book, Professor Jim Griffiths, started his career as a geomorphologist-cum-geologist in the geotechnical department of a large consulting civil engineering practice. I believe Jim was the first in Britain to have made a full-time career practising geomorphology within the commercial sector. His experience is similar to mine, although he brought a better and more up-to-date geomorphological understanding to problems of the time. Today, Jim is a leading practitioner in what has become the developing discipline of engineering geomorphology. Over the last decade he has also become a respected academic, specialising in natural hazards and surface materials. Importantly, Jim is training the next generation.

In Britain today there must be a dozen or more engineering geomorphologists within geotechnical departments with a background somewhat similar to Jim's. Having been around for a decade or two they are now starting to achieve fairly senior positions, as engineering geologists did three or four decades ago. They are learning, as engineering geologists had to, how the civil engineering world functions: the investigation, design and construction needs—operating within a contractual framework.

It was not until the late eighties that I met geomorphologist Dr Mark Lee—the other co-author of this book. Mark brings a different and very powerful insight into the developing world of engineering geomorphology. He is a physical geographer who, in the mid 1980s, found himself working with Jim at Geomorphological Services Ltd. In the late 1990s Mark had the good fortune to work with another leader in his field, Professor John Pethick, in a small university-based consultancy specialising in coastal and estuarine geomorphology. Working with John Pethick has given him an insight in the behaviour and management of coastal systems. Because it has proved of value he has also made geohazard risk assessment an authoritative speciality of his own, as you will see in the content of this book. This skill is yet another potent tool that geomorphology can bring to the civil engineer's work.

I used to think that it was an opportunity lost to geology that for the last hundred years or more geomorphologists in Britain (and I believe elsewhere) have been nurtured within schools of geography. In the last decade or so I have changed my views and now think that as geomorphology has become such a large and complex science, it is more than worthy of a degree in its own right, rather than being tacked on as part of a geology or, as happens now, as part of a geography degree. Albeit that geomorphology deals in the same concepts of time and space and materials, I believe it requires a somewhat different viewpoint to master from that of the traditional geologist. This is not to say that geologists should not be taught more and better geomorphology, especially the history of the Quaternary (this also applies to the teaching of geology for engineers) or that geomorphologists could not do with a better geological upbringing. The current situation has drawbacks, such as geologists not really understanding geomorphologists, and possibly vice-versa, but also geomorphological skills have become so powerful that they need a long training and considerable experience to be practised successfully. As with geology, observation and judgemental skills are an all-important part of such disciplines.

On reflection I have come to believe that it follows that engineering *geology* and engineering *geomorphology* are separate entities and should be taught as such, for example, at postgraduate level. A good engineering geologist must have a reasonable command of engineering geomorphology and vice-versa, but a skilled general practitioner in each discipline would have followed different, even separate pathways in gaining experience and authority. A polymath with equal skills in both disciplines would be very difficult to find. I say this because when listening to skilled engineering geomorphologists making field observations, or going through a train of thought on a particular subject, it is, perhaps not surprisingly, quite different from hearing a skilled geologist observing the same things. I see this as no different from a civil engineer having a different viewpoint from a geologist when discussing a geo-problem. Palaeontologists think differently from structural geologists and they can have difficulty understanding each other. Each has their own jargon. Each has their own background of considerable but different knowledge to draw upon to condition their own thoughts and judgements.

So what is engineering geomorphology? I still see it, as I saw it a decade ago in my Glossop lecture, as

'... similar to engineering geology, however specialising in geomorphology, i.e. surface and near-surface processes and characteristics of the earth's surface'.

I said then that

'... the practitioners of engineering geology have training and experience in ground problems that arise in civil engineering, and in the investigation, classification and performance of soils and rocks related to civil engineering situations; and a working knowledge of basic soil mechanics, rock mechanics and hydrogeology. Such practitioners provide engineering geology.'

This remains my view of engineering geology.

I, and I believe Jim Griffiths, think of engineering geomorphology as the relationship between materials, environment (especially climate) and the earth's surface processes. Mark's view is probably wider and more imaginative, from which I quote

'... the arena of engineering geomorphology (is) water, wind and gravity; complex system behaviour driven by variable environmental controls and uncertain forcing'.

His approach I believe has broadened the scope and strengthened our book. He often works with geomorphologists employed in industry to support river and coastal engineering, who, he suspects, greatly outnumber those geomorphologists currently working directly with geotechnical engineers and engineering geologists in foundation engineering. He does not think such geomorphologists look towards the Geological Society's Engineering Group as the spiritual home of engineering geomorphologists in the UK and I do feel some sympathy with this view. He goes further and thinks that such geomorphologists should be looking for inspiration from the various marine and water related groups within the ICE or CEIWEM, groups which do not necessarily immediately spring to mind as gathering places of engineering geologists.

We see the material in this book, perhaps for the first time, as covering this developing broad, even disparate, discipline of engineering geomorphology. Hopefully it will provide a framework that brings the needs of river, coastal and geotechnical engineering strands of 'engineering geomorphology' together. As such it may be a little immature in some areas, and in the future, with the wisdom of hindsight, our work may be shown to be somewhat disjointed and undoubtedly lacking in significant respects.

Whatever way engineering geomorphology develops, even if it is subsumed by engineering geology, or dare I say vice-versa, the book presents the subject as we see it now, albeit in a somewhat broad parsimonious and it is hoped a not too prejudicial style. We see it as being of value to a wide spectrum of engineering situations, particularly in providing a new set of judgemental skills to enable a better and more rigorous evaluation of situations than hitherto. Regrettably, in the last couple of decades, engineering geology has increasingly tended to be driven by the commercial needs of site investigation and, as with soil mechanics and geotechnics as a whole, bedevilled by clients who on occasion, short-sightedly, want the ostensibly cheapest version of everything. This has inevitably reflected on its drive and wonderment.

We are optimistic for the development of engineering geomorphology and its refreshing spirit and, we hope you like what has been written. I think it is a considerable improvement on my original, rather mundane outline for the book, and it has been improved beyond measure by Mark and Jim.

The colour block diagrams are again the inspired work of Geoff Pettifer – friend, engineering geologist and environmentalist.

Professor P. G. Fookes
F. R.Eng., Hon.F.R.G.S., Hon.D.Sc.(Plym)

Preface

Engineering geomorphology has developed in the last few decades to support a number of distinct areas of civil engineering:

- *River engineering*, complementing hydrologists, tends to concentrate on the nature and causes of alluvial river channel change (e.g. channel migration, bank erosion, bed scour).
- *Coastal engineering* provides an understanding of the occurrence and significance of shoreline changes, especially in response to changes in sea-level and sediment supply.
- *Ground (geotechnical) engineering*, where engineering geomorphology complements engineering geology, has been proven to be valuable for rapid site reconnaissance and slope instability studies.
- *Petroleum engineering* is working with other geo-specialists to support the routing of oil and gas pipelines, especially through remote regions.
- *Agricultural engineering*, linked closely with pedology, is involved with the investigation and management of soil erosion problems.

Although these applications of geomorphology developed separately in response to specific engineering needs, in recent years they have started to become integrated as a coherent discipline. Hence, we became interested in the idea of producing a book to set out the general principles that underpin the application of geomorphology to this diverse range of projects. By doing so we hope to show that engineering geomorphology is far more than terrain evaluation and site mapping.

PGF, EML, JSG

Acknowledgements

Writing the book has reminded us of the debt of gratitude owed to many others. We would like to acknowledge the huge influence of geomorphology professors Denys Brunsden, Ron Cooke and David Jones, and Doctor John Doornkamp – they were the first down this path in the UK.

The authors would also like to thank all those colleagues who have, over the years, provided them with stimulation when working on engineering geomorphological projects throughout the world, most especially Ken Ainscow, Professor John Atkinson, Dr Fred Baynes, Professor Eddie Bromhead, John Charman, Dr Alan Clark, Jim Clarke (BP), Professor Bill Dearman, Professor Andrew Goudie, Ken Head, Dr Gareth Hearn, Ian Higginbottom, Len Hinch, Professor John Hutchinson, Mike Kelly, Thomas Lyons, Chris Mamby, Dr Dick Martin, Peter Martin, Dr Anne Mather, Dr Roger Moore, Professor John Pethick, Dr Alan Poole, Saul Pollos, Mike Sanders, David Shilston, Professor Sir Alec Skempton, Doreen Smith, Mike Sweeney, Dr Martin Stokes, Professor Neil Taylor, Professor Peter Vaughan, Dr Tony Waltham, Dene Wilson and UT Cobley. Finally, we must thank Dr Keith Whittles of Whittles Publishing for his encouragement and patience during the long process of turning the book from a simple idea to the finished article.

Engineering geomorphology is an actively developing discipline of great relevance to those who work and build in all natural environments. Messrs. Fookes, Lee and Griffiths' ability to distil and convey the subject makes this book distinctive and potentially valuable to a wide audience of students and practitioners – in planning and environmental management, and in engineering design, construction and maintenance. Atkins' support of the book recognises the importance of this endeavour.

David Shilston CGeol CSci FGS FRSA
Technical Director – Engineering Geology
Atkins Ltd., Epsom, KT18 5BW, UK

Dedicated to our friend Michael Sweeney

Also shown is Len Hinch (to the rear), partner at Rendel Palmer and Tritton, for
whom a young Mike was working at the time of the Dharan Dhankuta Road
Investigation and Design, Nepal

1

Geomorphology and Engineering

Civil engineers apply the laws and principles of the physical sciences to the design and construction of structures that can resist or harness the forces of nature and improve the quality of life. This book is concerned with how the study of the earth's surface and the forces of nature that shape it i.e. *geomorphology* (in Greek, *geo* means earth and *morphe* means form) can help civil engineers develop their basis for rational design and technical management (see: http://www.ice.org.uk/ or http://www.asce.org/).

Introduction
The earth's surface is dynamic, not static, and *landforms* (features that make up the ground surface) change through time as a result of weathering and surface processes such as erosion, sediment transport and deposition. These changes have the potential to cause significant harm to civil engineering projects (i.e. *hazards* or *geohazards*). However, they may also lead to the formation of valuable *resources* that can be exploited, for example, deposits of sand and gravel needed in construction.

The application of geomorphology for engineering (*engineering geomorphology*) can help ensure that the technical problems faced by civil engineers are successfully identified and resolved. These include:
- the evaluation of the near surface ground for design and construction of man-made structures
- the risks to civil engineering projects from earth surface processes (both current processes and the legacy of past processes)
- the availability of resources for construction, especially aggregates
- the effects of civil engineering projects on the environment, notably on the operation of earth surface processes.

Geomorphology and Landform Change
A central theme of geomorphology is that surface processes and landform changes are the product of interactions between:
- the *physical setting*: the material characteristics of the ground, the abundance and properties of mobile sediments, and the current surface topography, climate and vegetation

- the *energy regime*: the availability of energy to 'drive' surface processes.

As a result geomorphology draws upon a wide range of sciences, including geology, geotechnics, pedology (soil science), hydrology, hydrodynamics (waves and tides), meteorology and ecology.

The unique way in which geomorphologists can help engineers is by providing a scientific basis for understanding landform change. Most landscape changes can be explained as responses to variations in energy inputs into *earth surface systems*.

Earth surface systems provide a spatial framework for explaining how the effects of surface processes and landform change in one location can be passed onto neighbouring parts of the natural landscape. Examples of earth surface systems include: slopes, catchments or coastal sediment transport cells. Variations in energy inputs might include changes in rainfall (intensity or total), temperature, river flows (discharge and sediment load) and wave/tidal energy arriving at the coast.

The earth surface system response can be difficult to predict because the energy inputs are *stochastic* (of random size, frequency and duration). In addition, the way in which an earth surface system responds is determined by the precise dimensions and properties of its individual components (i.e. landforms) at that particular moment. As these may change over time, often in subtle ways, occasionally dramatically, so the detailed effects of a particular level of energy input may vary over time. This so-called *state dependence* is a further source of uncertainty.

Understanding the past behaviour of earth surface systems can be useful in establishing and explaining the distribution of geologically recent deposits, such as fluvial sands and gravels, colluvium and landslide deposits. In this way geomorphology provides both a spatial and historical context for engineering geologists who are concerned with foundation conditions and the identification of construction resources.

Geomorphology and Engineering
Engineering geomorphology provides practical support for engineering decision making (project planning, design and construction)[1.1–2]. Engineering geomorphologists

need to work as part of an integrated team and provide information at many levels, ranging from crude qualitative approximations (e.g. at pre-feasibility stage) to quantitative analyses (e.g. to support detailed design and construction). The level of precision and understanding required needs to be sufficient for a particular problem or context, to enable an adequately informed decision to be made.

Typical questions that need to be addressed by an engineering geomorphologist include:

- What ground conditions could be expected in a particular area? e.g. the presence of relict solifluction sheets in temperate regions, the occurrence of running sands or buried peat horizons.
- Will the project be at risk from erosion or depositional processes over its design lifetime? e.g. will instability occur at the site, or is the proposed development set back sufficiently from a retreating cliff top?
- How could problems arise? e.g. from removal of support during excavation at the base of a slope.
- What is the likelihood of the project being affected by instability, erosion or deposition processes over its design lifetime? e.g. is there a very low chance of being affected by a channelised debris flow, or being engulfed by blown sand?
- What will be the effect of climate change or sea-level rise on the project risks? e.g. accelerated cliff recession, higher chance of landslide reactivation.
- Why has a problem arisen? e.g. leaking water pipes, disruption of sediment transport along a shoreline.
- What effects will the project have elsewhere? e.g. reduction in floodplain storage, changes in surface run-off and erosion potential.
- What magnitude event should be designed for to provide a particular standard of defence? e.g. what is the expected volume/depth and spatial coverage of the 1 in 50 year debris flow event?
- Where can suitable aggregates be found in sufficient quantities to satisfy the project demand for construction materials? e.g. are there sand and gravel deposits suitable for use in road pavement construction?

The answers to these questions can be both *qualitative* (e.g. recognition of pre-existing landslides with potential for reactivation) and/or *quantitative* (e.g. rates of change and the magnitude/frequency of events). However, all answers must be couched in practical terms and take account of uncertainty?

Landform Change and Engineering Time

Civil engineering project cycles are generally of the order of 10–100 years (engineering time), although occasionally longer. This imposes a limit to the types of landform change that are relevant to engineering:

- *High to relatively high probability events* that occur on timescales of <10 years. These are routinely incorporated into design, such as wind-blown sand, soil erosion, shallow hillside failures, flooding and river bank erosion, channel scour, coastal erosion and deposition.

- *Abrupt and dramatic changes* which are likely to be significant over a 10–100+ year timescale and have to be understood and taken into account in the design. Examples include establishment of gully systems, migration of sand dunes, river planform changes, coastal cliff recession, and the growth and breakdown of shingle barriers.
- *Low probability events* that would have a major impact on the project or development, such as flash floods, major first-time landslides, reactivation of pre-existing deep-seated landslides, excessive channel bed scour, neotectonic fault rupturing and tsunamis. The design needs to define an acceptable level of risk (e.g. build for a 1 in 100 year flood event, but not the 1 in 1000 year event).

Engineering geomorphology is directed towards understanding the way landforms or earth surface systems respond to relatively short to medium-term (<1 to 1000 years) changes in energy inputs (e.g. resulting from climatic variability, changes in sediment supply, land use change, neotectonics, the effects of man) rather than long-term landscape denudation and evolution. For any engineering works, however, there is a need to be aware of:

- the presence of potential problems inherited from the distant past, such as ancient landslides, periglacial solifluction sheets, vegetated sand dunes and karst features
- the finite nature of many aggregate resources, formed by surface processes operating under different environmental conditions, for example glacial sands and gravels in what are now temperate regions
- longer term trends, for example the Holocene decline in sediment availability experienced on many temperate coastlines.

Of significance for engineering geomorphology are the global oscillations in climate that occur in engineering time: the North Atlantic Oscillation, the El Niño Southern Oscillation and the Pacific Decadal Oscillation (see Chapter 9). These can influence the frequency and intensity of landscape changes and the generation of hazards to engineering projects. The environment is not constant even over engineering time; projects that do not allow for such fluctuations will be underestimating the level of risk that is being accepted.

Investigation Approaches

Problems related to geomorphology often arise on engineering projects because of:

- failure to correctly identify surface features e.g. pre-existing landslides
- failure to foresee the presence of problem ground conditions (e.g. soil pipes, aggressive or metastable soils) i.e. limited understanding of how systems operated in the past
- poor prediction of the rates of morphological change, based on extrapolation from past records i.e. based on a limited understanding of the geomorphological processes and their variability
- lack of appreciation of what might happen, as opposed to what has happened i.e. limited

understanding of the system behaviour and poor hazard models.

Ground-related problems are rarely unforeseeable, only unforeseen[1,2]. Sufficient effort should be given to ensuring that all potential situations and problems are anticipated, however remote their likelihood. There should be no surprises to be discovered during detailed design or construction.

Because each engineering project is unique it is not realistic to attempt to set out a precise methodology that is applicable to all situations. However, the basic principles of engineering geomorphology will be:

- Site conditions are the product of past geological and geomorphological processes (i.e. the *total geological and geomorphological history*[1,2–4]).
- The historical distribution and frequency of hazard events can be explained in terms of the system response to variations in the energy inputs or progressive changes in the system state (i.e. *contemporary system behaviour*[1,5]).
- Indications of the likelihood of hazards and future rates of change can be determined by considering the way in which systems could respond to changes in environmental controls (e.g. *future system behaviour*[1,6]).

Understanding of these issues has to be developed at the earliest opportunity in any project for it to be successfully engineered. Thus, most projects will follow a phased approach to investigation, involving:

- A *desk study review* of available information (e.g. maps, satellite imagery, aerial photography, documents, records) to develop an *initial terrain model*. The aims are to anticipate those processes or ground conditions that could influence the project and to plan subsequent investigations (see Chapter 38). These models make a key contribution to civil engineering; they comprise the starting point for predicting hazards, identifying material properties and establishing the extent of resources. They provide the framework for developing the geological and geomorphological basis for rational design, including the use of *Design Events* and *Reference Conditions*[1,7–8] to define the range of conditions that could reasonably be anticipated or foreseen (Table 1.1).
- *Field investigations* to evaluate the anticipated conditions (e.g. mapping landforms, describing materials and modelling processes; see Chapter 37). Models can be used to present the different kinds of information: *conceptual* models, *evolutionary* models and *observational* models, all of which have a place when working with engineers. For example, detailed site-specific terrain models are widely used.

Geomorphologists will often work side-by-side with engineering geologists to investigate ground conditions (Fig. 1.1). Typically this will require field investigations undertaken according to engineering geology standard practice[1,9–10] e.g. pitting, boring, geophysics, geotechnical testing; this represents a separate subject that is not covered in this book. Terrain models are an important component of the ground model used for designing and interpreting the engineering geological field investigations.

The models are not just static three-dimensional block diagrams. *Geographical Information Systems* (GIS) (Chapter 38) can be used to be fully integrate the models with a wealth of ground investigation information, risk assessment and engineering data.

- *On-going and post construction investigations.* During the construction phase the site-specific terrain models are checked, upgraded and refined to ensure all potential construction problems are taken into account. In the post-construction phase the model will be relevant to all issues arising from any claims associated with genuinely unforeseen ground conditions. If used correctly, these models will help form the basis of the correct application of the *Observational Method* (OM)[1,7–8, 1,11] for ground engineering where the Contract Documents allow this approach. The OM is a well established methodology that involves a continuous, managed, integrated process of design, construction control, monitoring and review. This enables previously defined appropriate modifications to be incorporated during or after construction as required by the ground conditions revealed during construction. (Table 1.2).

Table 1.1 The Function of Reference Conditions, adapted to incorporate geological and geomorphological information [1,8].

Aim: document the range of conditions that can be *reasonably be foreseen* for contract purposes, especially design and payment:

1. Formally define and describe the components of the terrain models (i.e. surface processes, materials and landforms). This will most likely involve erecting geological and geomorphological Reference Conditions by grouping together terrain units with similar engineering characteristics;

2. Simplify the geomorphological processes into a series of events that define the basis for design (e.g. the 1 in 100 year flood; a landslide event of a particular size and intensity).

Use of Reference Conditions can:

1. Allow a reduction in the overall laboratory testing schedule, as only representative samples from each Reference Condition have to be tested rather than testing all terrain units encountered;

2. Allow the incorporation of knowledge from similar terrain units that occur outside the project area that may be correlated with the Reference Conditions;

3. Be of practical help during construction anticipating ground conditions and predicting equipment performance and capability.

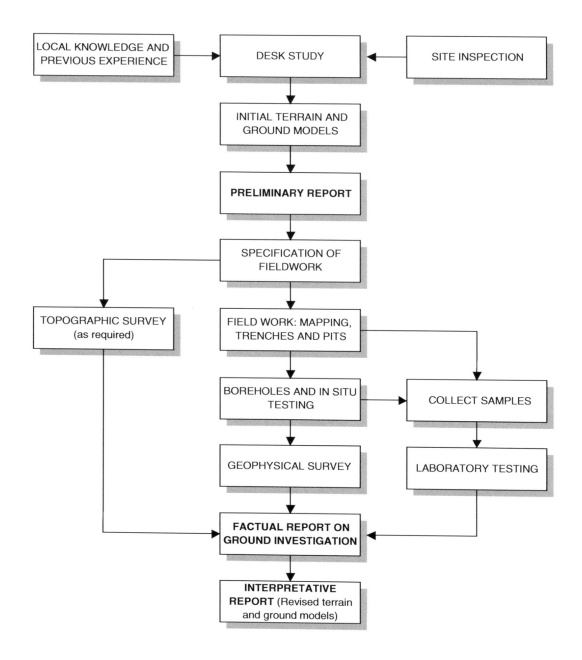

Figure 1.1 Flow chart for a simple one-stage ground investigation in support of geotechnical engineering design (after Fookes 1997).

Geomorphology for Engineers

The aim of this book is to provide undergraduate and post-graduate geomorphologists and geologists, as well as practising engineers, with a basic introduction to the developing subject of engineering geomorphology. In doing so, it is hoped that this will raise their awareness of the need to view the site-specific hazards generated by surface processes as the product of changes occurring throughout an earth surface system in response to variations in environmental conditions.

The book is divided into five parts:
- Part 1: the basic geomorphological concepts that underpin efforts to explain the causes, mecha-

nisms and consequences of landform change (Chapters 2 to 14).
- Parts 2-4: how the earth surface systems on hill-slopes and plains (i.e. *slopes*), rivers and the coast work (system behaviour), and, by doing so, generate hazards, define the ground conditions, and possibly provide resources for engineering projects. An indication is given as to how the hazards can be assessed and managed (Chapters 15 to 36).
- Part 5: common techniques available to the engineering geomorphologist to investigate geomorphological phenomena that might affect engineering works (Chapters 37 to 43).

It is a common complaint from engineers that *-ologists* never come straight to the point, preferring to wrap

Table 1.2 Important features of the 'Observational Method' to be employed during construction [1.7–8].

1. Exploration sufficient to establish at least the general nature, pattern, and properties of the deposits, but not necessarily the detail;
2. Assessment of the most probable conditions and the most unfavourable conceivable deviations from these conditions. In this assessment, geology often plays a major role;
3. Establishment of the design based on a working hypothesis of behaviour anticipated under the most probable conditions;
4. Selection of quantities to be observed as construction proceeds and calculation of their anticipated values on the basis of the working hypothesis;
5. Calculation of values of the same quantities under the most unfavourable conditions compatible with the available data concerning the sub-surface conditions;
6. Selection in advance of a course of action or modification of design for every foreseeable significant deviation of the observational findings from those predicted on the basis of the working hypothesis;
7. Measurement of quantities to be observed and an on-going evaluation of the actual conditions;
8. Modification of design to suit the actual conditions.

simple answers up in complex, rambling discussions. This perspective has been taken to heart in writing the text, and an attempt has been made to be brief and to the point. As a result, this book must be seen as a succinct introduction to the subject. If needed, flesh can be put on the bones through accessing the references or web sites listed in the text.

References

1.1 Fookes, P. G., Lee, E. M. and Milligan, G., (eds.) (2005) Geomorphology for Engineers. Whittles Publishing.

1.2 Fookes, P. G. (1997) First Glossop Lecture: Geology for engineers: the geological model, prediction and performance. *Quarterly Journal of Engineering Geology*, **30**, 290–424.

1.3 Fookes, P. G., Baynes, F. J. and Hutchinson, J. N. (2000) Total geological history: a model approach to the anticipation, observation and understanding of site conditions. *GeoEng 2000*, an International Conference on Geotechnical and Geological Engineering, Melbourne, Australia, **1**, 370–460.

1.4 Brunsden, D. (2002) Geomorphological roulette for engineers and planners: some insights into an old game.

Quarterly Journal of Engineering Geology and Hydrogeology, **35**, 101–142.

1.5 Brunsden, D. and Lee, E. M. (2004) Behaviour of coastal landslide systems: an inter-disciplinary view. *Zeitschrift fur Geomorphologie*, **134**, 1–112.

1.6 Lee, E. M. (2005) Coastal cliff recession risk: a simple judgement based model. *Quarterly Journal of Engineering Geology and Hydrogeology*, **38**, 89–104.

1.7 Nicholson, D., Tse, C-M. and Penny, C. (1999) The observational method in ground engineering: principles and applications. CIRIA Publication R185.

1.8 Baynes, F. J., Fookes, P. G. and Kennedy, J. F. (2005) The total engineering geology approach applied to railways in the Pilbara, Western Australia. *Bulletin of Engineering Geology and the Environment*, **64**, 67–94.

1.9 Waltham, T. (2002) *Foundations of Engineering Geology*. 2nd Edition, Spon Press, London.

1.10 BSI, (1999) BS 5930: *Code of Practice for Site Investigations*. British Standards Institute, London.

1.11 Fookes, P. G., Baynes, F. J. and Hutchinson, J. N. (2001) Total geological history: a model approach to understanding site conditions. *Ground Engineering* March, 42–47.

2

The Impetus for Change: Energy Inputs and Geomorphological Activity

Introduction

The surface processes that force landform changes are driven by energy from three sources:

1 *Solar radiation*: the upper atmosphere receives around 17.8×10^{16} W of radiant energy, of which around 30% is immediately reflected back. The remainder heats the atmosphere, forming thermal (heat) energy, and drives the hydrological cycle (Fig. 2.1).

2 *Tidal energy*: generated by the gravitational attraction of the Moon and the Sun (see Chapter 28) driving the movement of the ocean mass.

3 *Geothermal energy* (i.e. from heat deep inside the earth): provided by the long-term cooling of the earth and radioactive decay of unstable isotopes — largely uranium, thorium and potassium. This energy drives internal (*endogenic*) processes, such as plate tectonics, volcanic activity and seismicity, which shape surface and near-surface conditions.

The availability of energy to drive surface processes varies across the globe, reflecting differences in climate, seismic activity and relative relief. The influence of climate is the most pervasive, resulting in vast areas of broadly similar landscapes often with characteristic mate-

rials and surface processes (i.e. *morphoclimatic zones*[2.1–2]; Fig. 2.2; Table 2.1 and Chapter 38). The unequal distribution of energy across the world is the reason why some environments are more dynamically active in terms of surface processes (e.g. mountains, river channels, and coasts).

Potential and kinetic energy

Energy can be transformed into different forms; it is not destroyed (the first law of thermodynamics). Geothermal energy can result in tectonic uplift, increasing relief or elevation and *potential energy* (PE), a function of the height and weight of the material above a base level (often sea-level):

$$PE = m\,g\,h$$

where m = mass (kg); h = height above a datum (m) and g = gravitational acceleration ($9.81\ ms^{-2}$).

Potential energy can be released and converted into *kinetic energy* (KE), the energy of motion:

$$KE = 0.5\,m\,v^2$$

where v = velocity (ms^{-1}).

Potential energy is converted to kinetic energy as water passes through a channel section; this kinetic

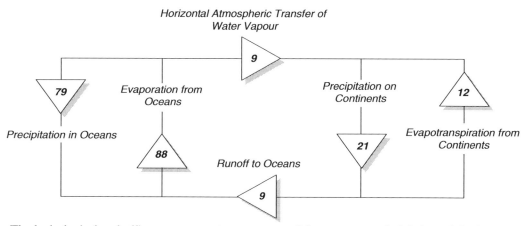

Figure 2.1 The hydrological cycle (figures represent percentages of the mean annual global precipitation of c.1000 mm).

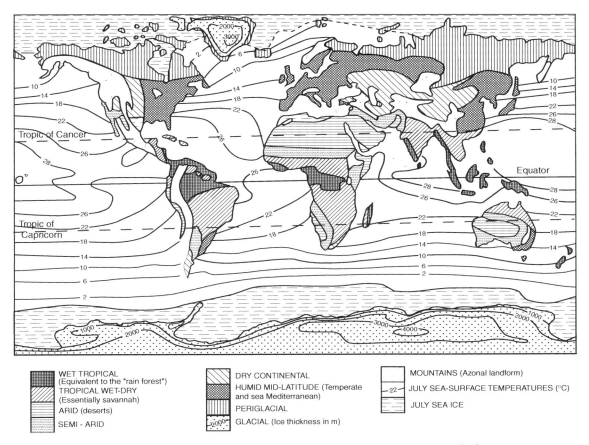

Figure 2.2 Global morphoclimatic zones (based on Tricart and Caileux 1972; Stoddart 1969)[2.1-2].

energy is available to overcome the frictional resistance of the channel bed and banks, generate heat and transport sediment. The potential energy loss (stream power, Js^{-1}) is related to the height difference (i.e. slope gradient) along the section:

$$\begin{aligned}
PE \ (loss) &= \rho_w \, g \, Q \ (h_1 - h_2) \\
&= \rho_w \, g \, Q \, s \\
&= 1000 \times 9.81 \times 100 \times 0.01 \\
&= 9810 \ Js^{-1}
\end{aligned}$$

where ρ_w = water density ($1000 \ kg \ m^{-3}$); Q = channel discharge (e.g. $100 \ m^3 s^{-1}$); h_1 = elevation at the start of the section; h_2 = elevation at the end of the section; s = slope gradient (e.g. 0.01).

Kinetic energy can be transferred to another body, causing it to move, or converted to a different form. For example, the kinetic energy of breaking waves is transferred to beach materials causing them to move (*sediment transport*) or is converted to heat and noise by the frictional drag of the water moving over the shore (*energy dissipation*).

Available energy is not evenly distributed across the world, it thus creates *energy gradients*, and work is performed as energy is redistributed along these gradients (the second law of thermodynamics). For example, the input of solar radiation varies between the equator and the poles, with the resulting energy gradients giving rise to the transfer of:

- *sensible heat* by the circulation of air in the atmosphere
- *latent heat* as water vapour in the atmosphere condenses
- *sensible heat* by the circulation of water in the ocean, creating ocean currents.

The warm air and ocean currents which flow towards the poles are matched by counter flows of cooler air and water. These energy flows result in the atmospheric circulation system and establishment of the climatic zones of the world.

Climatic zones

Regional variations in climate are the result of the unequal distribution of solar radiation, the global circulation of the atmosphere and the general pattern of the continents and oceans. Three major climate types occur, reflecting global variations in precipitation and temperature (Table 2.1, Fig. 2.2 and Chapter 38).

1. *Tropical*: dominated by Hadley cells (Fig. 2.3), comprising areas of low pressure near the equator (*equatorial trough*) towards which persistent winds (trade winds) blow; and areas of subsiding warm, dry air giving rise to major deserts (e.g. the Sahara). Intense rainfall, floods and/or drought events are associated with tropical storms (e.g. hurricanes and typhoons), the monsoon and the El Niño-Southern Oscillation (ENSO; see Chapter 9).

Table 2.1 Major morphoclimatic zones (see Fig. 2.2 and Chapter 38).

Climate zone	Morphoclimatic zone	Mean annual temperature (°C)	Mean annual precipitation (mm)	Relative importance of geomorphological processes
Azonal	Azonal mountain zone	Highly variable	Highly variable	Rates of all processes vary significantly with altitude; mechanical and glacial action become significant at high elevations.
Polar and tundra	Glacial	<0	0–1000	Mechanical weathering rates (especially frost action) high; chemical weathering rates low; mass movement rates low except locally; fluvial action confined to seasonal melt; glacial action at a maximum; wind action significant.
	Periglacial	–1 to +2	100–1000	Mechanical weathering very active with frost action at a maximum; chemical weathering rates low to moderate; mass movement very active; fluvial processes seasonally active; wind action rates locally high. Effects of repeated formation and decay of permafrost.
Temperate and Mediterranean	Temperate; wet mid-latitude	0–20	400–1800	Chemical weathering rates moderate, increasing to high at lower latitudes; mechanical weathering activity moderate with frost action important at higher latitudes; mass movement activity moderate to high; moderate rates of fluvial processes; wind action confined to coasts.
	Temperate; dry continental	0–10	100–400	Chemical weathering rates low to moderate; mechanical weathering, especially frost action, seasonally active; mass movement moderate and episodic; fluvial processes active in wet season; wind action locally moderate.
	Mediterranean; hot semi-dry (semi-arid tropical)	10–30	300–600	Chemical weathering rates moderate to low; mechanical weathering locally active especially on drier and cooler margins; mass movement locally active but sporadic; fluvial action rates high but episodic; wind action moderate to high.
Tropical	Hot drylands (arid tropical)	10–30	0–300	Mechanical weathering rates high (especially salt weathering); chemical weathering minimal; mass movement minimal; rates of fluvial activity generally very low but sporadically high; wind action at maximum.
	Savannah; hot wet-dry (humid-arid tropical)	20–30	600–1500	Chemical weathering active during wet season; rates of mechanical weathering low to moderate; mass movement fairly active; fluvial action high during wet season with overland and channel flow; wind action generally minimal but locally moderate in dry season.
	Hot wetlands (wet tropical, humid tropical)	20–30	>1500	High potential rates of chemical weathering; mechanical weathering limited; active, highly episodic mass movement; moderate to low rates of stream corrasion but locally high rates of dissolved and suspended load transport.

2. *Temperate and Mediterranean*: dominated by the westerly, upper atmosphere *jet stream* which controls the tracks of rotating, low pressure depressions. Variation in atmospheric circulation can lead to changes in the pattern of drought, wet years and flood frequency.

3. *Polar and tundra*: dominated by low solar radiation inputs and cold temperatures. The tundra region is associated with permanently frozen ground (*permafrost*). In the Arctic, Antarctic and high mountains, precipitation is very low and cold temperatures prevail throughout the year, seldom rising above 0°C.

Climate is a major influence on the rate, scale and significance of surface processes and weathering[2,3]:

• Water is the critical factor in many aspects of landscape change — from the erosive power of rainfall and running water (Chapters 16 and 22–27), to the effect of pore water pressures on slopes (Chapter 19) and the influence of groundwater throughflow on weathering (Chapter 21).

• Aridity in deserts results in severely restricted vegetation growth, leading to reduced surface stability and increased run-off when rainfall does occur i.e. increased *erosion potential* and susceptibility to *flash flooding* (see Chapter 26).

• The combination of high temperatures and low precipitation in deserts causes net evaporating conditions. Downward leaching of salts within the ground is limited; even highly soluble salts (e.g. gypsum and sodium chloride) can remain in the soil profile. *Capillary rise* is often very pronounced leading to the concentration of salts in the upper soil profile. This produces a highly aggressive environment in which *salt weathering* is an important factor in rock disintegration, ground heave and concrete attack (see Chapter 38).

• Frost action can only occur where ground temperatures fall below freezing, whereas *permafrost* will only develop where the mean annual temperature is at least below 0°C. This leads to frost splitting of rock, freeze-thaw ground movements, solifluction processes, and glacial activity (see Chapter 38).

• Wind action is most effective in dry regions with less than 200 mm mean annual rainfall (Chapter 18) resulting in the creation of mobile sand and dust hazards (e.g. *dune migration*) and wind-scoured erosional features.

In many regions it may be the extremes (e.g. high intensity rainfall events) that are responsible for initiating landscape changes.

Tectonic zones

The earth's surface has eight large, and numerous smaller plates (30–200 km thick), comprising both the thinner oceanic and the thicker continental crust[2,4]. The crust behaves as a ductile solid (see: http://pubs.usgs.gov/publications/text/dynamic.html). Geothermal heat energy from deep below the mantle drives *plate tectonics* — the ceaseless motion of the plates (typical rates: 10–100 mm year^{-1}; Fig. 2.4). Most of the world's volcanic, seismic and tectonic activity is associated with movement between plates and is concentrated around plate boundaries (Table 2.2):

• *Divergent boundaries*, where new crust is generated by *basaltic volcanoes* as plates move apart e.g. the mid-Atlantic Ridges (sea floor spreading rates are around 25 mm yr^{-1}). When the lava builds up above sea level it creates *basic shield volcanoes* (e.g. Iceland).

• *Transform boundaries*, where two plates move laterally past each other along a major *strike-slip* fault such as the San Andreas Fault zone in California.

• *Convergent boundaries*, where two oceanic plates are in motion towards each other. The thinner one slides beneath the other along a *subduction zone*, creating initially an intra oceanic island arc (e.g. the island chains in the western Pacific Ocean such as the Philippine Arc).

• A *collision margin*, where plate motion brings into contact two continents, a continent and an island arc or two island arcs. The Himalayas formed as a result of the collision of the Indian and Eurasian continental plates.

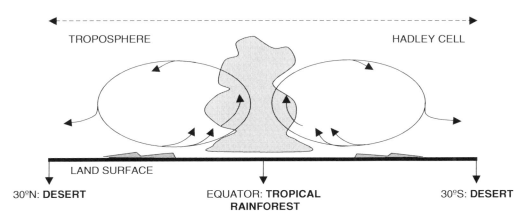

Figure 2.3 Inter-tropical convergence zones: Hadley cells.

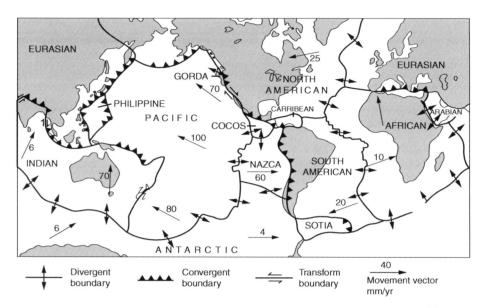

Figure 2.4 Global tectonic zones: plate boundaries and movement rates (based on Dewey 1972)[2.4].

Table 2.2 Simple characteristics of plate boundaries.

		Morphological and structural features		
Boundary type	**Stress**	**Oceanic-Oceanic lithosphere**	**Oceanic-Continental lithosphere**	**Continental-Continental lithosphere**
Divergent	Tensional	Mid-oceanic ridge, volcanic activity, sea floor spreading		Rift valley formation, volcanoes
Convergent	Compressional	Ocean trench and volcanic island arcs	Ocean trench and continental-margin mountain belt and volcanoes	Mountain belt
	Compressional	Complex island arc collision zone	Modified continental-margin mountain belt	Mountain belt, limited volcanic activity
Transform	Shear	Ridges and valleys normal to ridge axis		Fault zone, no volcanicity

- *Continental-margin orogen,* uplift created by the subduction of an oceanic plate beneath a plate carrying continental crust (e.g. on the western coast of South America resulting in the formation of the Andes).

Plate boundary earthquakes occur due to the concentration of stress along the edges of plates[2.5]. The size of an earthquake is related to the amount of stress released along the displacement. Subduction zones produce the world's largest earthquakes e.g. the magnitude (M_W, see Table 2.3) 9.5 1960 Chile earthquake, the M_W 9.2 1964 Alaska earthquake, and the M_W 9.0 December 26th 2004 earthquake off Sumatra and the Andaman Islands. These earthquakes can produce uplift and subsidence over large areas. For example the 1964 Alaska earthquake caused land-level changes of between 2–15 m in an area over 140 000 km^2.

Earthquakes also occur away from plate boundaries, because of strain accumulation and release within the plate e.g. the northwest India earthquake of 2001 (M_W 7.7) and the 1755 Lisbon earthquake (M_W c.8).

Earthquake magnitude is the most common measure of an earthquake's size. It is a measure of the size of the earthquake at its source i.e. the epicentre (Table 2.3).

Earthquake intensity is a qualitative measure of the shaking and damage caused by the earthquake. This value changes between locations depending on the distance from the earthquake and the local ground conditions (i.e. the Modified Mercalli Scale, see Table 2.4).

Explosive volcanoes are associated with convergent boundaries[2.6]. Active volcanic mountains such as Krakatoa, Mount St Helens, and Pinatubo all occur around the Pacific Ocean *Ring of Fire* (a series of contiguous plate boundaries). Eruptions, mainly from

Table 2.3 Earthquake magnitude measurements.

Measurement	Comment
Local magnitude, M_L	The Richter magnitude: a magnitude 0 event is an earthquake that would show a maximum combined horizontal displacement of 1 micrometre on a Wood-Anderson torsion seismometer 100 km from the earthquake epicenter.
	Because of the limitations of the Wood-Anderson torsion seismometer, the original M_L cannot be calculated for events larger than about $M_L = 6.8$.
Surface wave magnitude, M_s	Used for shallow (depth < 70 km) earthquakes at teleseismic distances (20–180°), employing the 20-second Rayleigh wave for the determination: $$M_s = \log \frac{A}{T} + S$$ where A = maximum displacement, T = period of displacement, and S = correction term for the distance of the station and the depth of the earthquake. M_s was developed by Gutenberg and Richter in 1936 as an extension to local magnitude at greater distances.
Body wave magnitude, m_b	m_b uses the amplitude of the P-wave train, the first arriving body wave, in its calculation. It is used at teleseismic distances from 16–100° $$m_b = \log \frac{A}{T} + Q(D, h)$$ where $Q(D, h)$ is a correction factor that depends on distance to the quake's epicenter D (°) and focal depth h (km).
Moment magnitude M_W	M_W is the most widely used measure of earthquake magnitude. The energy released or *moment* M_O of an earthquake is defined by: $$M_O = A D \mu$$ where A = rupture area, D = displacement, and μ = crustal rigidity. $$M_w = \log M_O - 9.1/1.5$$ Below magnitude 8, M_w matches M_s and below magnitude 6 matches m_b, which is close enough to the Richter M_L.

Table 2.4 The Modified Mercali Earthquake Intensity (MMI) scale.

MMI Value	Description
I	Not felt. Marginal and long period effects of large earthquakes.
II	Felt by persons at rest, on upper floors, or favourably placed.
III	Felt indoors. Hanging objects swing. Vibration like passing of light trucks. Duration estimated. May not be recognized as an earthquake.
IV	Hanging objects swing. Vibration like passing of heavy trucks; or sensation of a jolt like a heavy ball striking the walls. Standing motor cars rock. Windows, dishes, doors rattle. Glasses clink. Crockery clashes. In the upper range of IV, wooden walls and frames creak.
V	Felt outdoors; direction estimated. Sleepers wakened. Liquids disturbed, some spilled. Small unstable objects displaced or upset. Doors swing, close, open. Shutters, pictures move. Pendulum clocks stop, start, change rate.
VI	Felt by all. Many frightened and run outdoors. Persons walk unsteadily. Windows, dishes, glassware broken. Knick-knacks, books, etc., off shelves. Pictures off walls. Furniture moved or overturned. Weak plaster and masonry type D cracked. Small bells ring (church, school). Trees, bushes shaken (visibly, or heard to rustle).
VII	Difficult to stand. Noticed by drivers of motor cars. Hanging objects quiver. Furniture broken. Damage to masonry D, including cracks. Weak chimneys broken at roof line. Fall of plaster, loose bricks, stones, tiles, cornices (also unbraced parapets and architectural ornaments). Some cracks in masonry C. Waves on ponds; water turbid with mud. Small slides and caving in along sand or gravel banks. Large bells ring. Concrete irrigation ditches damaged.

(continued)

Table 2.4   *(continued)*

VIII	Steering of motor cars affected. Damage to masonry C; partial collapse. Some damage to masonry B; none to masonry A. Fall of stucco and some masonry walls. Twisting, fall of chimneys, factory stacks, monuments, towers, elevated tanks. Frame houses moved on foundations if not bolted down; loose panel walls thrown out. Decayed piling broken off. Branches broken from trees. Changes in flow or temperature of springs and wells. Cracks in wet ground and on steep slopes.
IX	General panic. Masonry D destroyed; masonry C heavily damaged, sometimes with complete collapse; masonry B seriously damaged. (General damage to foundations.) Frame structures, if not bolted, shifted off foundations. Frames racked. Serious damage to reservoirs. Underground pipes broken. Conspicuous cracks in ground. In alluvial areas sand and mud ejected, earthquake fountains, sand craters.
X	Most masonry and frame structures destroyed with their foundations. Some well-built wooden structures and bridges destroyed. Serious damage to dams, dikes, embankments. Large landslides. Water thrown on banks of canals, rivers, lakes, etc. Sand and mud shifted horizontally on beaches and flat land. Rails bent slightly.
XI	Rails bent greatly. Underground pipelines completely out of service.
XII	Damage nearly total. Large rock masses displaced. Lines of sight and level distorted. Objects thrown into the air.

Masonry A: Good workmanship, mortar, and design; reinforced, especially laterally, and bound together by using steel, concrete, etc.; designed to resist lateral forces.

Masonry B: Good workmanship and mortar; reinforced, but not designed in detail to resist lateral forces.

Masonry C: Ordinary workmanship and mortar; no extreme weaknesses such as failing to tie in at corners, but neither reinforced nor designed against horizontal forces.

Masonry D: Weak materials, such as adobe; poor mortar; low standards of workmanship; weak horizontally.

explosive volcanoes, are dominated by pyroclastic and ash fluidised flows and airfall tephra, whereas lava flows are viscous but relatively minor and short.

Some oceanic volcanoes such as the Hawaiian Islands or the Canary Isles are not associated with a plate boundary. They have formed over anomalous *hot spots* within the mantle, where it is thought rising thermal plumes have burned holes through the overlying plates. Movement of the plates relative to the hotspots over time has created lines of volcanoes.

Mountain zones
There are major variations in potential energy around the world. Potential energy increases as the height of material above a base level; therefore it is greatest in upland and mountainous areas. Mountains are extremely dynamic environments; they are probably the most landslide and erosion prone landscapes in the world[2.7–8].

Many the world's major mountain belts (Fig. 2.5) are the product of the Mesozoic to present-day plate collisions (the Alpine orogeny, began around 100 Ma ago). Today, the main tectonically active mountain belts are:
- *Alpine-Himalayan* belt, running from Borneo through northern India into Iran, Turkey and through southern Europe
- *Circum-Pacific* belt, including the Andes of South America and the Rocky Mountains of the USA, and encompassing the island arcs of the west Pacific.

Climate varies considerably in mountainous regions because of the effects of altitude and slope orientation, and the fall in temperature with elevation (*adiabatic lapse rates*) of around 1–2°C per 300 m rise. As a result, several climatic zones can occur within an area of high relative relief and glaciers can occur in the tropics (e.g. Puncak Jaya, Irian Jaya); the climate type is known as *azonal* because it is not related to distance from the equator.

Mountains have a major impact on global airflow patterns that affect the climates of adjacent regions[2.9]. For example, the presence of the Himalayas prevent the movement of warm air north into central Asia and cold air south into India, permitting tropical climates to extend farther north in India and southeast Asia than elsewhere in the world.

Geomorphological zones and time
Most landscapes are the product of many millions of years of weathering and surface processes[2.10–11]. These processes operate with different intensities in different climatic and tectonic regimes. However, over geological timescales the boundaries of these environments are dynamic, drifting across the continents in response to major climatic changes or plate movements. As these environments have shifted, so the intensity and character of geomorphological processes will have changed.

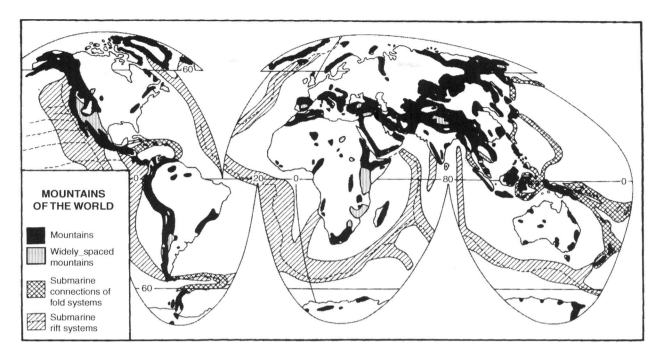

Figure 2.5 Global mountain regions (after Gerrard 1990)[2.7].

In some areas, current processes may have behaved in a similar manner for extremely long periods of time (e.g. the ancient cratonic regions of Central Africa). In other areas, current processes are actively reshaping an *inherited landscape*, destroying it or simply having no effect at all. As a result:

- Landscapes are *composite forms*, containing a range of features of varying antiquity inherited from different environments as major climatic or tectonic changes left their imprint.
- The landscape rarely reflects any one climate or period of geomorphological change; it is usually an assemblage of landforms that have superimposed histories (i.e. a *palimpsest* — like a surface which has been written on many times after previous inscriptions have only been partially erased).
- Present day conditions and morphoclimatic zones do not necessarily give a good guide to the processes that created particular landforms or left behind unusual deposits.

The implications of this inheritance for engineering projects can include:

- A legacy of pre-existing hazards: although they may have been created under very different conditions, these features may continue to present significant risks to engineering projects. Examples include pre-existing landslides (prone to reactivation; see Chapter 19), relict periglacial features in temperate zones[2.12–13] (Fig. 2.6 see colour section) and karst terrain[2.14] (potential for collapse; see Chapter 21).
- The presence of near-surface materials that have accumulated under former environmental conditions.

For example, spreads of wind-blown loess derived from glacial outwash plains that bordered Quaternary ice sheets (reaching thicknesses of > 200 m in Central Asia[2.15]), glacial tills and solifluction deposits.

- Inherited resources, essentially non-renewable under contemporary conditions and processes, such as fluvio-glacial or relict floodplain terrace sands and gravels[2.16–17].

References

2.1 Tricart, J. and Cailleux, A. (1972) *Introduction to Climatic Geomorphology*. Longmans, London.

2.2 Stoddart, D. R. (1969) Climatic geomorphology: review and reassessment. *Geography*, **1**, 159–222.

2.3 Lee, E. M., and Fookes, P. G. (2005) Climate and weathering. In P. G. Fookes, E. M. Lee and G. Milligan (eds.) *Geomorphology for Engineers*. Whittles Publishing, 31–56.

2.4 Dewey, J. F. (1972) Plate tectonics. *Scientific American*, **226**, 5.

2.5 Hengesh, J. V. and Lettis, W. R. (2005) Active tectonic environments and seismic hazards. In P. G. Fookes, E. M. Lee and G. Milligan (eds.) *Geomorphology for Engineers*. Whittles Publishing, 218–262.

2.6 Waltham, T. (2005) Volcanic landscapes. In P. G. Fookes, E. M. Lee and G. Milligan (eds.) *Geomorphology for Engineers*. Whittles Publishing, 633–662.

2.7 Gerrard, A. J. (1990) *Mountain Environments: an Examination of the Physical Geography of Mountains*. Belhaven, London.

2.8 Charman, J. and Lee, E. M. (2005) Mountain environments. In P. G. Fookes, E. M. Lee and G. Milligan (eds.) *Geomorphology for Engineers*. Whittles Publishing, 501–534.

2.9 Barry, R. G. (1992) *Mountain Weather and Climate*. 2nd edition. Routledge, London.

2.10 Brunsden, D. (1990) Tablets of Stone: Toward the ten commandments of Geomorphology. *Zeitschrift für Geomorphologie*. Supplementband **79**, 1–37.

2.11 Brunsden, D. (1993) The persistence of landforms. *Zeitschift fur Geomorphologie*, **93**, 13–28.

2.12 Walker, H. J. (2005) Periglacial forms and processes. In P. G. Fookes, E. M. Lee and G. Milligan (eds.) *Geomorphology for Engineers*. Whittles Publishing, 376–399.

2.13 Fookes, P. G. 1997. First Glossop Lecture: Geology for engineers: the geological model, prediction and performance. *Quarterly Journal of Engineering Geology*, 30, 290–424.

2.14 Waltham, T. (2005) Karst terrains. In P. G. Fookes, E. M. Lee and G. Milligan (eds.) *Geomorphology for Engineers*. Whittles Publishing, 662–687.

2.15 Derbyshire, E. and Meng, X. (2005) Loess. In P. G. Fookes, E. M. Lee and G. Milligan (eds.) *Geomorphology for Engineers*. Whittles Publishing, 688–728.

2.16 Owen, L. A. and Derbyshire, E. (2005) Glacial environments. In P. G. Fookes, E. M. Lee and G. Milligan (eds.) *Geomorphology for Engineers*. Whittles Publishing, 345–375.

2.17 Gregory, K. J. (2005) Temperate environments. In P. G. Fookes, E. M. Lee and G. Milligan (eds.) *Geomorphology for Engineers*. Whittles Publishing, 400–418.

3

The Basics of Change: Stress, Strain and Strength

Introduction

Landform change is driven by the forces (*stress*) imposed on the surface and near-surface materials. The fundamental properties, in particular the *strength*, of these materials govern the nature and rate that landscapes evolve over time.

Weathering (see Chapter 6) physically and chemically changes near surface soils and rocks. The weathered materials are then transported by gravity, water, ice or wind and deposited elsewhere as new landforms. The combination of weathering and transportation is known as *erosion*. From an engineering perspective, the most significant landform changes resulting from erosion and deposition are:

- Removal of material from hillslopes through surface erosion (water and wind) and mass movement (landsliding).
- Migration of sand dunes in dune fields and sand seas.
- Bank erosion, bed scour, sedimentation and channel changes along river systems, deltas and estuaries.
- Cliff recession, erosion and deposition along shorelines.

Despite the variety of processes, the basic principles remain the same:

- *Detachment* of material through surface erosion or mass movement occurs when the imposed *stresses* are greater than the *strength* of the material.
- *Deposition* of material which occurs when the *stresses* imposed on the materials in transport fall below a *critical* or *threshold level*.

Stress

When a force is applied to a solid body, stresses (force per unit area; Pascal, Pa) are transmitted within the material. Examples include *tensile* stress (a stretching force), *compressive* stress (the force imposed by a load) and *shear* stress (a tangential force).

Gravity imposes a shear stress τ on all slopes (Fig. 3.1):

$$\tau = W \sin\beta$$

where W = weight of slope materials ($W = \gamma z \cos\beta$); β = slope angle (degrees); γ = unit weight (kN m^{-3}) and z = depth (m).

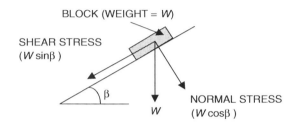

Figure 3.1 The forces acting on a slope.

As a slope becomes steeper, so the shear stress increases:

Unit Weight γ (kNm^{-3})	Depth z (m)	Angle $\beta°$	Cos β	Sin β	Shear stress (kN m^{-2})
19	3	10	0.985	0.174	9.74
19	3	15	0.966	0.259	14.25
19	3	20	0.940	0.342	18.31
19	3	25	0.906	0.423	21.83
19	3	40	0.766	0.643	28.06

Under seismic loading (i.e. earthquakes) the shear stress is supplemented by a seismic force:

$$\tau = (W - F_v) \sin\beta + F_h \cos\beta$$

where F_v and F_h are the vertical and horizontal seismic forces, defined as follows:

$$F_v = \frac{a_v W}{g}$$

and

$$F_h = \frac{a_h W}{g}$$

where a_v and a_h are the vertical and horizontal acceleration.

Moving fluids impart a fluid drag (a shear stress) on sediment particles. For moving water, the force along the channel bed is:

$$\tau = \rho\, g\, R\, S$$

where ρ = fluid density; g = gravitational acceleration (9.81 m s^{-2}); R = hydraulic radius of the channel (channel cross-section divided by the wetted perimeter) and S = bed and water surface slope gradient.

Shear stress increases linearly with flow depth and slope:

Fluid density (kg m⁻³)	Flow depth (m)*	Slope Gradient S	Shear stress τ (MPa)
1000	0.2	0.001	1.96
1000	2	0.001	19.6
1000	20	0.001	196
1000	2	0.001	1.96
1000	2	0.01	19.6
1000	2	0.1	196

* The hydraulic radius (R) is approximately equal to flow depth for channels that are wide relative to their depths.

For wind, the shear stress (mean drag per unit surface area) is:

$$\tau = \rho_a u*^2$$

where ρ_a = air density (0.00125 g cm⁻³) and $u*$ = drag (shear) velocity of the wind, or air resistance.

Drag velocity $u*$ is proportional to the slope of the wind velocity profile i.e. the change in wind speed with height above the ground surface. Small changes in drag velocity can result in significant increases in the shear stress:

Air density (kg m⁻³)	Drag Velocity (m sec⁻¹)	Shear stress (MPa)
1.25	0.1	0.012
1.25	0.25	0.078
1.25	0.5	0.312

Strength

The resistance to the forces imposed by gravity or fluid motion is a function of the strength of the materials: *tensile* strength (resistance to stretching), *compressive* strength (resistance to crushing) and *shear* strength (resistance to tangential forces). The shear strength s determines resistance to erosion:

$$s = c + \sigma \tan\phi$$

where c = actual cohesion; σ = normal stress and ϕ = angle of internal friction.

Actual cohesion c is the result of electro-chemical bonding or cementation of particles. This is a major component of rock strength, and can be a factor in the short-term shear strength of soils.

Friction is the result of the compressive forces that hold particles together. It is derived from two components:
- The angle of internal friction ϕ, a measure of the frictional resistance of the material.
- The normal stress σ, i.e. the effect of gravity operating at right angles to the slope or shear surface (see Fig. 3.1), where $\sigma = W \cos\beta$.

As slope increases, the normal stress decreases:

Unit Weight γ (kN m⁻³)	Depth z (m)	Angle β°	Cos β	Normal stress (kN m⁻²)
19	3	10	0.985	55.28
19	3	15	0.966	53.18
19	3	20	0.940	50.33
19	3	25	0.906	46.82
19	3	40	0.766	33.45

Under seismic loading the resisting force becomes:

$$s = c + ((W - F_v) \cos\beta - F_h \sin\beta) \tan\phi$$

where F_v and F_h are the vertical and horizontal seismic forces.

The presence of water within a slope reduces shear strength because inter-particle water exerts a pressure u (*pore water pressure*). This reduces the particle-to-particle contact and hence the frictional component of strength.

The frictional resistance depends on the difference between the applied total normal stress σ and the pore water pressure u. That part of the normal stress that is effective in generating shear resistance is the *effective stress* σ':

$$\sigma' = \sigma - u$$

Thus the shear strength becomes what is referred to as the *Coulomb equation* (see Chapter 15):

$$s = c' + (\sigma - u) \tan\phi'$$

where c' and ϕ' are modified parameters with respect to *effective* stress. Pore water pressure can be negative, in which case a soil can be considered to be held partly together by suction.

For quartz *sands and gravels*, $c' \geq 0$ and $\phi' \geq 30°$. ϕ' increases with density and particle packing. Soils with angular particles and/or a range of particle sizes (poorly sorted), which achieve greater densities, have higher ϕ' values.

Clays tend to exhibit ϕ' values in the range 20–25°. Their cohesion is strongly dependent on the amount of diagenesis the soil has experienced (see Chapter 7). Over-consolidation or cementation (e.g. with calcite or iron compounds) increases the cohesion c'.

The Onset of Erosion

For *slopes* the balance between stress and strength can be defined as a *limiting equilibrium* relationship represented by the *factor of safety F*:

$$F = \text{shear strength } s \,/\, \text{shear stress } \tau$$

A stable slope is one where the resisting forces are greater than the destabilising forces and, therefore, can be considered to have a *margin of stability* ($F > 1.0$). A slope at the point of failure has no margin of stability, as the resisting and destabilising forces are approximately equal i.e. $F = 1.0$. At this point the slope will fail through landsliding (see Chapter 19).

For *mobile fluids* (e.g. water or air) the threshold between failure and stability (i.e. the onset of erosion) is often represented by a critical shear stress or velocity of flow. For *water*, the critical bed shear τ_c on a rough channel bed increases with particle size of the bed material according to Shield's function:

$$\tau_c = 0.06 \, (\rho_p - \rho_w) \, g \, d$$

where ρ_p and ρ_w are the specific weight of the particles and water respectively; g is the gravitational acceleration and d is the particle diameter (mm).

Simpler versions of this relationship are Baker and Ritter's function[3.1]:

$$\tau_c = 0.03 \, d^{1.49}$$

and Costa's formula[3.2]:

$$V_c = 0.18 \, d^{0.49}$$

where V_c = critical flow velocity (ms^{-1}). There is a non-linear relationship between threshold flow velocity and grain size[3.3-4] (Fig. 3.2).

For *wind*, the initial movement of loose particles begins at a critical shear velocity[3.5]:

$$u*_t = A\sqrt{\frac{\rho_p - \rho_a}{\rho_a} \, g \, d}$$

where ρ_a, ρ_p = density of air and the particle respectively and A is a constant which at the start of sand movement = 0.1 (see Chapter 18; Table 18.2).

Figure 3.3 shows the relationship between particle diameter and the critical shear velocity. The first movement occurs at the *fluid* or *static* threshold. Once movement has started the threshold under particle bombardment (the *dynamic* or *impact* threshold) is around 80% of the static threshold (see Chapter 18).

Deposition

Eroded material is carried by the transporting agents (ice, water or wind) until the velocity of the flow drops below a critical velocity, causing the material to be deposited (i.e. a *depositional threshold*). This threshold can be as low as 35% of the erosional threshold (Fig. 3.2).

Mass movement features (landslides) move down the slope under the influence of gravity until they reach a new stable position with a factor of safety F of 1.0 or higher. Some landslides result in considerable *run-out*, often because of the fluidity of the debris[3.6].

Variations in Stress and Strength

The balance between stress and strength is not static. Landform changes are generally associated with variations in either or both of the stress or strength. Examples include:

- A sudden occurrence of high shear stresses associated with extreme events such as high velocity water or wind flows, or storm waves.
- A sudden but temporary increase in stress caused by earthquakes, producing a horizontal acceleration on

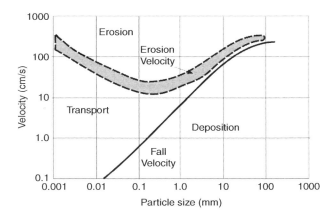

Figure 3.2 Critical threshold velocities for erosion, transport and deposition by water (based on Hjulström 1935)[3.3].

a static body of material, exceeding the gravitational acceleration.
- A gradual increase in shear stress on a slope as it naturally steepens (e.g. basal erosion by a river).
- A sudden reduction in effective shear strength associated with, for example, the rapid increase in pore water pressures during an intense rainstorm.
- A gradual reduction in shear strength associated with the slow recovery over time (possibly decades) of pore water pressures in fine-grained materials.
- A gradual reduction in shear strength associated with weathering of materials (see Chapter 6).

Landform change events are often the product of the variable interaction between a range of factors that influence shear stress and shear strength:
- *Preparatory factors*, which make a site susceptible to change without actually initiating change e.g. the reduction in beach volume at the base of a cliff associated with disruption of longshore sediment transport by a groyne field.
- *Triggering factors,* which initiate an event e.g. an intense rainstorm or earthquake.

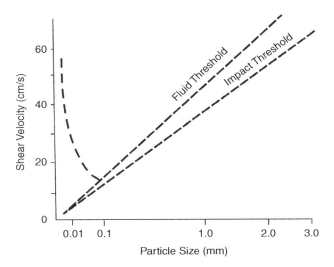

Figure 3.3 Wind erosion: threshold wind velocities (after Bagnold 1941)[3.5].

Weathering often acts as a preparatory factor in the generation of events. In many mountainous areas, soil development proceeds until a critical soil depth and a landslide-triggering event (e.g. intense rainfall) occur in combination, generating shallow hillside failures. After the initial failure, the rate of weathering and soil formation will influence the timing of future failures at a particular site[3.7].

References

3.1 Williams, G. P. (1984) Palaeohydrologic equations for rivers. In J. E. Costa and P. J. Fleisher (eds.) *Developments and Applications of Geomorphology*, Springer Verlag, Berlin, 343–367.

3.2 Costa, J. E. (1983) Palaeohydraulic reconstruction of flash-flood peaks from boulder deposits in the Colorado Front range. *Geological Society of America Bulletin*, **94**, 986–1004.

3.3 Hjulström, F. (1935) Studies of the morphological activity of rivers as illustrated by the River Fyries. *Bulletin Geological Institute*, University of Uppsala, **25**, 221–527.

3.4 Sundborg, A. (1956) The River Klarälven, a study of fluvial processes. *Geografiska Annaler*, **38**, 127–316

3.5 Bagnold, R. A. (1941) *The Physics of Blown Sand and Desert Dunes*. Chapman and Hall, London.

3.6 Wong, H. N., Ho, K. K. S. and Chan, Y. C. (1997) Assessment of consequence of landslides. In D. Cruden and R. Fell (eds.) *Landslide Risk Assessment*. Balkema, Rotterdam, 111–149.

3.7 Crozier, M. J. (1986) *Landslides: Causes, Consequences and Environment*. Croom Helm.

4

Earth Surface Systems

Introduction

Earth surface systems are conceptual models that can be used to describe how sediment and energy transfers (e.g. down slopes or along stream channels) provide the *inter-linkages* between landforms in an area. They provide the basis for explaining landscape change (see Chapter 2). Although landscape changes can be explained as the cumulative product of site specific interactions between the imposed stresses and the strength of the materials (a 'bottom-up' approach), this understanding can be enhanced by considering the changes in the context of the behaviour of earth surface systems (a 'top-down' approach). For example, although the workings of a carburettor can be explained in isolation, its function and performance can only be understood within the context of a larger system — the car engine. In the same way, the erosion and accretion of mudflats needs to considered within the context of an estuary.

Systems form a useful spatial framework for evaluating:

• How *hazards* and *risks* to a particular site can arise as a result of processes operating elsewhere within a system. For example, changes in land use in the headwaters of a catchment can lead to changes in flood frequency and river channel erosion or deposition downstream (see Chapter 12).

• The potential *impacts* of an engineering project on other landforms within a system. For example, reclamation of an intertidal wetland can have significant effects on the whole estuary, through the resulting changes to the tidal prism and mean depth (see Chapter 31).

Types of System

Earth surface systems can be defined at a range of scales, from river drainage basins (watersheds or catchments) and coastal cells (sediment transport cells) to individual hillslopes, dunes or cliffs. Irrespective of the scale, each system comprises:

• An assemblage of individual *morphological* components (landforms)

• Transfers (*cascades*) of energy (e.g. kinetic energy of flowing water or wind) and sediment.

Rivers are systems that carry water and sediment. At the scale of a drainage basin (landscape level), the overall system comprises three broad zones, each with a characteristic form and function (Fig. 4.1):

• Zone 1: the water and sediment *source area*, generally mountain or upland areas around the margins of the drainage basin.

• Zone 2: the sediment transport zone, usually the main river *channel*.

• Zone 3: the sediment store or *depositional zone*, comprising the river floodplain, delta or estuary.

A similar arrangement of components can be found within a mudslide system (Fig. 4.1):

• Zone 1: the *source area*, comprising individual landsides around the mudslide head.

• Zone 2: the *track*, a relatively steep chute along which mudslide debris is transported from the head to the accumulation zone.

• Zone 3: the *accumulation zone*, comprising sequences of overlapping lobes of debris.

Large systems can be divided into a series of smaller sub-systems, each with their own arrangement of source areas (inputs from adjacent sub-systems), transport channels or pathways, storage and outputs (transfers to adjacent sub-systems).

Figure 4.2 presents a simple model of part of a river channel system (a single stretch of the river or *reach* can be considered to be a sub-system). Changes in the bed and banks occur in response to changes in the energy (water discharge) and sediment inputs and outputs. The interaction between these inputs and the existing channel bed (the bed form and available sediments) control the sediment transport rate which, through erosion and deposition, controls the channel geometry.

System Specification

The influences on an earth surface system can be organised into a generalised hierarchy (Fig. 4.3):

1. *System environmental controls* (independent variables or boundary conditions): these include the climate, gravitational attraction, eustatic sea-level, tectonic

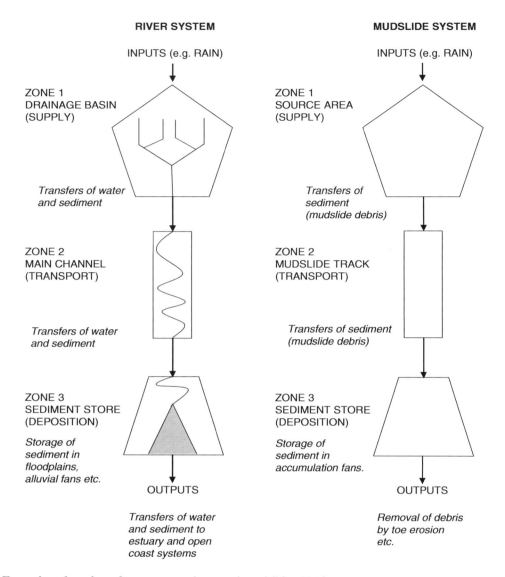

Figure 4.1   Examples of earth surface systems: rivers and mudslides. Each system comprises an assemblage of landforms and transfers of energy and sediment.

history and the post-glacial effects of isostatic re-adjustment, and the existing geology, topography and sediment availability (i.e. the *inheritance* from past processes).

2. *Energy regime factors*: these factors vary in response to changes in the system controls and include rainfall, wind regime, temperature, tidal range, wave climate and relative sea-level. These are the drivers of change, referred to as *forcing factors*.

3. *System state*: defined by the dimensions and characteristics of the landforms that form the system. It is the product of past change and sets the framework for future evolution.

A model of a coastal cliff system is shown in Fig. 4.4; this highlights the interactions between terrestrial (cliff) and marine (shoreline) sub-systems. Each of these two linked systems has its own environmental controls. The energy regime factors are identified as a *primary*

response to the environmental controls. Morphological changes to the shoreline and cliff represent the *local response* to these factors.

System environmental controls should not be regarded as being constant over engineering time, especially climate and sea-level (see Chapters 9 and 10). The magnitude and frequency of the energy inputs tend to be random (*stochastic*), introducing uncertainty into system behaviour (see Chapter 5).

System Change

System change can occur in response to:
- *Changing energy inputs (perturbations)*
  For example, more than 9000 landslides were triggered in Guatemala when Hurricane Mitch stalled over Central America in October 1998 and dropped 2 m of rain over mountainous terrain[4.1]. The growth of Romney Marsh and the Dungeness spits in the UK around 6500 BP led to the

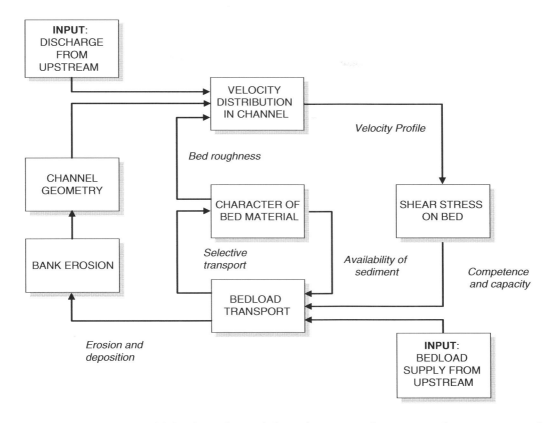

Figure 4.2 A simple river system model: bank erosion and channel geometry changes occur in response to variations in input.

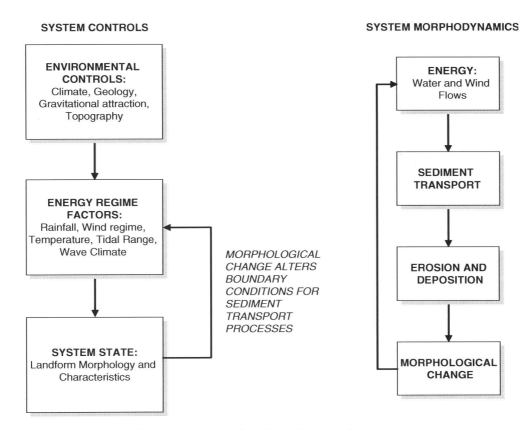

Figure 4.3 Organisation of earth surface systems: controls and morphodynamics.

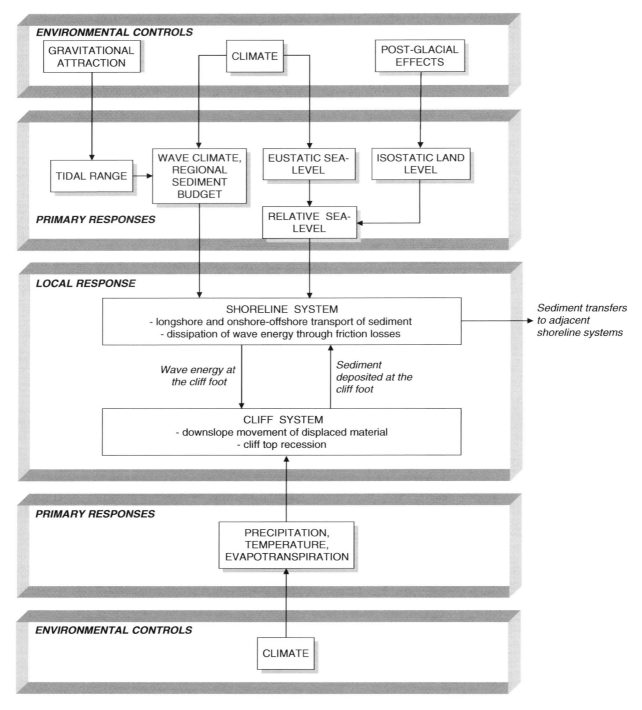

Figure 4.4   Coastal cliff system: generalised hierarchy of influences between shoreline and cliff systems.

reduction in wave attack at the foot of the cliff-line between Hythe and Lympne, Kent[4.2]. As a result the cliff-line became *abandoned*.

- *Changing sediment inputs*
  For example, 19th Century dumping of hydraulic gold mining waste along the Sacramento River in the Sierra Nevada caused aggradation, braiding and channel switching[4.3]. However, as the input of waste decreased the river began to incise through the recent deposits.
- *Modification to the existing system components*
  The 1993 Holbeck Hall landslide in Scarborough, UK, was the result of the gradual decline in strength

of the cliff materials (glacial till and mudstone) due to a combination of progressive failure and strain softening[4.4]. The periodic exposure of resistant clay (*clay plugs*) in the Lower Mississippi valley have a significant effect on the local rate of bank erosion and influence the form and evolution of meanders in the reach[4.5].

- *Incipient instability* (chaos) *within the system components*
  Incipient instability is common in alluvial river channels, where channel sinuosity and meander amplitude increase until a cut-off or channel avulsion occurs (e.g. the River Dane, UK[4.6]). This is

due to channel lengthening and gradient reduction and is not necessarily related to external changes.

- *Changes in the energy flows through the system*
  For example, changes in run-off from hillslopes can lead to significant river channel changes because of the increased flow velocity. Around 1865 to 1915 trench-like gulleys (*arroyos*) developed on the plains in the arid and semi-arid south-west USA (e.g. Kaneb Creek in south Utah developed a 25 m deep and 80–120 m wide channel between 1880 and 1914)[4.7]. The changes were blamed on over-grazing by cattle, and the resultant increase in run-off.

- *Changes in the sediment flows through the system*
  For example, a major landslide on the east Devon coast (UK) in 1840 disrupted the eastwards longshore transport of shingle and caused the decline of the beach in Pinhay Bay, 2 km to the east[4.8].

These changes can be immediate (e.g. the impact of Hurricane Mitch) or may take many years to become apparent (e.g. the beach changes on the east Devon coast). The *lag time* is the time taken for the system to begin reacting to a change. The *relaxation time* is the time taken by the system to fully respond to the change.

System Dynamics and Landscape Stability

Systems include elements of different age and *stability* under the present day environmental controls including:

- *Actively dynamic landforms*: subject to almost continuous adjustment to changes in energy inputs (e.g. rainfall, river discharges, wave attack) or sediment availability such as alluvial river channels and the coastline.

- *Episodically active landforms*: subject to occasional adjustments to extreme events, such as hillsides prone to shallow debris slides in tropical regions, sand dune fields, steep mountain catchments prone to debris flows and flash floods.

- *Stable, relict landforms*: although they are stable under present conditions, they would not have evolved under them i.e. they are relicts of a former set of conditions. For example, periods of extreme aridity and high wind speeds in the Quaternary led to the development of the 100–200 m high primary dunes of the Saharan sand seas (*ergs*). In the current hyper-arid climate, sand dune mobility is restricted by relatively low wind speeds and, hence, the high primary dunes are effectively stable.

- *Unstable, relict landforms*: landforms that developed under past conditions, but are being progressively destabilised by present conditions. Many shingle beaches in high latitudes of the northern hemisphere (e.g. Slapton Sands, UK), were created from material moved onshore as sea levels rose early in the Holocene and, thus, are relict features with no significant contemporary source of sediment (see Chapters 29 and 34).

- *Fossil landforms*: essentially unchanging elements of the landscape, such as low angled pediment slopes in deserts, ancient surfaces in cratonic areas.

Landscape Sensitivity

There can be considerable variation in the susceptibility of landforms and systems to change. This is known as *landscape sensitivity* and ranges from[4.9]:

- fast responding systems that are very sensitive to the effects of human activity e.g. badlands, alluvial river channels, shorelines
- slowly responding systems, such as hard rock cliffs, plains and gentle slopes.

In addition to the material strength and potential energy available at a site, the main controls on sensitivity are the *linkages between landforms* within a system (especially the flows of water and sediment).

References

4.1 Bucknam, R. C. (2001) Landslides triggered by Hurricane Mitch in Guatemala: Inventory and Discussion. US Geological Survey Open File Report **01** 443.

4.2 Jones, D. K. C. and Lee, E. M. (1994) *Landsliding in Great Britain*. HMSO.

4.3 Gilbert, G. K. (1917) Hydraulic mining debris in the Sierra Nevada. Professional Paper USGS 105.

4.4 Lee, E. M. (1999) Coastal Planning and Management: The impact of the 1993 Holbeck Hall landslide, Scarborough. *East Midlands Geographer*, **21**, 78–91.

4.5 Schumm, S. A. and Thorne, C. R. (1989) Geologic and geomorphic controls of bank erosion. In M. A. Potts (ed.) *Hydraulic Engineering*. ASCE, New Orleans, 106–111.

4.6 Hooke, J. M., Harvey, A. M., Miller, S. Y., and Redmond, C. E. (1990) The chronology and stratigraphy of the alluvial terraces of the River Dane, Cheshire, NW England. *Earth Surface Processes and Landforms*, **15**, 717–737.

4.7 Cooke, R. U. and Reeves, R. W. (1976) *Arroyos and Environmental Changes in the American Southwest*. Oxford University Press, Oxford.

4.8 Lee, E. M. (2001) Estuaries and Coasts: Morphological adjustment and process domains. In D. Higgett and E. M. Lee (eds.) *Geomorphological Processes and Landscape Change: Britain in the Last 1000 years*. Blackwell, 147–189.

4.9 Brunsden, D. and Thornes, J. B. (1979) Landscape sensitivity and change. *Transactions of the Institute of British Geographers*, **4**, 463–484.

5

The Behaviour of Earth Surface Systems

Introduction

Many earth surface systems can be extremely complex and, at times, chaotic. Others will appear to be unchanging over time (e.g. stable, relict and fossil landforms). System complexity is, in part, due to the fact that the components and surface processes are interlinked, as changes to one will influence the other. This complexity is most apparent in dynamic systems:

- The activity of surface processes results in *morphological changes* and the release of sediment leading to deposition.
- *Morphological changes* alter the boundary conditions (topography) for the surface processes.
- *Sediment properties and abundance* (sediment supply) affect surface processes and the extent to which potential sediment transport and deposition is realised.

As a result of this inter-linkage between form and process (termed *morphodynamic behaviour*; see Fig. 4.3), sequences of landform change events are not independent of each other, but are influenced by the existing topography, along with the size and location of previous events (*state dependence*). Thus landform change is a process with a 'memory' in that the effects of past events influence current and future behaviour, known as sensitive dependence upon the initial conditions[5.1]:

- Past *landslide events* can influence the location and style of future events. For example, the size of channelised debris flows in tropical mountain regions will depend on the volume of materials supplied by past incidents of debris slide activity stored along stream valley floors.
- The position of past overwashing and crest lowering along a *shingle barrier* will provide the focus for future events unless the beach is able to build up during the period between storms (see Chapter 34). The extent to which the beach can recover from a particular storm is conditional on the precise timing and sequence of subsequent storm events and the availability of sediment.

System Complexity

Systems can be classified in terms of their degree of complexity and randomness of behaviour:

- *Highly organised systems* that can be modelled and analysed using deterministic mathematical functions (e.g. the prediction of tides).
- *Unorganised complex systems* that exhibit a high degree of random behaviour and hence can be modelled by stochastic methods (e.g. the probability of waves of particular heights or flood discharges).
- *Organised complex systems*, where the behaviour involves both deterministic and random components (e.g. sand grain movement follows well-defined physical laws but actual dune migration is controlled by random variations in wind speed and direction).

Most dynamic earth surface systems are characterised by organised complexity. This is because they exhibit *non-linear behaviour* (i.e. a change in one variable will not result in a proportional change in another), involving:

- *Negative feedback* (i.e. self-regulation) that tends to maintain the balance within a system, limiting the extent to which system morphology can change e.g. the recovery of a channel form after the impact of a large flood.
- *Positive feedback* that tends to generate short periods of accelerated activity e.g. the expansion of a quick clay landslide after an initial failure (see Chapter 19) or the rapid development of a blow-out in a coastal dune system (see Chapter 35).

There is uncertainty in system behaviour because of the stochastic nature of forcing events (*randomness*) and the fact that the system response can be very sensitive to the initial system state, which is often only known in broad terms (*deterministic uncertainty* i.e. even if we knew the future forcing events we could not be sure of the system response). Average states may be predictable but the *precise* conditions at a particular time in the future are not. We cannot expect to make completely reliable predictions about future changes. It is often more appropriate to consider a range of possibilities (*scenarios*) and express the predictions as probability distributions rather than single values.

Thresholds

System change occurs when thresholds are exceeded; at the simplest level the balance between stress and strength on a slope represents a threshold (see Chapter 3). Other thresholds are more complex. The system response to a particular level of energy input is the product of the variable interaction between *preparatory* and *triggering* factors (see Chapter 3). If a potential triggering event (e.g. an intense rainstorm) occurs before the system has been sufficiently 'prepared' (e.g. by weathering) then no change will occur. Often it is possible to define:

- *Minimum threshold* for change: above this threshold the conditions are sufficient to trigger change whereas below it, there is insufficient impetus for change. However, conditions above the minimum threshold are *not always* sufficient to initiate change, as some events will be *redundant* or *ineffective* because of the recent event history.
- *Maximum threshold* for change: above this the conditions are *always* sufficient to initiate change.

Figure 5.1 shows the minimum and maximum daily rainfall thresholds associated with landslide activity in Wellington, New Zealand[5.2]. The minimum threshold (20 mm day^{-1}) corresponds with the daily rainfall category below which no landslide activity has been recorded and above which it may occur under certain conditions (i.e. the 24 hour rainfall *sometimes* causes landsliding). The maximum threshold (>140 mm day^{-1}) is the 24 hour rainfall category above which landsliding has *always* occurred (probability = 1).

Major landslide events at Fairlight on the Sussex coast, UK are associated with gradual profile (cliff top to cliff foot) steepening by basal erosion until a critical threshold slope angle is reached. Once this overall slope angle has been reached (c. 32°), then the cliff is susceptible to deep-seated failure. The timing of major failures is then associated with the occurrence of wet winters (see: http://www.fairlightcove.com/).

Some changes may be an inherent (*intrinsic*) feature of the system behaviour and not necessarily the product of external factors:

- In semi-arid areas, the accumulation of sediment in the upper parts of a valley floor gradually increases the valley slope angle (long profile). As the valley floor gradient increases, the ability to withstand the effects of potentially erosive flood events declines[5.3]. Eventually a flood of a particular magnitude will exceed the threshold of valley stability and erosion will occur (Fig. 5.2).
- Many estuaries oscillate between *flood dominant* (sediment sink) and *ebb dominant* (sediment source) forms (see Chapter 31); the switching of form is controlled by changes in mean channel depth (an intrinsic threshold related to the progressive effects of deposition or erosion).

The implications of thresholds to engineers are self-evident; works undertaken at some sites may result in dramatic and unexpected changes. For example, the 6 Mm^3 Rissa quick clay landslide in Norway[5.4] followed an excavation of about 700 m^3 on the edge of a lake (http://www.ngi.no/).

Modes of System Behaviour

Although changes may be easy to observe, record or measure, there can be problems in establishing what the

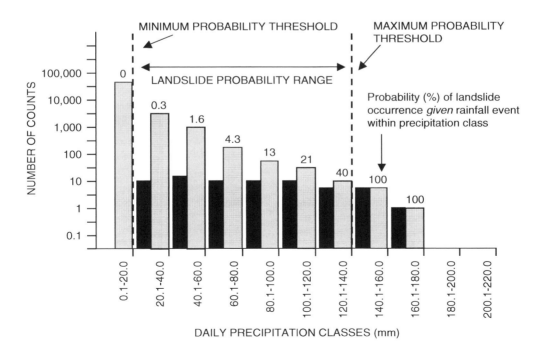

Figure 5.1 Minimum and maximum rainfall thresholds for landsliding, Wellington, New Zealand (after Glade 1998)[5.2]; black bars represent recorded landslide-triggering daily rainfall events, grey bars indicate actual daily rainfall events.

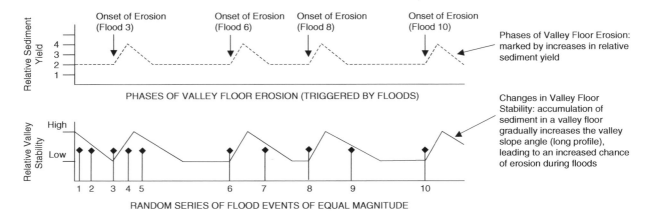

Figure 5.2   The complexity of system response. The relationship between flood events (vertical lines) and phases of valley floor erosion. All 10 floods were the same size but only events 3, 6, 8, and 10 caused erosion (after Schumm 1977)[5.3].

changes actually mean for an engineering project. Change can be a reflection of a number of distinctive modes of system behaviour:

1. *Episodic, progressive change* where no recovery occurs. For example, cliff erosion involves sequences of landslide events that result in loss of cliff-top land. Cliffs do not recover by advancing forward.

2. *Periodic or cyclical change*, where the landform responds to an event (e.g. a storm or large flood) by altering the morphology, but then gradually recovers (either partially or completely) its original pre-event form. For example, variations in wave energy can cause major changes to beach form. Sediment is eroded from the upper shore and transported offshore under high wave energy conditions and then brought back onshore in lower-energy conditions, leading to beach accumulation (see Chapter 33).

3. *Continuous change*, where landform change is an inherent part of the natural evolution of the system. *Chaos* is known to be present within turbulent flows[5.1], and this has implications for river and coastal system behaviour, e.g. meander development across floodplains[5.5] or the saltmarsh response to sea-level rise[5.6].

4. *State change*, where the landform responds to an event by changing its overall form because threshold levels have been exceeded. Examples include the change from meandering to braided channels (see Chapter 25) or the breaching of barrier beaches during storms to create barrier island and tidal inlet systems (see Chapter 34).

5. *Progressive change*, where there is a longer-term trend in the evolution of the land component or system. For example, in southern England there has been an increasing potential for landsliding over the last 2 million years[5.7]. This is because tectonic uplift of the Weald (over 100 m in 2 Ma) has caused an increase in relative relief, leading to gradual river incision and the slow development of scarps and vales. Landslide activity became widespread along the scarps once the slope angle or height had reached a limiting value, or the landscape dissection had exposed a landslide-prone stratum.

Equilibrium Concepts

Despite the changes associated with erosion and deposition, many landforms appear to maintain stable average forms that persist over time. For example, as coastal cliffs retreat, the individual components of the cliff (the distribution and activity state of landslide features on the cliff face) will be continuously changing. However, the overall form will tend to remain the same. This average form can be viewed as the most likely stable or robust configuration for the prevailing conditions. In the language of chaos, these robust forms are *attractors*[5.8].

The persistence of robust forms reflects a balance between the imposed stresses, the processes operating within the system and the materials. This balance is termed *equilibrium*[5.9]. The concept is not as straightforward as it initially appears, raising a number of important questions:

- Are the landforms adjusted to the current conditions or to conditions that operated at some time in the past?
- Is the equilibrium reflected in a constant average form and rate of processes, or in changing forms and rates?
- Has there been sufficient time between major process events (e.g. a storm) for a system to achieve an equilibrium form, or are events so frequent that an equilibrium form is an unachievable target?
- Could changes in the imposed stresses or materials lead to the establishment of different equilibrium forms?
- Do different processes acting on different landforms composed of different materials produce the same average form (*equi-finality*)?

Steady State Equilibrium

This describes the condition where the landform or process rate (e.g. recession rate) oscillates around a stable mean value (Fig. 5.3). This condition is associated with constant (i.e. stationary) boundary conditions (see Chapter 4) and is the product of negative feedback mechanisms that re-establish the equilibrium form after an event.

For example, alluvial river channel geometry has been seen as being in steady state equilibrium (i.e. in *regime*) with the prevailing flows. Channel cross-sectional form adjusts by erosion and deposition to accommodate the sediment load and all but the highest flows. As mean discharge increases so width, mean depth and velocity all increase (see Chapter 25).

Dynamic Equilibrium

This describes the condition where the landform or process rate oscillates around a mean value that has been gradually changing through time (Fig. 5.3). The condition is associated with variable (non-stationary) boundary conditions.

For example, the Holderness cliffs in the UK are prone to severe marine erosion. Erosion posts have been used to record annual cliff recession since 1951. The average recession rate has increased over this period from around 0.5 m year[-1] in the 1950s, to over 3.5 m year[-1] by the 1990s (Fig. 5.4). However, the average cliff form has remained the same (see: http://www.hull.ac.uk/coastalobs/easington/erosionand-flooding/).

Metastable Equilibrium

This describes the condition where periods of oscillating changes in form or process rate (i.e. dynamic equilibrium) are separated by periods of abrupt change to new forms i.e. *states* (Fig. 5.3). This condition is associated with non-stationary boundary conditions, positive feedback mechanisms and the exceedence of threshold levels.

An example is the way in which shingle barrier beaches switch from stable forms being prone to *overtopping* to unstable forms affected by *roll-over* and *breaching* (see Chapter 34). The shingle barrier at Porlock, in North Devon, UK, failed during a storm in October 1996; this was the result of a progressive reduction in sediment supply to the beach i.e. beach volume gradually declined (dynamic equilibrium) until a critical threshold was reached below which breaching was inevitable[5.10] (Fig. 5.5).

Engineering Geomorphological Implications

- Earth surface systems are complex, sensitive and have different thresholds for change.
- Equilibrium does not necessarily imply stability over engineering timescales.
- It is tempting to view the landscape as being in steady state equilibrium with the prevailing conditions (e.g. the equilibrium beach[5.11]). However, climate and sea-level has been variable over engineering time and will continue to be variable in the future. It is unlikely that true equilibrium forms would have been able to develop under variable conditions.
- In some cases it is more realistic to assume that many forms are transient or unstable unless it can be demonstrated otherwise. Many rivers and shorelines appear to adjust their form more frequently and rapidly than had previously been appreciated both by geomorphologists and engineers.

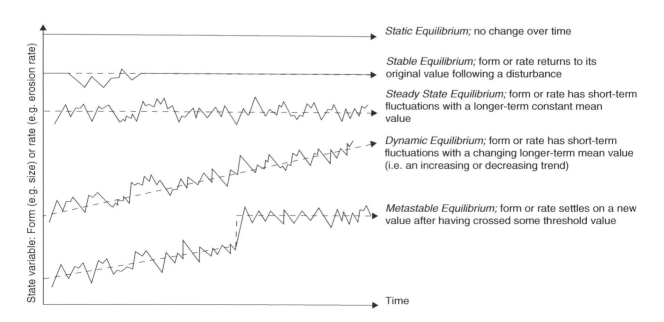

Figure 5.3 Types of system equilibrium (after Chorley and Kennedy 1971)[5.9].

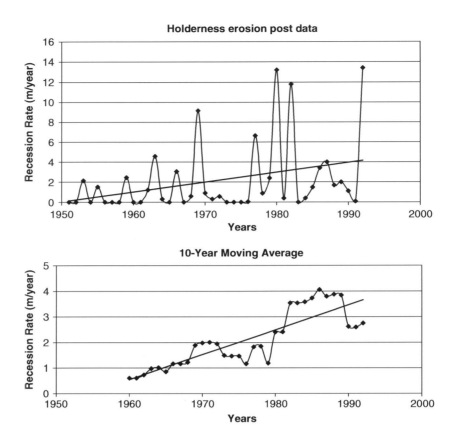

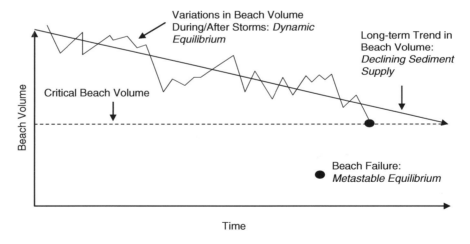

Figure 5.4   Erosion post data from the Holderness coast, UK.

Figure 5.5   Metastable equilibrium response of a shingle barrier beach to storm events and declining sediment supply (after Jennings and Orford 1999)[5.10].

• Where stable forms exist it is possible that they are relict forms, adjusted to suit previous climatic conditions. Although the forms may be stable under current conditions they might never have evolved under them (*passive disequilibrium*). However, human interference could lead to them becoming *reactivated* (e.g. the landslides that disrupted the construction of the Sevenoaks by-pass, UK)[5.12].

• Many systems contain evidence of past changes (e.g. debris flow and flood deposits, landslide events, stabilised gully channels; Fig. 5.6). This evidence can be used to reconstruct the processes that have operated at a site in the past and might be expected in the future. For example, flood magnitude can be estimated from slackwater flood deposits or the largest boulder sizes found along a stream channel (see Chapter 27).

Figure 5.6   Relict debris flow deposit in late Tertiary deposits from southeast Spain. Beds have been tilted to 80°. Base of flow is the fine-grained deposit on LHS of photograph. Width of section displayed is c. 2.5 m.

- Some of inherited features may be the product of very different environmental conditions to those that operate today, having survived many thousands of years.

References

5.1 Gleick, J. (1988) *Chaos: Making a New Science*. London, Cardinal.

5.2 Glade, T. (1998) Establishing the frequency and magnitude of landslide-triggering rainstorm events in New Zealand. *Environmental Geology*, **35**, 160–174.

5.3 Schumm, S. A. (1977) *The Fluvial System*. Wiley, New York.

5.4 Bjerrum, L., Loken, T., Heiberg, S. and Foster, R. (1969) A field study of factors responsible for quick clay slides. Seventh International Congress on *Soil Mechanics & Foundation Engineering*, **1**, 531–540.

5.5 Hooke, J. M., Harvey, A. M., Miller, S. Y. and Redmond, C. E. (1990) The chronology and stratigraphy of the alluvial terraces of the River Dane, Cheshire, NW England. *Earth Surface Processes and Landforms*, **15**, 717–737.

5.6 Phillips, J. D. (1992) Qualitative chaos in geomorphic systems, with an example from wetland response to sea level rise. *Journal of Geology*, **100**, 365–374.

5.7 Jones, D. K. C. and Lee, E. M. (1994) *Landsliding in Great Britain*. HMSO.

5.8 Cowell, P. J. and Thom, B. G. (1994) Morphodynamics of coastal evolution. In R. W. G. Carter and C. D. Woodroffe (eds.) *Coastal Evolution: Late Quaternary Shoreline Morphodynamics*. Cambridge University Press, 33–86.

5.9 Chorley, R. J. and Kennedy, B. A. (1971) *Physical Geography: a Systems Approach*. Prentice-Hall, London.

5.10 Jennings, S. C. and Orford, J. D. (1999) The Holocene inheritance embedded within contemporary coastal management problems. Proceedings of the 34th *MAFF Conference of River and Coastal Engineers*, **9.2.1–9.2.15**. MAFF, London.

5.11 Dean, R. G. (1991) Equilibrium beach profiles: characteristics and applications. *Journal of Coastal Research*, **7**, 53–84.

5.12 Skempton, A. W. and Weeks, A. G. (1976) The Quaternary history of the Lower Greensand escarpment and the Weald Clay vale near Sevenoaks. *Philosophical Transactions of the Royal Society*, London, **A283**, 493–526.

6

System Controls: Geology

Introduction

Ground materials comprise unweathered rock (*bedrock*), in-situ weathered rock (*saprolite*) or soils. Engineers consider any non-lithified materials overlying solid rock to be a soil (*regolith* or *overburden*; Fig. 6.1; see http://www.geolsoc.org.uk/). Regolith may comprise saprolite, alluvium, till, loess, dune sand, volcanic dust, etc.

Geological setting and materials exert a primary control on earth surface systems. The geological character (rocks and structure) determine:

- the broad-scale form of systems, relief and slope gradients
- the materials that are available to be eroded, transported and deposited by surface processes.

The rate and nature of landscape change is dependent on the mass characteristics of the rocks and soils (i.e. intact strength, discontinuities and solubility).

The nature of bedrock underlying an area is the product of its geological history i.e. the formation of the component rocks, along with the *diagentic* (post deposition changes, which includes compaction, lithification and cementation), *tectonic* and *weathering* disturbances it has been subject to. The extent to which the bedrock geology of an area is expressed in the landscape depends on whether the rocks are covered by regolith.

The enormous range of combinations of rock types, geological structure and denudation history means that, at a detailed level, every landscape and earth surface system will be unique. This inherent variability is also a source of deterministic uncertainty in system behaviour (see Chapter 5) i.e. we cannot expect to know *precisely* the materials within a system or the ground conditions at a particular site.

Rocks

The main rock types (*lithologies*) are:

- *Igneous*: originating from the earth's crust or *lithosphere*. *Intrusive* rocks have slowly solidified from magma bodies before they reach the surface

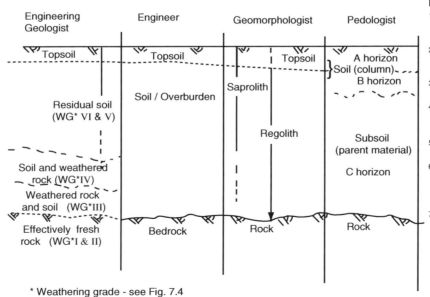

Figure 6.1 Delimitations of soil from different perspectives.

(e.g. the coarser crystalline rocks such as dolerite or deep-seated *plutonic* intrusions such as granite); *extrusive* rocks cool quickly from surface lava flows (e.g. finer crystalline rocks such as basalt), *pyroclastic* rocks form from ejected volcanic ash and debris (tephra).

- *Metamorphic*: rocks altered by the effects of high confining pressures and/or high temperatures. *Regional metamorphism* involves high temperature and pressure, typically associated with mountain chains along continental plate collision margins (e.g. the Himalayas). *Thermal* or *contact metamorphism* involves high temperature around igneous intrusions. The metamorphic rock type varies with the nature of the original rock and the temperature/pressure conditions to which they were subjected (e.g. clays can produce shales, slates, phyllites, schists and gneiss depending on the conditions). Metamorphic rocks are divided into two main divisions: *foliated/banded*, where the texture is layered and the minerals have a preferred orientation (e.g. schists) or are separated into distinctive bands of different composition (e.g. gneiss) and *non-foliated* (e.g. hornfels).

- *Sedimentary*: formed at or close to the earth surface from material derived from the weathering and erosion of pre-existing rocks (*clastic* sedimentary rocks, e.g. quartz sandstone), the hard parts of animals or plants (*organic* sedimentary rocks, e.g., shelly limestone, coal), or from the precipitation out of solution of dissolved minerals (*chemical* sedimentary rocks, e.g. rock salt, gypsum). A distinctive feature of most sedimentary rocks is their stratification or bedding[6.1].

Rocks of similar types occur in *suites* or *associations*[6.2]; examples of common rock-landform associations are presented in Figs 6.2 to 6.4.

In general, in a particular area the older rocks (Table 6.1) tend to be stronger, more lithified and with more complex structure than younger rocks. However, there are often major differences in the character of rocks of the same age in different tectonic settings (e.g. the contrast between rocks of the east and west coasts of the USA), due to the contrasting geological histories to which they have been subjected.

The mineral composition, fabric and porosity of the rock determine the mechanical strength and resistance to chemical weathering. Shale is mechanically weak, but resistant to chemical weathering; limestones are often strong, but readily soluble resulting in the formation of karst landscapes (see Chapter 21). The long-term differential weathering and erosion of different rocks is reflected in the various landscapes that have developed across the globe.

Rock structure

Stresses generated by plate tectonics can be accommodated by the extremely slow flow and deformation of rocks, or by fractures (*discontinuities* or *defects*). The rock structure is related to the regional/local geological stress histories; an understanding of this can help anticipate the local/site structural conditions[6.3].

Joints are minor fractures within a rock mass along which no movement has occurred, developed in nearly all rocks. The density and length of joints can be extremely variable. They often occur in distinct patterns related to the stress history of the region. Master joints can have a persistent or through-going form. Planar weaknesses develop independent of the bedding in foliated metamorphic rocks, causing *rock cleavage* and *schistocity*. Discontinuities and bedding planes tend to be the dominant control on rock slope stability. Geomechanical survey of the discontinuities along a rock slope is an important component of rockfall hazard assessment, especially the development of a potential fall frequency/volume distribution.

Faults are fractures along which movements have occurred[6.4] (Fig. 6.5). Common fault types are (Figs 6.6 and 6.7, see http://earthquake.usgs.gov/learning):

- *Strike-slip faults*: movement is parallel to the fault strike. Faults are either *right lateral* (dextral) or *left lateral* (sinistral) depending on the relative motion of the block on the opposite side of the fault. Active surface ruptures can cause a zone of ground cracking, bulging and tearing of near surface materials up to around 15 m wide.

- *Normal faults*: these show dip-slip displacement in which the hanging wall has moved downward relative to the footwall (Fig. 6.6). Ground rupture can occur along multiple fault splays across zones 5–10 km wide e.g. the 1983 Borah Peak earthquake, Idaho, USA (see: http://nsmp.wr.usgs.gov/GEOS/IDO/idaho.html).

- *Reverse faults*: these show dip-slip displacement in which the hanging wall block has moved up relative to the footwall block (Fig. 6.6). Fault traces typically form a sinuous, discontinuous trace across the ground surface. Earthquakes can form a relatively broad zone of permanent ground deformation up to 1 km wide. Secondary deformation such as backthrusts resulting from surface rupture on reverse faults may occur several kilometres from the main fault trace.

Sustained stress under high confining pressures can lead to a range of bedrock structures, from gentle tilting and doming to intense deformation and *folding* (Figs 6.7 and 6.8). Most folds originate at some depth; simpler forms include *monoclines*, *anticlines* (oldest beds in the centre), *synclines* (youngest beds in the centre) and *periclines* (dipping anticlines). *Recumbent folds* occur where rocks are overturned and are the product of horizontal compression; this may lead to shearing in the upper part of the fold along a *thrust fault*. Folds near the ground surface (exposed by erosion) can dominate the landscape with forms that mimic the fold pattern

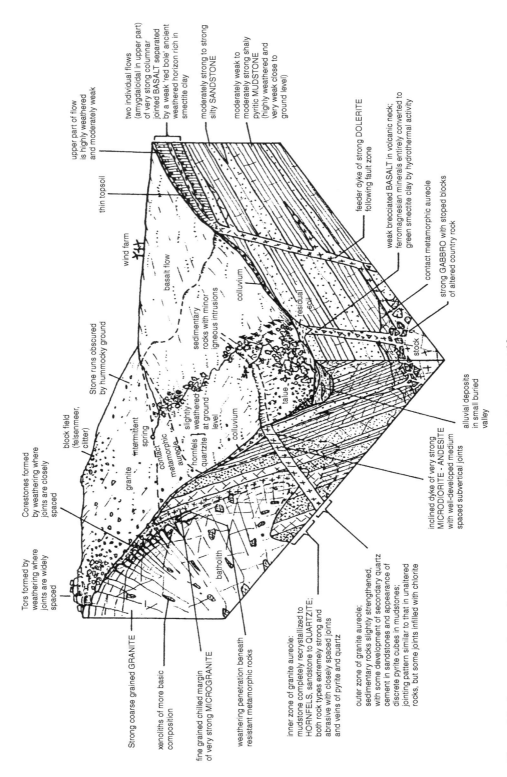

Figure 6.2 Igneous rock associations (wet temperate climate) (after Fookes 1997)[6.2].

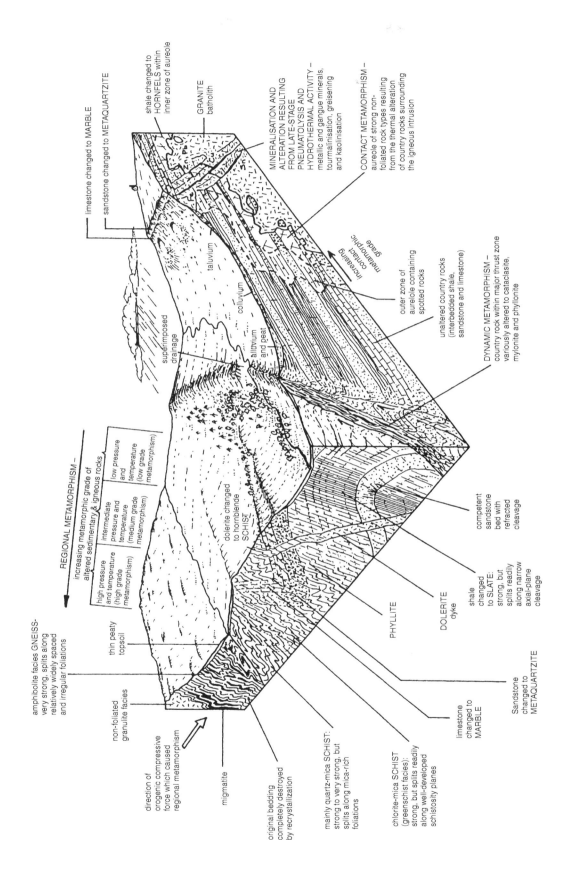

Figure 6.3 Metamorphic rock associations (wet temperate climate) (after Fookes 1997)[6.2].

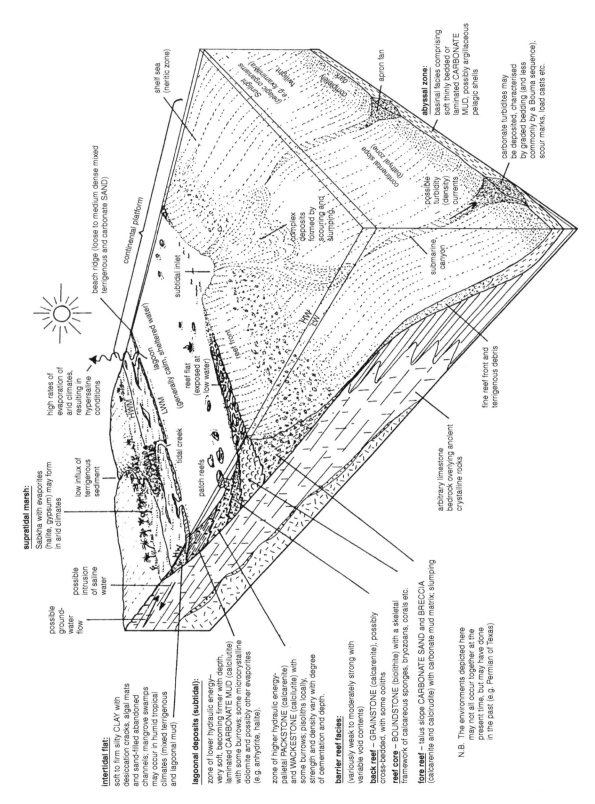

Figure 6.4 Tropical and sub-tropical carbonate depositional environments (after Fookes 1997)[6.2].

Table 6.1 Simple geological age (stratigraphic) column.

Era	Period	Age range (Ma)
CENOZOIC	Quaternary	2–0
	Tertiary	66–2
MESOZOIC	Cretaceous	144–66
	Jurassic	208–144
	Triassic	245–208
LATE PALAEOZOIC	Permian	286–245
	Carboniferous	360–286
	Devonian	408–360
EARLY PALAEOZOIC	Silurian	438–408
	Ordovician	495–438
	Cambrian	545–495
PROTEROZOIC	Precambrian	2500–545
ARCHAEAN		3800–2500
PRE-ARCHAEAN	No rock record	3800–4600

Figure 6.5 Fault/fissure created by tension developed at the North-American and European divergent constructive plate boundary exposed in Iceland.

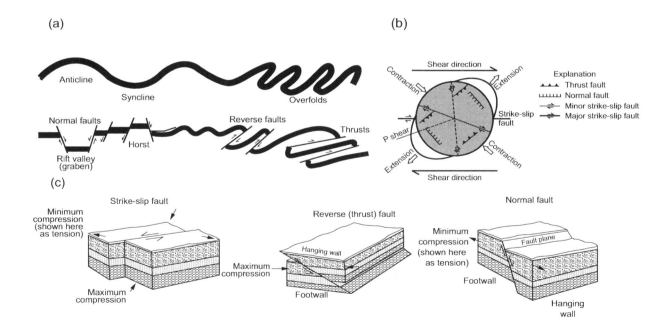

Figure 6.6 Geological structures: folding and faulting (a) types of fold; (b) faulting and (c) types of fault (after Hengesh and Lettis 2005)[6.4].

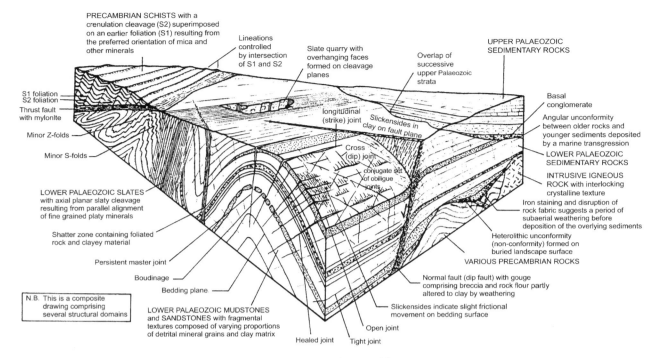

Figure 6.7   Examples of rock mass discontinuities (after Fookes 1997)[6.2].

(e.g. Zagros Mountains, Iran). Inverted relief occurs where long-term weathering and erosion has preferentially removed the more fractured fold units (e.g. in a syncline), leaving what had been the core as high ground.

Weathering

This is a combination of mechanical, chemical and biological processes that cause the breakdown of rocks[6.5–6] (Table 6.2):

- *Disintegration*: physical or mechanical weathering involving the breakdown of the rock into smaller fragments without any change in their chemical properties.
- *Decomposition*: chemical weathering involving a change in the chemical composition of the rocks or their component minerals.

The relative significance of weathering process depends on the environment (e.g. climate), the nature of the geological materials and biological conditions (Table 6.3).

Weathering normally results in the loss of rock strength and coherence (see Table 7.1 and Fig. 7.1); it 'prepares' debris for removal and transportation by mass

Figure 6.8   Chevron folds in Carboniferous rocks at Millook Haven, North Devon, UK (photograph courtesy of Dr M Anderson, University of Plymouth).

movement, water, ice, or wind (see Chapter 3). In some circumstances, chemical processes (e.g. precipitation of salts) can create localised stronger materials (case-hardening, rock varnish) or more extensive cemented weathering crusts (*duricrusts*; see Chapter 7), although in other situations salt crystal growth will cause a rock to disintegrate.

Table 6.2 Main weathering processes.

Principal processes of Disintegration	Principal processes of Decomposition
Crystallisation processes	Hydration and hydrolysis
• Salt weathering (crystal growth, hydration, differential thermal expansion)	
• Frost weathering	
Temperature/pressure change processes	Oxidation and reduction
• Insolation weathering	
• Sheeting, unloading	
Weathering by wetting and drying	Solution, carbonation, sulphation
• Moisture swelling	
• Alternate wetting and drying	
• Water-layer weathering	
Organic/biological processes	Chelation
• Root wedging	Bio-chemical changes
• Colloidal plucking	• Micro-organism decay
• Lichen activity	• Bacteria
	• Lichens

Table 6.3 Rock properties and resistance to weathering[6.6].

Rock properties	Physical weathering (disintegration)		Chemical weathering (decomposition)	
	Resistant	Non-resistant	Resistant	Non-resistant
Mineral Composition	High feldspar content	High quartz content	Uniform mineral composition	Mixes/variable mineral composition
	Calcium plagioclase	Sodium plagioclase	High silica content (quartz, stable feldspars)	High $CaCO_3$ content
	Low quartz content	Heterogeneous composition	Low metal ion content (Fe-Mg)	Low quartz content
	$CaCO_3$		Low biotite	High calcic plagioclase
	Homogeneous composition		High aluminium ion content	High olivine
				Unstable primary igneous minerals
Texture	Fine-grained	Coarse-grained	Fine-grained dense rock	Coarse-grained igneous
	Uniform texture	Variable texture	Uniform texture	Variable texture (porphyritic)
	Crystalline or tightly packed clastics	Schistose	Crystalline	Schistose
	Gneissic	Coarse-grained silicates	Clastics	
	Fine-grained silicates		Gneissic	
Porosity	Low porosity	High porosity	Large pore size	Small pore size
	Free-draining	Poorly draining	Low permeability	High permeability
	Low internal surface area	High internal surface area	Free-draining	Poorly draining
	Large pore diameter permitting free drainage after saturation	Small pore diameter hindering free drainage after saturation	Low internal surface area	High internal surface area

(continued)

Table 6.3 (*continued*).

Rock properties	Physical weathering (disintegration)		Chemical weathering (decomposition)	
	Resistant	**Non-resistant**	**Resistant**	**Non-resistant**
Bulk properties	Low absorption	High absorption	Low absorption	High absorption
	High strength, elasticity	Low strength	High compressive, tensile strength	Low strength
	Fresh rock	Partially weathered rock	Fresh rock	Partially weathered rock
	Hard	Soft	Hard	Soft
Structure	Minimal foliation	Foliated	Strongly cemented	Poorly cemented
	Clastics	Fractured, cracked	Dense grain packing	Calcareous cement
	Massive formations	Mixed soluble, insoluble mineral component	Siliceous cement	Thin-bedded
	Thick-bedded sediments	Thin-bedded sediments	Massive	Fractured, cracked
				Mixed soluble, insoluble mineral component
Representative Rocks	Fine-grained granites	Coarse-grained granites	Acidic igneous varieties	Basic igneous varieties
	Some limestones	Dolomites, marbles	Crystalline rocks	Limestones
	Dolerites (Diabases), gabbros	Many basalts	Rhyolites, granites	Marbles, dolomites
	Coarse-grained granites	Soft sedimentary rocks	Quartzite	Poorly cemented sandstones
	Rhyolites	Schists	Granitic gneisses	Slates
	Quartzites		Metamorphic rocks	Carbonates
	Strongly cemented sandstones			Schists
	Slates			
	Granitic gneisses			

References

6.1 Braithwaite C. J. R. (2005) Stratigraphy. In P. G. Fookes, E. M. Lee and G. Milligan (eds.) *Geomorphology for Engineers*. Whittles Publishing, Caithness, 107–123.

6.2 Fookes, P. G. (1997) First Glossop Lecture: Geology for engineers: the geological model, prediction and performance. *Quarterly Journal of Engineering Geology*, **30**, 290–424.

6.3 Waltham, T. (2005) Tectonics. In P. G. Fookes, E. M. Lee and G. Milligan (eds.) *Geomorphology for Engineers*. Whittles Publishing, Caithness, 85–106.

6.4 Hengesh, J. V. and Lettis, W. R. (2005) Active tectonic environments and seismic hazards. In P. G. Fookes, E. M.

Lee and G. Milligan (eds.) *Geomorphology for Engineers*. Whittles Publishing, Caithness, 218–262.

6.5 Lee, E. M., and Fookes, P. G. (2005) Climate and weathering. In P. G. Fookes, E. M. Lee and G. Milligan (eds.) *Geomorphology for Engineers*. Whittles Publishing, Caithness, 31–56.

6.6 Charman, J. and Lee, E. M. (2005) Mountain environments. In P. G. Fookes, E. M. Lee and G. Milligan (eds.) *Geomorphology for Engineers*. Whittles Publishing, Caithness, 501–534.

7

System Controls: Engineering Soils

Engineering Soil Types

Engineering soils (unlithified material; Fig. 7.1) can be described according to their *dominant particle size*, for example using the Unified Soil Classification System[7.1] or BS 5930[7.2].

Three main types of soil are recognised:

1. *Residual soils*, the product of in situ weathering of bedrock[7.3]. Soil thickness and type are broadly associated with climate and intensity of weathering (Figs 7.2 and 7.3). The thickness of the weathering profile reflects the relative balance between the rate of bedrock weathering and the rate of removal by soil erosion or landsliding.

   A weathering profile is created over the unweathered rock. Distinctive zones (horizons) can develop in response to variations in the intensity of weathering and movement of minerals. The upper layers will contain rock debris that has been completely weathered to a soil. Lower down in the profile there will be an increasing amount of unweathered or partly weathered rock. Figure 7.4 presents a common weathering grade model used by engineers; the base of Grade III is often regarded as the soil/rock boundary (rockhead – typically a highly irregular surface).

   Depths of weathering may exceed 100 m, especially in humid tropical environments and will usually be highly irregular (e.g. the Jos Plateau, Nigeria). Irregular profiles and variable weathering depths make for difficult foundation and excavation conditions. Relict discontinuities and faults are important in controlling engineering behaviour.

   Tropical residual soils usually exhibit distinctive engineering properties and characteristics[7.4–5]. Three main phases are recognised within weathering Grades IV to VI (saprolite and solum), forming a *continuum* of residual soil mineral development in tropical areas, ranging from *fersiallitic* to *ferruginous* and *ferrallitic* (Table 7.1).

Fresh rock (Grade I) strength decreases with increased weathering to Grade III (which may contain rock *corestones*).

Figure 7.1 Poorly stratified fluvial sediments, mainly cobbles and gravel, in an alluvial fan within a glaciated valley near Stavanger, Norway (photograph courtesy of Tony Waltham, Geophotos).

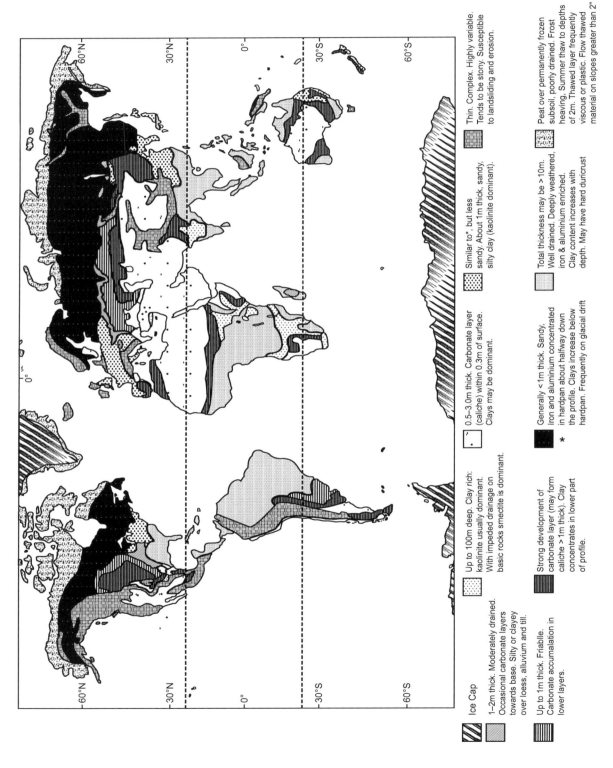

Figure 7.2 Simplified global soils map (after Fookes 1997)[7.3].

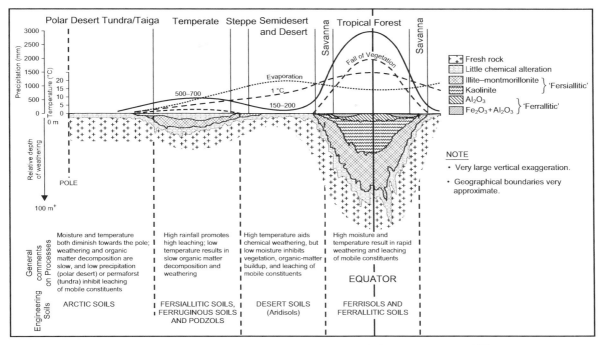

Figure 7.3   Climate zones, weathering characteristics and pedological soils (after Fookes 1997)[7.3].

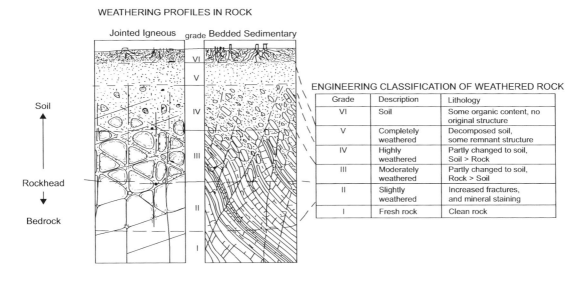

Figure 7.4   Examples of weathering zones (see Fookes 1997)[7.4].

2. *Organic soils*, formed *in situ* by the growth and decay of vegetation (e.g. peat). Peat is an accumulation of partly decomposed plant material (often Sphagnum moss). Around 15% of Ireland is covered by blanket peat (1.2 Mha). Peat forms where the ground is waterlogged i.e. in *anaerobic conditions*. This restricts microbial activity within the soil, slowing down the breakdown of plant litter. Peat formation is favoured in areas of high rainfall and low temperatures, which reduce evapotranspiration, stimulating further waterlogging.

3. *Transported soils*, the product of erosion of residual soils or bedrock, transport of this material and its deposition (simplified in Table 7.2).

The engineering behaviour of these soils and the associated implications for excavation and foundation conditions are the main concern of engineering geologists and geotechnical engineers (Table 7.3). Understanding of landform-soil associations can be important in developing models to support the design of ground investigations (see Chapters 37 and 38). Soil behaviour also exerts a major influence on the operation of earth surface systems, especially the rate and frequency of detachment

Table 7.1   Summary of tropical soil phases, location and climate[7.3, 7.5].

Soil phase	Mineralogy	Climate needed to reach the phase	Typical locations of the phase
1. Fersiallitic	Upper soils undergo decalcification and weathering of primary minerals. Quartz, alkali feldspars and muscovite not affected. Free iron usually >60% of total iron. Main clay mineral formed is 2:1 smectite; 1:1 kaolinite may appear in older well-drained surfaces. With recent volcanic ash porous andosol soils formed which are eventually replaced by 1:1 halloysites.	Cool Tropical • mean annual temperature 13–20°C • annual rainfall 0.5–1.0 m • dry season – yes	Mediterranean climates; higher altitudes in tropics.
2. Ferruginous (ferrisols-transitional)	More strongly weathered soils form but orthoclase and muscovite typically remain unaltered. Kaolinite is the dominant clay mineral; 2:1 minerals are subordinate and gibbsite usually absent. On older land surfaces and more permeable and base rich parent material, ferrisols transitional to phase 3. Partial alteration to gibbsite may occur.	Dry Tropical • mean annual temperature 20–25°C • annual rainfall 1.0–1.5 m • dry season – usually	Savannah
3. Ferrallitic	All primary minerals except quartz are weathered by hydrolysis and much of the silica and bases are removed by solution. Remaining silica combines to form kaolinite but usually excess aluminium gibbsite is formed. Depending on the balance between iron and aluminium, iron oxide or aluminium oxide will predominate. Soils take a long time to form and currently take 10^4 or more years.	Wet Tropical • mean annual temperature >25°C • annual rainfall >1.5 m • dry season – no	Can occur in modern savannah from previous wetter climate. Conversely, some currently hot wet areas are still only in the ferruginous phase (e.g. by climate change or by rejuvenation of slopes).

Table 7.2   Main transported soil types.

Soil Type	Formation	Nature of Deposit
Alluvium	Transport and deposition by rivers	Fine clay to coarse gravels. Coarse particles usually rounded. Soils usually sorted and often show pronounced stratification
Solifluction*	Slow downslope movement of water-logged soil material	Variable. Characteristic of cold regions, but can occur in tropics
Colluvium	Transport by mass movement and surface water flow (e.g. soil erosion)	Variable, from clay to gravel and boulders, but generally fine
Taluvium	Transport by mass movement i.e. landslide deposits and screes	Variable coarse grading
Glacial	Transport and deposition by ice	Tills** of various types and moraines, usually highly variable. Tills are often heavily over-consolidated
Fluvio-glacial	Transport and deposition by ice meltwater	Outwash materials, becoming finer with distance from the meltwater source. Fine material usually laminated and varved (glacial lake deposits)
Aeolian	Transport and deposition by wind	Usually silts (loess) and fine sands with uniform grading
Volcanic	Ash and pumice deposited during eruptions	Silt size particles with larger volcanic debris. Highly angular, often vesicular. Weathering generally produces highly plastic clays

  * There are a number of regional names in the UK for solifluction deposits e.g. *head* in south England; *coombe rock* in the chalklands of southeast England.
** Older UK geological maps may refer to till as *boulder clay*.

Table 7.3   Typical properties of engineering soils[7.14].

Soil type	Properties/characteristics	Applications
Gravels	Densities vary widely according to deposition. Strength high and compressibility low. Permeability variable, depending on grading and packing, often very high and can only be determined by pumping tests. Can be gap-graded with voids only partially filled with fines, which may then migrate if hydraulic gradient increased	Generally good foundation, full consolidation occurs during construction. Large flows into excavations. Generally good fill material; single size gravels are self-compacting when deposited in water. However, fills which are 'choked' with silt and clay matrix material can be difficult to compact when wet, and may not lose water readily under gravity drainage
Sands and silts	Properties usually improve with geological age. Densities vary widely according to deposition. Loose sands and silts very susceptible to liquefaction during earthquakes, can develop flow slides, large settlements when subject to vibration. Permeability moderate to high. Very erodible, piping risks when subject to internal water flow. Surface erosion by water flow and wind. Possibility of collapse of dry sands on wetting, particularly if weathered. Silts subject to frost heave	Generally good foundation, consolidation during construction. In excavations, beware base failures by piping, loss of soil through sheet pile clutches etc. Danger of soil and water inflows into tunnels, high abrasion of tunnelling machines. Generally good fill material, but dry single size sands and silts have poor trafficking characteristics. Silty soils prone to rapid deterioration and poor trafficking during wet weather, and may not lose water under drainage by gravity. Silty soils may 'bounce' when trafficked, creating difficulties in forming graded surfaces prior to laying road bases etc, and increasing fuel consumption of plant. If loose dumped, moist sands and silts may have low density, and collapse may occur on inundation
Alluvial clays (*general properties*) Note: There exist significant differences between over-, normally- and under-consolidated clays listed below	Properties vary widely with mineralogy and grading (proportion of silt and sand). Drained strength decreases and compressibility increases with increasing clay content and plasticity. In clayey materials with more than 30–40% platy clay minerals present, shearing produces discontinuities of low residual strength. Permeability low unless sand or silt layers are present. Strength likely to be anisotropic. Density and *in situ* strength depend on stress history due to burial, desiccation etc	For saturated clays, short-term undrained strength governs stability in loading cases (foundations). Long-term drained strength governs unloading cases (excavations). Little consolidation or swelling during construction except near drainage boundaries; post-construction consolidation and swelling occurs.
Normally-consolidated clays	Low undrained strength, high compressibility and secondary compression (creep), which increase with plasticity. Exposed surfaces usually overconsolidated by desiccation	Low allowable loading pressures under structures and embankments. Large post-construction settlements. Base heave and failure in strutted excavations, and high strut loads. Down-drag on piles. Low strength and difficult working conditions for plant during excavation
Over-consolidated clays	If effective stresses due to engineering work exceed pre-consolidation pressure, behaviour reverts to that of normally-consolidated clay; otherwise compressibility much less. Undrained strengths higher but difficult to predict. Permeability may be controlled by flow through fissures	Mass strength of foundations often affected by fissuring. Pre-existing shear surfaces, particularly in highly plastic clays, may control stability of slopes. Probable high *in situ* horizontal stress in heavily over-consolidated clays, large horizontal movements during and after excavation, high lateral stresses on buried structures
Under-consolidated clays	Excess pore pressures still present. Very low undrained strength and high compressibility relative to depth	Ground surface still settling. Stability of excavated slopes may be controlled by undrained strength

(continued)

Table 7.3 (continued)

Soil type	Properties/characteristics	Applications
Colluvial soils	Formation often involves shearing. Low strength shears may be present, often continuous and at base of soil, on slopes currently too flat for slope movement. Properties may differ from those measured at time of site investigation due to seasonal variations	Problems for embankment foundations on sidelong ground and in excavations
Talluvial materials	Landslides are often at limiting equilibrium, on major shear surfaces. Strata disturbed in massive landslides; open fissures and fracture zones formed. Soil more porous and wetter than parent soil. Perched water table effects	Excavation and filling likely to initiate new landslide movements. Screes may develop avalanche/flow slide behaviour when disturbed
Hot desert soils	Generally have little or no fines. Usually granular, uniformly graded and of low density when wind blown or coastal. Often coarse, well-graded and with angular particle shape when deposited by ephemeral spate flow in wadis or fans. Water table near to the surface leads to precipitation of evaporite salts.	Soil may be highly erodible once thin protective pavement removed or disturbed. Duricrusts and densely packed boulders (in wadis) may cause excavation difficulty. Aggregate may be in short supply or contaminated by salts. Problems of sediment movement by water in sudden storms and by wind. Deposites of wind-blown sand liable to collapse on wetting. Salty ground highly aggressive to structures and road pavements
Glacial soils	Often variable and heterogeneous, horizontally and vertically, with complex ground water conditions, including artesian. Density and strength of tills depend mainly on density of deposition, not on stress history. Grading curve may be almost straight line over wide range of particle size. Nature of fine matrix material and presence of clay minerals control properties. Dense materials may be very strong and stiff. Often contain local inclusions of water-deposited laminated sands silts and clays. Often have discontinuities as for alluvial clays	Problems with water-deposited soils as for alluvial soils. Tills generally good foundation material. Boulders cause problems in piling, tunnelling, excavation and filling. Drag structures at base of tills on weak rocks etc. cause errors in rock level estimation, problems with piles etc. Layers or lenses of high permeability may heavily influence water flows into excavations and stability of slopes in short term Problems with fills as for alluvial soils
Periglacial soils	Permafrost in active conditions, seasonal volume changes. In relic conditions, past ground freezing is likely to have produced extensive colluvium deposits on slopes. Ground contorted by freeze/thaw features e.g. cryoturbation, frost wedges, pipes. Parent soils and rocks will have been fractured, brecciated and uncemented by ground freezing. Such effects may be present generally or locally	Thaw stable coarse, thaw unstable fine materials. Construction may damage frozen environment; special techniques required. Effects of relict cambering and valley bulging may cause problems with foundations, excavations and tunnels
Organic soils	Highly compressible and subject to severe long-term creep. Often of very low unit weight. Methane gas may be present	Very large settlements of foundations and embankments underlain by organic soils, requiring lightweight fills or precompression by surcharging. Problems in slope stability due to low passive resistance. Non-saturated peat deposits may float when flooded. Wastage of peats when exposed and subject to drying. Usually difficult to run plant on and to handle as spoil. Acid soils, may be sulphates.

(continued)

Table 7.3 (continued)

Soil type	Properties/characteristics	Applications
Volcanic soils	Properties significantly different from those of sedimentary soils due to porosity and crushability of silt and sand size particles; *in situ* moisture content higher than usual; greater reduction in strength, but smaller reduction in compressibility, with increase in stress; no clear peak in compaction curve	Low particle density makes earthwork fills very susceptible to erosion. May soften with compaction, and easily damaged by earth-moving machinery, leading to loss of trafficability. Drying produces non-reversible improvement; addition of quicklime effective
	Fine soils often of high plasticity (with smectite, allophane), but strength higher and less dependent on plasticity than with sedimentary soils	
Residual soils	Wide range of grading, plasticity, mineralogy, etc., depending primarily on weathering processes. Grading often depends on unweathered quartz particles present. Re-cementation of soils may occur. 'Black' soils, usually formed with poor drainage have high plasticity, expansive soils with large volume changes on wetting and drying. 'Red' soils, usually formed with good drainage, but with pronounced structure due to weathering process. Engineering properties depend more on structure than on grading, mineralogy etc. Usually behave as if bonded, with structure yielding at a certain stress level. Strength and compressibility depend on this yield as much as on density. Mineralogy and properties can be changed by drying. Structure usually gives high *in situ* permeability. Porous soils with a high degree of saturation may be sensitive, giving low undrained strength on remoulding and destruction of structure, most likely in soils from volcanic rocks	Core stones cause problems in piling, can influence excavation methods in open cut. Relic low-strength discontinuities i.e. slope instability if of critical extent and angle. Erodible soils; severe gullying during heavy rain.
		Rapid consolidation during construction. Porous soils existing in dry conditions may collapse on wetting. Porous soils with high degree of saturation become weak with reworking and cause problems with operating plant during excavation, etc
		Strength due to structure lost in fills, then strength and compressibility depend on density achieved by compaction, which depends on water content at source. Compaction and loss of structure usually gives substantial reduction in permeability. General properties of fill are similar to those of alluvial clays of similar grading and mineralogy, but mineralogy often differs from that of sedimentary clays

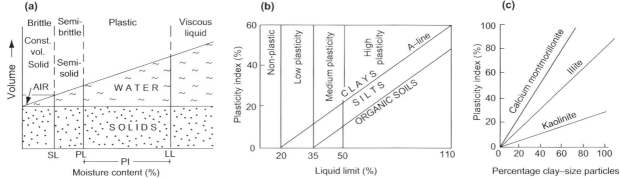

LL: Liquid Limit; PL: Plastic Limit; PI: Plasticity Index; SL: Shrinkage Limit

Figure 7.5 Atterberg Limits (in part after Skempton 1953)[7.6] (a) Three-phase system of soils and the influence of water-content on the Atterberg Limits and the Volume of a soil; (b) A simplified classification of fine-grained soils based on Atterberg Limits and (c) The influence of the amount and species of clay in a soil on its plasticity index.

Table 7.4 Clay minerals and their indicative properties.

Clay mineral	Occurrence	Liquid Limit (%)	Plastic Limit (%)	Activity	Residual shear strength ϕ'°_r	Cation Exchange Capacity (me/100g clay)
Smectites (e.g. bentonite and montmorillonite)	Widespread, but not too common	100–1300	50–100	Calcium montmorillonite: 1.5 Sodium montmorillonite: 6–13	4–10	80–150
Allophane	Volcanic areas	120–250	100–140	> 3	2–40	25–70
Illite	Common, esp. marine clays	60–120	35–60	0.5–0.9	c.10	10–40
Halloysite (hydrated)	Not common	50–70	47–60	0.1–0.4	25–40	40–50
Halloysite (dehydrated)	Not common	35–55	30–45	0.5	25–40	5–20
Chlorite	Not common	44–47	36–40	0.3–0.5	–	10–40
Kaolinite	Common	30–110	25–40	0.3–0.5	12–18	3–15

by surface erosion (see Chapters 17 and 18) and mass movement (see Chapter 19).

Engineering Soil Behaviour: Clays

The behaviour of clays (BS 5930 grain size <2 μm) will depend on their mineralogy and the grading. *Atterberg Limits* (Fig. 7.5) define the boundaries between *brittle, plastic* and *liquid* behaviour[7.6]:

- *Plastic limit* PL: the minimum water content at which a soil can be rolled into a 3 mm diameter cylinder, it marks the transition between brittle and plastic states.
- *Liquid limit* LL: the minimum water content at which soil flows under its own weight, it marks the transition between plastic and liquid states.
- *Plasticity index* PI = LL – PL: a measure of plasticity i.e. the property of non-returnable deformation in response to an applied force.
- *Liquidity index* LI = (w – PL) / PI: w is the weight of water (% of dry weight). It is a measure of consistency and strength. At the liquid limit, LI = 100%, at the plastic limit LI = 0%.
- *Activity* PI/%Clay: activity correlates with the cation exchange capacity and indicates the *reactiveness* of the clay (see Table 7.4 for common clay minerals and their indicative range of properties).
- *Undrained shear strength* s_u (where there is no drainage of pore water) is directly related to the liquidity index; s_u at PL (i.e. LI = 0) is always 100 times s_u at LL (i.e. LI = 1).
- *Drained shear strength* s decreases with increasing clay content and plasticity while the *compressibility* increases. For saturated clays the short-term *undrained strength* governs stability when loads are applied (e.g. placing fill), but when material is removed by erosion or excavation the long-term *drained strength* is applicable.
- *Permeability*: clays are low in permeability (the rate of water flow through a soil or rock),

with homogenous intact clays being practically impermeable ($<10^{-9}$ ms^{-1}).

Normally-consolidated clays: the current *effective stress* (see Chapter 3) is the maximum to which the soil has been subjected. These soils are water-rich; they have low undrained shear strength and are highly compressible. When remoulded, these soils display a reduction in their undrained shear strength (*sensitive* soils).

Under-consolidated clays: normally-consolidated clays that have yet to reach their fully consolidated state under the current *vertical stress* (see Chapter 3). These occur where there has been rapid deposition (e.g. in deltaic or estuarine environments) or where they have been recently loaded by new material. These clays exhibit high pore-water pressures (which will not have had time to dissipate), very low undrained shear strengths and high compressibilities.

Over-consolidated clays: the current *effective stress* is less than a previous maximum (the *pre-consolidation pressure*). Examples include older *geological clays* (*engineering clays* such as the London Clay or Oxford Clay, UK), which have been exposed after the removal of hundreds of metres of overburden by natural erosion processes and *glacial tills* deposited beneath thick ice. These clays are usually fissured, and often contain shear surfaces created by processes such as denudational unloading, tectonic movements or landsliding. The discontinuities lead to bulk permeabilities in the range 10^{-4}–10^{-9} ms^{-1}. Undrained and drained strengths are higher than normally-consolidated clays but are difficult to predict, and compressibilities are much lower. Over-consolidated clays are extremely susceptible to landsliding e.g. the Isle of Wight Undercliff, UK, where basal shear surfaces occur in the Cretaceous Gault Clay and thin clay beds within the Sandrock [7.7].

Laminated clays with sand and silt layers: usually associated with *lacustrine, lagoonal* (*varves* are annual lake bed accumulations, generally light grey in the summer and dark grey in the winter), *estuarine* or *near-shore* deposition. They are highly anistropic in strength, with mass behaviour determined by the most plastic layer present. Higher permeability layers (the sands and silts with permeabilities of 10^{-4}–10^{-5} ms^{-1}) can lead to more rapid consolidation or swelling in the horizontal direction, resulting in increasing drained shear strength over time. Movement of water on exposed faces is likely to lead to seepage erosion.

Quick clays are very sensitive marine clays that have been subject to leaching by percolating groundwater resulting in an open-texture soil skeleton that can collapse. Failure of a sensitive clay slope can result in the material remoulding to form a heavy liquid that flows out of the original slide area, often supporting rafts of intact clay. For example, the quick clay slide that occurred in 1971 at St Jean Vianney, Quebec, involved 6.9 Mm^3 of material, destroyed 40 homes and killed 31 people[7.8].

Engineering Soil Behaviour: Silts, Sands and Gravels

Sands (BS 5930 grain size 0.06–2 mm) and silts (grain size 2–60 μm): densities vary widely according to the environment of deposition, and generally the engineering properties improve with geological age. Intact permeabilities range from 10^{-5}–10^{-9} ms^{-1} in silts and fine sands (grain size 0.06–0.2 mm) to 10^{-5}–10^{-2} ms^{-1} in clean sands or sand and gravel mixtures. Because of the high permeabilities these materials tend to consolidate rapidly. The particles can be weakly bonded/cemented or interlock where angular particles dominate; both create *apparent cohesion* in an unconsolidated deposit. Loose sands and silts are very susceptible to liquefaction (see Chapter 19). Silts are very susceptible to natural piping and erosion by flowing water and wind (see Chapters 17 and 18), as well as frost heave.

Sandy soils (e.g. weathered granitic soils, volcanic deposits and sandy river terrace deposits) can be prone to dramatic large run-out landslides. Failure involves *sliding-surface liquefaction* and occurs when a shear surface develops in sandy soils and the grains are crushed or comminuted in the shear zone[7.9]. The resulting volume reduction causes excess pore pressure generation which continues until the effective stress becomes small enough that no further grain crushing occurs. This mechanism is common in earthquake-triggered debris slides and flow. For example, the 1995 Nikawa landslide in the Kobe area, triggered by the Hyogoken-Nanbu earthquake (M_L 7.2), destroyed 11 houses and buried 34 people[7.10].

Gravels (grain size 2–60 mm): densities vary according to the environment of deposition, but frictional strengths are generally high and compressibility is low. Clean gravels are effectively free-draining (permeability $> 10^{-5}$ ms^{-1}) but because of the large voids, fines can migrate in to pore spaces reducing permeability.

Loess is structureless windblown silt, common in the interiors of the northern continents. Much of it is derived by wind deflation from Pleistocene glacial outwash plains. Dry or moist loess will stand in a vertical face, but it is easily gullied and piped by running water. It can suffer *metastable collapse* when ground is wetted under its new load (e.g. a building), liquefaction and landsliding. In 1920 around 200,000 people were killed when a magnitude 8.5 earthquake triggered numerous loess landslides in China[7.11].

Engineering Soil Behaviour: Organic Soils

They are usually highly compressible, with a low unit weight and low shear strength. Methane gas may be present and acid sulphate conditions can exist (i.e. aggressive ground conditions). When dry they can be subject to wind erosion (see Chapter 18) and peat wastage. For example, in the Fens of UK, significant peat wastage has occurred because of land drainage works; at the Holme Post near Peterborough there has been around 4 m of surface lowering since 1850[7.12].

Duricrusts

Duricrusts are indurated horizons at or near the ground surface[7.13]. *Ferricretes* and *alcretes* form by the *relative accumulation* of iron and aluminium oxides in the soil as more mobile compounds are leached out of the weathering profile. They tend to be associated with hot, high rainfall climates.

Silica (*silcrete*), calcium carbonate (*calcrete*), magnesium carbonate (*dolocrete*) or gypsum (*gypcrust*) duricrusts form through their *accumulation* in the weathering profile. Accumulation may occur as a result of capillary rise (*per ascensum* model), downward percolation (*per descensum* model) or throughflow of solute-rich groundwater. These duricrusts are frequently associated with hot desert environments.

Duricrusts are generally stronger than the underlying materials (which may be porous due to leaching) and, hence, tend to armour the landscape by forming a hard cap to flat-topped hills and plateaux surfaces; the *cretes* can be > 2 m thick and massive, although joints may be widened by solution weathering; *crusts* tend to be thinner, weaker and are often transient features.

The presence of these indurated horizons may have significant implications for excavation operations. Often they are sources of construction stone and aggregates, although potential salt contamination and solution voids need to be considered.

References

7.1 NAVFAC (1986) *Soil Mechanics: Design Manual 7.01*. US Naval Facilities Engineering Command, Alexandria, Virginia.

7.2 British Standards Institute (BSI) (1999) *Code of Practice for Site Investigations*. BS 5930. HMSO. London.

7.3 Fookes, P. G. (1997) First Glossop Lecture: Geology for engineers: the geological model, prediction and performance. *Quarterly Journal of Engineering Geology*, **30**, 290–424.

7.4 Fookes, P. G. (ed.) (1997) *Tropical Residual Soils*. Geological Society Engineering Group Working Party Revised Report. The Geological Society, 184.

7.5 Duchaufour, P. (1982) *Pedology, Pedogenesis and Classification*. George Allen and Unwin, London.

7.6   Skempton, A. W. (1953) The colloidal activity of clays. Proceedings of the 3rd International Conference on *Soil Mechanics and Foundation Engineering*, **1**, 57.

7.7   Jones, D. K. C. and Lee, E. M. (1994) *Landsliding in Great Britain*. HMSO.

7.8   Tavenas, F., Chagnon, J. Y. and LaRochelle, P. (1971) The Saint-Jean Vianney landslide: observations and eyewitness accounts. *Canadian Geotechnical Journal*, **8**, 463–478.

7.9   Sassa, K. (1996) Prediction of earthquake induced landslides. In K. Senneset (ed.) *Landslides*. Balkema, Rotterdam, 115–132.

7.10 Sassa, K., Fukuoka, H. and Sakamoto, T. (1999) The rapid and disastrous Nikawa landslide. In K. Sassa (ed.) *Landslides of the World*. Kyoto University Press, 27–31.

7.11 Close, U. and McCormick, E. (1922) Where the mountains walked. *National Geographic Magazine*, **41**(5), 445–464.

7.12 Higgitt, D. and Lee, E. M. (eds.) (2001) *Geomorphological processes and landscape change: Britain in the last 1000 years*. Blackwell.

7.13 Goudie, A. S. (1973) *Duricrusts in Tropical and Subtropical Landscapes*. Oxford Research Series in Geography.

7.14 Milligan, G., Fookes, P. G. and Lee, E. M. (2005) Engineering behaviour of soils and rocks. In P. G. Fookes, E. M. Lee and G. Milligan (eds.) *Geomorphology for Engineers*. Whittles Publishing, 137–172.

8

System Controls: Mobile Sediments

Introduction

Sediments are mobilised by the surface processes of weathering and erosion before accumulating in temporary '*stores*' (e.g. channel bars, sand dunes) or long-term '*sinks*' (e.g. floodplain alluvium, sand seas). The availability and nature of mobile sediment has a major control on the evolution and behaviour of dynamic systems (e.g. rivers and coasts). In a river the kinetic energy (KE) of flowing water (see Chapter 2) is available to overcome the frictional resistance of the channel bed and banks, generate heat and transport sediment. If the sediment transport potential is greater than the amount of sediment available to be moved, the excess kinetic energy may be used in eroding the channel bed and banks.

The inputs (I) and outputs (O) from a channel reach with a particular discharge, slope, sediment input from upstream and bed sediment size can be described in terms of the excess/deficit in KE available for sediment transport:

- KE < sediment input $\Rightarrow$ net deposition (i.e. I > O)
- KE > sediment input $\Rightarrow$ net erosion of bed and banks (i.e. I < O)
- KE = sediment input $\Rightarrow$ net transport (i.e. I = O)

Similar responses occur on the coast (see Chapter 29) and in wind-dominated environments (see Chapter 18).

Relative sediment availability (i.e. sediment budget) determines the erosion/deposition status of a dynamic system:

- *Positive* sediment budget (I > O): net deposition e.g. channel and estuary infill, shoreline advance, dune formation, beach growth and cliff abandonment
- *Negative* sediment budget (I < O): net erosion e.g. channel incision, shoreline retreat, cliff recession

The *nature* of the sediment within a system influences the amount of sediment transport that occurs:

- *Friction losses* increase with the surface area/shape of the sediment over which the water is flowing i.e. KE available for sediment transport decreases with increased bed *roughness* (see Chapter 23).
- *Resistance to entrainment* varies with sediment size/mass i.e. the critical bed shear/velocity increases with particle size (see Chapter 3, Fig. 3.2).

- *Potential for deposition* varies with sediment size/mass i.e. the threshold velocity decreases with particle size (see Chapter 3, Fig. 3.2)
- The *erodibility* of the bed/bank/shoreline materials varies with particle size (see Chapter 17).

Important features of dynamic systems are:

- Variations in the availability and size of sediment inputs will result in morphological adjustments.
- The adjustments in morphology (i.e. erosion and deposition) can cause changes in the bed/bank/shoreline roughness and, hence, the KE available for sediment transport.
- Sediment transport provides a link between adjacent systems and sub-systems.

Sediment Sources

Common sediment types include:

- *Clastic sediments*, derived from the breakdown of rocks, and
- *Carbonate sediments*, derived from the skeletal remains of organisms or direct precipitation of calcium carbonates (e.g. coral reefs and carbonate rich lagoons).

River and coastal systems contain sediments derived primarily from the rocks and superficial deposits in their source areas (e.g. the hillslopes and channel banks within a catchment). The range of grain size and mineral composition depends, to a large extent, on the mechanical and chemical weathering processes operating in these source areas.

The most important sediment sources are *transported soils*, especially glacial deposits in mid-high latitudes and loess (see Chapter 7, Table 7.2). Quaternary ice sheets caused widespread erosion, remobilising unconsolidated materials and scouring bedrock (see Chapter 9). The eroded material was re-deposited in a variety of landforms (e.g. drumlins, moraines, eskers, and till sheets[8.1]; see Fig. 38.2). These features are often highly susceptible to erosion and therefore are a major source of sand, gravel and boulders for rivers and the shoreline. *Paraglacial systems* are those where the current processes and forms are still influenced by the availability of sediments released from glacial deposits[8.2].

Variations in Sediment Supply

The amount of sediment available for transport and deposition within a system does not remain constant over time. Important trends include:

- *Progressive decrease in availability* since the early-mid Holocene. For example, many mid-high latitude river and coastal systems developed in the early Holocene (see Chapter 9), in response to the abundant supply of glacially-derived sediment. Over time, however, the sediment sources have gradually been exhausted (removed by erosion) or stabilised by vegetation. In other areas (e.g. the coast of Northern Territory, Australia) seabed sources have gradually been depleted by the long-term movement of fine sediments to floodplain 'sinks' (e.g. the South Alligator River estuary)[8.3].

- *Rapid increase in sediment supply* over the last few centuries. Human activity (such as deforestation, mining operations, agricultural development) has led to a major influx of sediment from many catchments (e.g. the massive increase in sedimentation in the Fly River, Papua New Guinea, as a result of the development of the Ok Tedi gold and copper mine[8.4]; Fig. 8.1). The timing of this influx varies around the world, but has led to aggradation of rivers (i.e. channel instability), estuary infill, rapid delta growth and the formation of dunes and beach ridges.

The balance between these trends varies around the world. In the UK, for example, the influx of human-triggered sediments has declined because of soil conservation measures and river engineering works; current landform changes may simply be the adjustment to long-term sediment depletion.

Sediment Properties (fabric)

The key properties of clastic sediments include:

- *Grain size*: particle size distributions can be measured with callipers, sieving, hydrometer and pipette analysis. Geotechnical engineers describe soil grading according to Codes of Practice (e.g. BS 5930 in the UK[8.5], Table 8.1; ASI 1726 in Australia[8.6]).

River and coastal engineers often describe sediment size using the Udden-Wentworth scale (Table 8.2). Sediment sizes are usually converted to a logarithmic scale before statistical analysis – the phi ϕ grain size. Standard measures include the mean (x), standard deviation (sd) and skewness (sk), calculated from percentiles that can be derived from the particle size distribution curve (50% of particles are equal to or less than the $\phi50$ particle size):

$$x = \frac{\phi16 + \phi50 + \phi84}{3}$$

$$sd = \frac{\phi84 - \phi16}{4} + \frac{\phi95 - \phi5}{6.6}$$

$$sk = \frac{\phi16 + \phi84 - 2\phi50}{2(\phi84 - \phi16)} + \frac{\phi5 + \phi95 - 2\phi50}{2(\phi95 - \phi5)}$$

Mean size gives an indication of the average forces required to move the grains. For coarse sediments (> 0.2 mm) there is essentially a linear relationship between the velocity required for entrainment and transport and grain size (see Fig. 3.2).

The standard deviation (a measure of the variability around the mean value) is an indication of the range of the processes that produced the sediment. For example,

Figure 8.1 Ok Tedi river near Tabubil, Papua New Guinea. The river is filled with silt and clay tailings from the Ok Tedi gold and copper mine at Mount Fubilan resulting from the 'run-of-the-river' tailings mine waste disposal system that had been adopted (photograph taken in November 1989).

Table 8.1 Grain size according to BS 5930: 1999.

Particle size mm	Principle soil type	Particle shape	Composite soil types		
> 200	Boulders				
> 60	Cobbles	Angular	*Term **		*Approx %*
		Sub-angular	*Coarse soil*		*secondary*
		Sub-rounded			
> 20	Coarse gravel	Rounded	Slightly (sandy)		<5%
> 6	Medium gravel	Flat			5 to 20%
> 2	Fine gravel	Tabular	Very (sandy)		> 20%
> 0.6	Coarse sand	Elongated	SAND and GRAVEL		50% each
> 0.2	Medium sand				
> 0.06	Fine sand				
			Fine soil		
			Slightly (sandy)		< 35%
> 0.02	Coarse silt				
> 0.006	Medium silt		(sandy)		35 to 65%
> 0.002	Fine silt		Very (sandy)		> 65%
< 0.002	Clay				

* 'sandy' is an example and this system can also be used for 'gravelly', silty or clayey.

Table 8.2 The Udden-Wentworth sediment size classification[8.7].

Size mm	μm	φ*	Wentworth Size Class	Friedman Size Class
2048		−11	Cobbles	Very large boulders
1024		−10		Large boulders
512		−9		Medium boulders
256		−8		Small boulders
128		−7		Large cobbles
64		−6		Small cobbles
32		−5	Pebbles	Very coarse pebbles
16		−4		Coarse pebbles
8		−3		Medium pebbles
4		−2		Fine pebbles
2	2000	−1	Granules	Very fine pebbles
1	1000	0	Very coarse sand	Very coarse sand
0.5	500	1	Coarse sand	Coarse sand
0.25	250	2	Medium sand	Medium sand
0.125	125	3	Fine sand	Fine sand
0.063	63	4	Very fine sand	Very fine sand
0.031	31	5	Silt	Very coarse silt
0.016	16	6		Coarse silt
0.008	8	7		Fine silt
0.004	4	8	Clay	Very fine silt
0.002	2	9		Clay

Note: $\phi = -\log_2$ mm.

a large standard deviation can indicate little sorting of grains during transport or deposition (e.g. glacial tills). A well-sorted material with a narrow grain size distribution and small standard deviation is likely to be indicative of fluvial or aeolian activity.

Skewness (a measure of the separation between the modal class and the mean values) provides an indication of the history of the sediment. A positive skew indicates an excess of fine grains (e.g. addition of fines, such as wind blown material), a negative skew indicates selective removal of fine grains (e.g. beach sand).

• *Grain mass and density*: mass m is a measure of the inertia of a grain i.e. the resistance that it offers to having its position changed by an applied force:

$$m = w / g$$

Table 8.3 Settling velocity for various particle sizes.

Particle radius (m)	Particle density ρ_p (kg m^{-3})	Density of water ρ_w (kg m^{-3})	Dynamic viscosity (N s m^{-2})	Settling velocity (m s^{-1})
0.00006	1280	1000	0.0007	0.003
0.00012	1280	1000	0.0007	0.013
0.00024	1280	1000	0.0007	0.050
0.0005	1280	1000	0.0007	0.218
0.001	1280	1000	0.0007	0.872

Where w is the weight, which varies with the composition of the material, and g is the gravitational acceleration.

Density D is the mass per unit volume V:

$$m = V \times D$$

For spheriodal particles, the mass varies with the cube of the radius. Thus, 125 times more shear force is required to initiate the movement of a 10 mm diameter particle than a 2 mm diameter particle (diameter increases by 5, mass increases by 5^3).

- *Grain shape*: form can be described by measuring the long a, intermediate b and short c axes of the particles. Roundness can be determined from standard charts[8.8]. Formulae are available for calculating a range of detailed form and roundness indices[8.9–11].

- *Packing*: the tighter the packing arrangement the greater the shear stress required to entrain the grains. Grain shape will have an effect on how tightly packed a sediment can get. The degree of *sorting* is an important factor; poorly sorted sediments tend to pack together better than well sorted ones, since smaller grains can rest in the spaces between larger ones, confusingly such deposits are called *well-graded* in geotechnical engineering. Deposits will become more tightly packed over time as increasing overburden causes grains to realign themselves and water is extruded (*consolidation*).

Fine-grained Sediments

Fine-grained sediments (silts and clays) exhibit an *apparent cohesive strength* in addition to the frictional shear strength. This is believed to be a function of the attractive forces between particles, short-term cementation, or the development of negative pressures in the pores between particles. Hence silt and clay rich sediments have tended to be called *cohesive soils*. The effect of this short-term apparent cohesion is that, in fluid environments, the resistance to erosion of fine-grained particles increases progressively with decreasing grain size (Fig. 3.2); it requires the same water flow velocity (c 5 ms^{-1}) to erode a clay surface (0.001 mm size grains) as it does to move a 0.1 m size pebble.

Fine-grained sediments are transported in suspension; turbulent eddies in the flow prevent the grains from falling downwards i.e. the flow velocity must exceed the fall velocity (*settling velocity*) for the grains

Table 8.4 Settlement velocities of sediment grains (particle density = 2000 kg m^{-3}; fluid viscosity at 25° = 0.001).

Grain size millimetres	Microns (i.e. 10^{-6} m)	Terminal settling velocity (cm s^{-1})
2	2000	240
0.2	200	2.4
0.02	20	2.4×10^{-2}
0.002	2	2.4×10^{-4}
0.0002	0.2	2.4×10^{-6}

to be held in suspension. The settling velocity is expressed by Stokes' law:

$$v = (2 r^2 g \Delta\rho) / 9 \eta$$

Where v is the settling velocity (ms^{-1}), g is the gravitational acceleration, r is the radius of the particle (m), ($\Delta\rho$ is the difference in density between the particle ρ_p and fluid ρ_w (kg m^{-3}) and η is the dynamic viscosity of the fluid (Nsm^{-2}); Table 8.3. The settling velocity of fine-grained sediments is proportional to the square of the grain diameter i.e. small decreases in grain size results in dramatic decreases in settlement velocity.

In *salt water* clay particles have a tendency to stick together forming aggregates due to *flocculation*. These larger *flocs* enable clays to settle out in estuaries and inlets, otherwise the required settling velocities would be too low to allow significant deposition. From Table 8.4, a 0.002 mm grain settles out at 0.00024 cm s^{-1} i.e. it would take nearly 60 hours for the grain to fall 0.5 m onto a mudflat. As such low velocities only occur for a few minutes at low and high tides, negligible deposition would occur. However, a 0.5 mm floc has a settling velocity of 0.5 cm s^{-1} i.e. it can fall 0.5 m in 100 seconds.

References

8.1 Owen, L. A. and Derbyshire, E. (2005) Glacial environments. In P. G. Fookes, E. M. Lee and G. Milligan (eds.) *Geomorphology for Engineers*. Whittles Publishing, 345–375.

8.2 Church, M. and Ryder, J. M. (1972) Paraglacial sedimentation: a consideration of fluvial processes conditioned by glaciation. *Geological Society of America Bulletin*, **83**, 3059–3072.

8.3 Woodroffe, C. D., Thom, B. G. and Chappell, J. (1985) Development of widespread mangrove swamps in mid-Holocene times in northern Australia. *Nature*, **317**, 711–713.

8.4 Murray, L., Thompson, M., Voigt, K. and Jeffrey, J. (2000) Mine waste management at Ok Tedi mine, Papua New Guinea: a case study. *Tailings and Mine Waste '00*, Balkema, Rotterdam, 507–515.

8.5 British Standards Institute (BSI), 1999. *BS 5930 The Code of Practice for Site Investigations.* HMSO.

8.6 Standards Association of Australia, 1993. *AS 1726 Geotechnical Site Investigations.* Sydney, NSW.

8.7 Wentworth, C. K. (1922) A scale of grade and class terms for clastic sediments. *Journal of Geology*, **30**, 377–392.

8.8 Krumbeim, W. C. (1941) Measurement and geological significance of shape and roundness of sedimentary particles. *Journal of Sedimentary Petrology*, **11**, 64–72.

8.9 Wadell, H. (1933) Sphericity and roundness of rock particles. *Journal of Geology*, **41**, 310–331.

8.10 Powers, M. C. (1953) A new roundness scale for sedimentary particles. *Journal of Sedimentary Petrology*, **27**, 355–372.

8.11 Folk, R. L. (1955) Student operator error in determination of roundness, sphericity and grain size. *Journal of Sedimentary Petrology*, **25**, 297–301.

9

System Controls: Climate Variation

Introduction

Climate has varied throughout the history of the earth over geological timescales, primarily due to combination of variations in the orbit around the Sun and solar radiation (*Milankovitch cycles*) and to the slowly changing global distribution of land masses (i.e. plate tectonics). Examples of these variations are the repeated 'icehouse' glacial and 'greenhouse' interglacial episodes during the Quaternary Period. There were also periods in earth history, such as during the Cretaceous Period (144–66 Ma ago), when the planet was in a greenhouse cycle, with no ice sheets over the poles, higher global temperatures and high levels of carbon dioxide in the atmosphere.

In addition to the climatic changes that occur over geological time, there are important climatic variations that influence geomorphological processes over shorter periods, such as the Little Ice Age during the Middle Ages (around 1300–1900 AD).

The Quaternary Period

The legacy of the climatic instability over the last c. 2.5 Ma (the Quaternary period; Tables 9.1 and 9.2) can

Table 9.1 Glacial, simplified interglacial and postglacial stages.

General Description	Britain	NW Europe	Alps	North America
Holocene	Flandrian			
Last Glacial	Devensian	Weichselian	Würm	Wisconsin
Last Interglacial	Ipswichian	Eemian	Riss- Würm	Sangamon
Penultimate Glaciation	Wolstonian*	Saalian	Riss	Illinoian
Penultimate Interglacial	Hoxnian	Holsteinian	Pre-Riss	Yarmouth
Older glacials	Anglian	Elsterian	Mindel	Kansan
	Cromerian	Cromerian		Aftonian
	Beestonian	Menapian	Günz	Nebraskan

* The Wolstonian did not produce an ice cap in Britain.

Table 9.2 Quaternary stratigraphy in Great Britain.

Stage	Sub-stage	Stadial and Interstadials	Boundary age BP (Start date)
Flandrian			11 500
Devensian Glacial	Late Devensian	Loch Lomond Stadial	13 000
		Windermere Interstadial	15 000
		Dimlington Stadial	26 000
	Middle Devensian	Upton Warren Interstadial Complex	?
	Early Devensian	Brimpton Interstadial	?
		Chelford Interstadial	122 000
Ipswichian Interglacial			132 000
Wolstonian Glacial	Paviland Glaciation		352 000
Hoxnian Interglacial			428 000
Anglian Glacial			480 000
Cromerian Interglacial Complex			810 000

1. The transition between the Devensian Glacial and Flandrian marks the Pleistocene/Holocene boundary.

2. The transition between the Wolstonian Glacial and Ipswichian marks the Middle to Late Pleistocene boundary.

3. The beginning of the Cromerian may mark the Early to Middle Pleistocene boundary.

4. Dates may vary depending on the authority.

have a major influence on engineering projects. Key features are:

- *Marked global temperature fluctuations*, from values similar to present day during *interglacials*, to levels that were sufficiently cold to treble the volume of land ice during *glacial* periods. There have been at least 17 major glacial/interglacial cycles in the last 1.6 Ma (Fig. 9.1) based on O[16/18] ratios (see Chapter 10).
- *The build up of ice sheets* up to 4 km thick over mid to high latitudes, especially in the northern hemisphere[9.2] (Fig. 9.2). Beyond the ice limits, permafrost and periglacial conditions had a profound impact on slope stability (e.g. widespread landsliding and solifluction in the Northern Hemisphere). The last glacial advance had its peak around 18 000 BP, with deglaciation starting around 15 000 BP.
- *Marked fluctuations in global sea level* (see Chapter 10), which resulted in buried valleys below modern rivers; complex river terrace sequences along valleys; onshore relict cliffs and raised beaches; dead coral reefs and submarine canyons extending from continental shelves to the deep sea.
- Extensive covering of much of the land and sea bed surface by complex *till sequences* (i.e. morainic material, or *boulder clay*) both from valley and continental glaciers (Table 9.3; Fig. 38.2).
- Vast volumes of *granular glaciofluvial debris* issuing from the margins and snouts of glaciers, often covering the tills laid down by the glaciers themselves.
- *Formation of numerous lakes* of all sizes near the glaciated regions, often containing laminated clays and silts.
- Enormous volumes of silt carried away by the wind from valleys that drained the melting glaciers and deposited as *loess* (predominantly silt and fine sand) over vast areas of the Northern Hemisphere continents.
- *Abrupt changes between warm and cold periods*. For example, at the end of the last glaciation, the transition from the Younger Dryas (near-glacial wetter conditions around 13 000 to 11 500 BP) to the Holocene period (warmer conditions) is believed to have occurred in less than a decade[9.3]; in Greenland, temperatures rose 10°C in a decade[9.4].
- *Rapid retreat and decay of the ice sheets* during the interglacial periods, with a replacement of tundra landscapes by forest in the mid-latitudes.
- In low latitudes the growth and contraction of the mid to high latitude ice sheets generally corresponded with periods of greater moisture (*pluvials*) and periods of less moisture (*interpluvials*). Sand seas developed and advanced during dry periods.
- During the current interglacial period, the Holocene, there have been notable climatic changes superimposed on the overall glacial/interglacial cycle[9.5] (Table 9.4). For example, around 7000 BP the present day Sahara experienced a prolonged humid period (Fig. 9.3)[9.6].
- Over the last millennium, climate has continued to change (Fig. 9.4). In the UK a Medieval Warm Period (around 1100–1300 AD) was followed by a deterioration to a period of colder and stormier conditions known as the Little Ice Age (around 1300–1900 AD). Evidence suggests that these changes were related to sunspot activity – the 'Grand Maximum' and the 'Spörer' and the 'Maunder Minima', respectively[9.7].

The Thermohaline Circulation

The thermohaline circulation (THC; see http://www.ncdc.noaa.gov/paleo/abrupt/story3.html) is the ocean circulation in the Atlantic, driven by density (i.e. temperature and salinity) differences between different water bodies. The deep outflow of cold North Atlantic Deep Water is matched by a warm northward surface flow. This effectively transports heat into the North Atlantic; Europe is up to 8°C warmer than other regions at the same latitude, with the largest effect in winter.

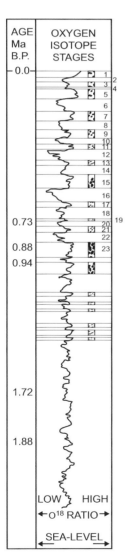

Figure 9.1 Climate change over the last 2 million years: the oxygen isotope record (after Summerfield 1991)[9.1].
Note: Lower O[16/18] ratios correspond with colder climate and low eustatic sea-levels.

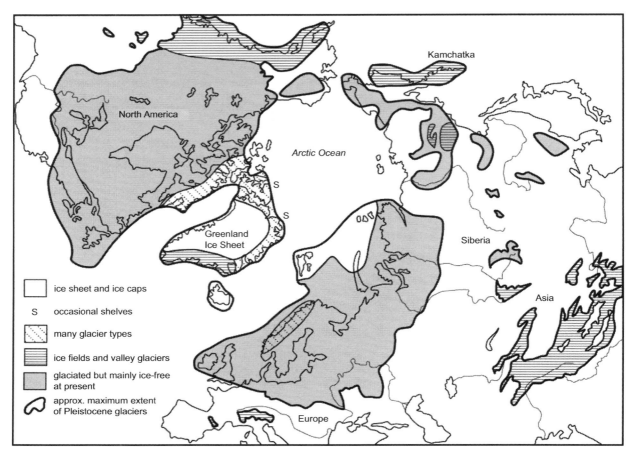

Figure 9.2 The former extent of northern hemisphere glaciated areas during the Pleistocene (modified from Denton and Hughes 1981)[9.2].

Table 9.3 Characteristics of glacial deposits.

Generic Type	Soil Types
Glacioterrestrial (tills or boulder clay)	Over-consolidated, low to medium plasticity clays, often containing sand pockets, varying gravel content, occasional cobbles and boulders (lowland areas).
	Gravel and cobble mixtures, with only a small proportion of fines (upland areas). Coarse material usually angular or sub-angular. The fines are often rock flour produced by the grinding of rock materials rather than genuine clay minerals.
Glaciofluvial (outwash deposits)	Sands and gravels of widely varying particle sizes; silt pockets frequent.
Glaciolacustrine (glacial lake clays)	Laminated silty clays.

Table 9.4 The climatic periods of the European Late Devensian and Holocene (the Blytt-Sernander scheme)[9.5].

Date (calendar years BP)	Period	Pollen Zone	Climate	Archaeological Period
18 000–15 000	Oldest Dryas	Ia	Cold	Later Upper Palaeolithic
15 000–13 500	Bølling Oscillation	Ib	Warm/wet	Later Upper Palaeolithic
	Older Dryas	Ic	Cold/dry	Later Upper Palaeolithic
	Allerød Oscilaltion	II	Warm/wet	Later Upper Palaeolithic
13 500–11 500	Younger Dryas	III	Cold	Later Upper Palaeolithic
11 500–10 500	Pre-Boreal	IV	Cool/dry	Early-Mid Mesolithic
10 500–7800	Boreal	V/VI	Warm/dry	Mesolithic
7800–5700	Atlantic	VIIa	Warm/wet	Neolithic and Bronze Age
5700–2600	Sub-Boreal	VIIb	Warm/dry	Bronze Age and Iron Age
2600–Present	Sub-Atlantic	VIII	Cool/wet	

Note: Pollen zones are a system of subdividing the Late Devensian and Holocene paleoclimate using the data from pollen cores.

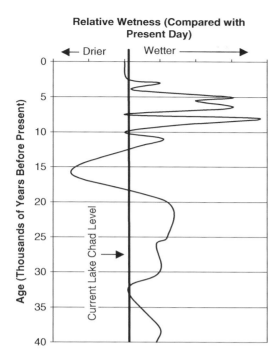

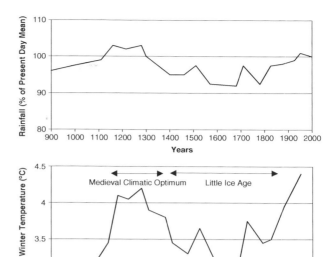

Figure 9.4 Climatic variability in England over the last Millennium (after Adams *et al.* 1991)[9.7].

Figure 9.3 Climatic changes in the Sahara over the last 40 000 years: variations in water levels in the Lake Chad basin compared to current levels (vertical line; after Servant and Servant-Vildary 1980)[9.6].

It is believed that the THC operates in 2 modes:
- A warm switched-on mode similar to the present-day Atlantic, and
- A cold switched-off mode with cold water forming south of Iceland. This appears be associated with major inputs of freshwater, either from surging glacial ice sheets (*Heinrich events*) or meltwater floods (e.g. *Younger Dryas event* around 13 000 BP when temperatures in Europe were 9°C colder than today).

The cold Younger Dryas event (the Loch Lomond Stadial in Great Britain) occurred due to changes in the flow of meltwater from the retreating Laurentide ice sheet on North America. During the initial stages of deglaciation the meltwater had flowed to the Gulf of Mexico via the Mississippi River; as the ice sheet retreated further it flowed out the St Lawrence waterway to the North Atlantic, freshening it sufficiently to halt the THC circulation[9.8].

A THC collapse due to global warming is now considered to be a low probability-high impact risk. More likely is a weakening of the THC by 20–50%, as indicated by many climate models[9.9].

The North Atlantic Oscillation

This involves the north-south shift (or vice versa) in storm and depression tracks across the North Atlantic Ocean and into Europe. The *NAO index* can vary from year to year, but tends to remain in one phase for several years.
- *Positive NAO index* (e.g. winter/springs of 1989, 1990, and 1995): a strong subtropical high pressure centre and a deep Icelandic low. The increased pressure difference results in stronger and more frequent winter storms crossing the Atlantic Ocean on a more northerly track i.e. warm and wet winters in Europe and cold and dry winters in northern Canada and Greenland. The eastern USA experiences mild and wet winter conditions.
- *Negative NAO index* (e.g. winter/springs of 1917, 1936, 1963, and 1969): a weak subtropical high and a weak Icelandic low. This results in fewer and weaker winter storms crossing the Atlantic on a more west-east pathway, bringing moist air into the Mediterranean and cold air to the UK and northern Europe. The USA east coast experiences more cold air and snow conditions.

El Niño Southern Oscillation

The El Niño Southern Oscillation (ENSO) is associated with strong fluctuations in ocean currents and surface temperatures within the Pacific Basin (see http://www.elnino.noaa.gov/). It causes abnormal atmospheric and environmental conditions, primarily in equatorial regions. There are two components:
- El Niño ('Christ Child'): associated with unusually warm ocean surface temperatures in the Equatorial region of the Pacific. During typical El Niño conditions, warmer sea surface temperatures spread further east. This coincides with a weakening of the atmospheric circulation (the *Walker circulation*, an east-west atmospheric circulation pattern – rising air above Indonesia and the western Pacific and sinking air above the eastern Pacific). It may cause lower rainfall over the western Pacific, and excessive rain on parts of Peru and Ecuador. The most intense El Niño of the 20th century occurred in the period 1997–98; it followed the longest recorded event, from 1991–1995[9.10].
- La Niña ('Little Girl'): associated with abnormal cold ocean surface temperatures in the Equatorial Pacific.

PDO Phase	North Pacific Sea Surface Pressure	North Pacific Sea Surface Temperature	Influence on El Niño Conditions	Influence on La Niña Conditions
Positive	Low	Cold	Enhance	Weaken
Negative	High	Warm	Weaken	Enhance

La Niña events are associated with cooler sea surface temperatures extending further west. Strengthening of the Walker circulation causes an increase in precipitation, particularly over Indonesia, and drier conditions are experienced over Peru and Ecuador.

The Southern Oscillation Index (SOI) is used to quantify the strength of an ENSO event. It is calculated from the difference between the sea level pressure at Tahiti and Darwin, Australia. The frequency of El Niño events has been increasing; since 1970 there have been 5 events (1972–73, 1982–83, 1986–88, 1991–95 and 1997–98); the same number occurred in the preceding 70 years[9.10].

The Pacific Decadal Oscillation

The Pacific Decadal Oscillation (PDO see http://tao.atmos.washington.edu/pdo/) is an oscillation in northern Pacific sea surface temperatures, between normal and below-normal conditions. The pattern operates on a 20–30 year time scale. Shifts in the PDO regime occurred in 1925, 1947, 1977 and, possibly, 1995[9.11]. PDO phases can combine with El Niño/La Niña conditions to affect climate, particularly in winter.

Volcanic Eruptions

Explosive volcanic eruptions can have a major influence on global and regional climate[9.12]. Particles are ejected into the lower stratosphere and spread to form a veil over the whole planet, reducing the amount of solar energy that reaches the earth's surface (Fig. 9.5). An individual eruption may generate global cooling of 0.2–0.3°C for around 1–2 years. For example, the explosion of Santorini (Thera) in the Eastern Mediterranean around 3628 BP led to a period of cooler, wetter climate. After Tambora, Indonesia, exploded in 1815 the following year became known as the "the year without a summer" in many parts of the Northern Hemisphere (see http://vulcan.wr.usgs.gov/Volcanoes/Indonesia/).

Major eruptions in lower latitudes have the greatest impact; particles can reach the higher latitudes of both hemispheres, because of the nature of the atmospheric circulation. Material from major eruptions in the mid to high latitudes of each hemisphere tends to remain poleward of the eruption latitude. Major Icelandic or Alaskan/Aleutian/Kamchatkan eruptions

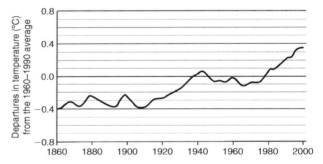

Figure 9.6 Combined annual global land surface air and sea surface temperature anomalies (°C) 1861–2000, relative to 1960–90 overall average (after IPCC 2001)[9.13].

Figure 9.5 Mount Pinatubo erupting in June 1991 (image courtesy of the United States Geological Survey's Cascades Volcano Observatory).

only influence the higher latitudes of the Northern Hemisphere.

Evidence for Future Climate Change

Global climatic changes are occurring as the result of human-induced accumulation of greenhouse gases such as carbon dioxide (CO_2) in the atmosphere (see http://www.ipcc.ch/). Geochemical evidence from ice cores, supplemented by direct measurements since the mid-1950s, reveals a steady rise in greenhouse gas concentrations from the late 1700s changing to a rapid rise post 1950. Atmospheric concentrations of CO_2, the main man-related greenhouse gas, have risen from about 270 ppm in pre-industrial times to over 360 ppm (the current concentrations have probably not been exceeded in the last 20 Ma) [9.13].

Global temperature has risen by about 0.6 ± 0.2°C since the beginning of the 20th century, with about 0.4°C of this warming occurring since the 1970s (Fig. 9.6). In the 20th century precipitation increased by 0.5–1% per decade over the Northern Hemisphere continents; there was also a 2–4% increase in the frequency of heavy precipitation events. Warm El Niño episodes have become more frequent, persistent and intense since the 1970s.

Climate Change Predictions

The Inter-governmental Panel on Climate Change (IPCC) was established by the World Meteorological Organisation (WMO) and the United Nations Environment Programme (UNEP) in 1988. Climate predictions have been made in a series of Assessment Reports; the most recent (the Third Assessment Report) was published in 2001[9.13].

The Third Assessment Report results provide an indication of the scale of changes that could be expected by 2100[9.13]:

- A mean global surface temperature rise of 1.4–5.8°C is projected. Note that warming at the higher end of this range would shift climatic zones towards the poles by about 550 km.
- Glaciers and ice caps will continue to retreat (they have retreated throughout the 20th century) and Northern Hemisphere snow cover and sea ice will

decrease. There remains the possibility of a collapse of the West Antarctic Ice Sheet; this could cause sea-level to rise rapidly by around 5–6 m (see Chapter 10).

- At latitudes of 45° or greater (i.e. northern Europe, Russia, China, northern and central USA, Canada and the southern extremes of South America) annual precipitation will increase by 100–300 mm.
- In the lower latitudes of 5–45° (i.e. Australia, southern Africa, southern USA, western South America, Central America, the Caribbean, north Africa, the Mediterranean region, the Middle East and India) annual precipitation will decrease by 100–700 mm. It is also likely that the strength and duration of the Asian summer monsoon will become more variable.
- Around the Equator annual precipitation changes are expected to be complex, with a decrease of 100–600 mm predicted for the Americas and Southeast Asia, but an increase of 100–300 mm expected in central Africa.
- Global mean sea-levels are expected to rise by around 0.4 m, in response to thermal expansion of sea water (see Chapter 10).

It is not just the climate that could change, but also the weather. It is expected that the frequency of climate extremes (droughts, hurricanes, intense rainstorms, periods of extreme heat and cold) will increase.

Engineering Geomorphological Implications

The predicted changes in climate and weather are expected to lead to significant changes in the behaviour of many earth surface systems i.e. changing patterns of landsliding, erosion, flooding, sediment transport and deposition. However, the response to climate change is likely to be complex (see Chapter 5).

In southern England, for example, over the next century there could be up to a 25% increase in mean monthly effective rainfall (see http://www.ukcip.org.uk/)[9.14]. Figure 9.7 presents the return period of different monthly *effective rainfalls* totals (i.e. the precipitation minus evapotranspiration) for the Isle of Wight Undercliff (a large pre-existing landslide complex)[9.15]. The return period (annual probability) of a monthly effective rainfall total of 150 mm (sufficient to trigger landslide reactivation) is predicted to change from the current 1:200 years to 1:40 years, under a *high emissions* scenario. This change is expected to cause an increase in the frequency of landslide activity in the Undercliff.

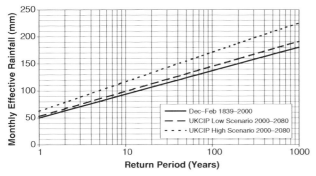

Figure 9.7 Predicted changes to winter effective rainfall in the Isle of Wight Undercliff, UK (after Halcrow 2001)[9.15].

References

9.1 Summerfield, M. A. (1991) *Global Geomorphology.* Longman Group Ltd., Harlow.

9.2 Denton, G. H. and Hughes, T. J. (eds.) (1981) *The Last Great Ice Sheets.* Wiley, New York.

9.3 Alley, R. B., Meese, D. A., Shunam, C. A., Gow, A. J., Taylor, K. C., Grootes, P. M., White, J. W. C., Ram, M., Waddington, E. D., Mayewski, P. A. and Zielinski, G. A. (1993) Abrupt increase in Greenland snow accumulation at the end of the Younger Dryas event. *Nature*, **362**, 527–529.

9.4   Cuffey, K. M. and Clow, G. D. (1997) Temperature, accumulation, and ice sheet elevation in central Greenland through the last deglacial transition. *Journal of Geophysical Research*, **102**, 26383–26396.

9.5   Roberts, N. (1998) *The Holocene*. 2nd edition, Blackwell, Oxford.

9.6   Servant, M. and Servant-Vildary, S. (1980) L'environnement Quaternaire du Bassin du Tchad. In M. A. J. Williams and H. Faure (eds.) *The Sahara and the Nile*. Balkema, Rotterdam, 133–162.

9.7   Adams, J., Maslin, M. and Thomas, E. (1999) Sudden climatic transitions during the Quaternary. *Progress in Physical Geography*, **23**, 1–36.

9.8   Clark, P. U., Marshall, S. J., Clarke, G. K. C., Hostetler, S. W., Licciardi, J. M. and Teller, J. T. (2001) Freshwater forcing of abrupt climate change during the last glaciation. *Science*, **293**, 283–287.

9.9   Stocker, T. F. and Schmittner, A. (1997) Influence of CO2 emission rates on the stability of the thermohaline circulation. *Nature*, **388**, 862–865.

9.10 WMO (1999) *WMO Statement on the Status of the Global Climate in 1998*. WMO-No. 896, World Meteorological Organization, Geneva.

9.11 Minobe, S. (1997) A 50–70 year climatic oscillation over the North Pacific and North America. *Geophysical Research Letters*, **24**, 683–686.

9.12 Kelly, P. M., Jones, P. D. and Jia Pengqun (1996) The spatial response of the climate system to explosive volcanic eruptions. *International Journal of Climatology*, **16**(5), 537–550.

9.13 Inter-Governmental Panel on Climate Change (IPCC) (2001) The IPCC third assessment report. *Summary for Policy Makers*. http://www.ipcc.ch/

9.14 Hulme, M., Barrow, E. M., Jenkins, G. J., New, M., Osborn, T. J. and Viner, D. (1998) *Climate change scenarios for the UK Climate Impacts programme*. UKCIP Technical Report. Norwich: Climatic Research Unit. http://www.ukcip.org.uk/

9.15 Halcrow Group Ltd. (2001) *Preparing for the Impacts of Climate Change*. Report to the Standing Conference on Problems Associated with the Coast (SCOPAC), Ventnor, Isle of Wight. http://www.scopac.org.uk/publications.html.

10

System Controls: Sea-level Change

Introduction

Sea-level provides a datum for all coastal processes and determines the *base level* (i.e. the limit on vertical erosion) to which most earth surface systems operate (base level in internal basins is the lowest elevation in that basin e.g. 133 m below sea-level in the Qattara Depression, Egypt). Changes in sea-level alter the position of the shoreline and the potential energy available in the landscape (see Chapter 2).

Sea-level has been extremely variable in the past (Fig. 10.1)[10.1]. In the last 20 000 years there have been changes in excess of 100 m (Fig. 10.2). Such sea-level changes can involve:

- Changes in the absolute volume of seawater (*eustatic changes*). These are worldwide because of the interconnections between the oceans and are associated with the growth and decay of land based ice sheets during the Quaternary.
- Changes in the absolute land level due to *tectonic activity* (e.g. uplift and subsidence) or as a result of *isostatic adjustments* to the loads imposed on the land surface (e.g. ice or sedimentation). These tend to be local/regional effects.

Relative sea level change is the change in the level of the sea relative to the land, taking account of both eustatic and tectonic/isostatic changes. Eustatic sea-level rise or land subsidence produces a *positive* change, whereas eustatic sea-level fall or uplift results in a *negative* change.

Mean sea-level is the average elevation of the water surface, taking account of short-term variations in tides but excluding waves and atmospheric effects. Mean sea-level on a coast is normally related to a fixed datum e.g. Ordnance Datum (UK) or American Sea Level Datum. There is no globally accepted base datum for sea level, which causes problems during the construction of tunnels and bridges between some countries.

The movement of the shoreline in response to sea-level change can be:

- *transgressive*: movement up the shore profile i.e. inland or
- *regressive*: movement down the shore profile i.e. seawards, exposing the former sea floor.

Quaternary Sea-level Changes

The record of global sea-level changes can be established from:

1. *Oxygen isotope records* in deep marine sediments: the lighter oxygen isotope O^{16} is preferentially evaporated compared with the heavier O^{18}, resulting in

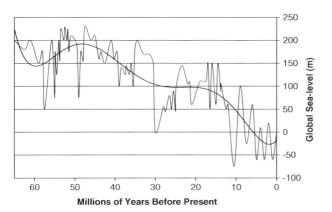

Figure 10.1 Global eustatic sea-level curve since the Cretaceous (after Haq *et al*. 1987)[10.1].

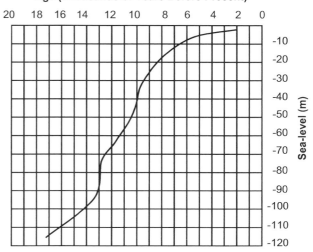

Figure 10.2 General eustatic sea-level curve for Barbados (after Fairbanks 1989)[10.3].

O^{16} rich precipitation. This leads to markedly different isotope compositions for ice sheets and ocean water. As ice builds up during glacial periods, large quantities of O^{16} become "trapped" in the ice, leaving the oceans enriched in O^{18}. The shells of marine organisms reflect the isotopic composition of the water at the time that they were formed; a 0.01% increase in the O^{18} concentration indicates around 10 m decrease in global sea-level. Deep sea ocean cores have revealed regular climatic fluctuations during the last 2.5–3 ka, with glacial periods of around 80–100 Ma length alternating with warmer interglacial periods lasting 10–20 ka (Fig. 9.1). High sea-levels caused by major deglaciation events are associated with oxygen isotope stages 1, 5, 7, 9, 11, 13, 15, 17 and 19 (stage 3 was a partial deglaciation event with moderately high sea-levels)[10.2].

2. *Ancient shoreline elevation*: uplifted coral terraces (e.g. in Barbados and the Huon Peninsula, Papua New Guinea) represent the peaks of each phase of sea-level rise over the last 250 000 years[10.3–4]. Sea-level calculations for these sites are based on the assumption that uplift rates have remained constant.

Previous high eustatic sea-levels during the Quaternary were similar to present day levels (Fig. 10.3).

Post-glacial and Holocene Sea-level Changes

Sea level has risen by over 100 m during the past 18 000–15 000 years, i.e. since the end of the last period of significant ice advance. Evidence of the low-stand position comes from oxygen isotope records, shallow and inter-tidal water mollusc shells. Local sea-level curves for the Holocene vary because of the effects of isostatic recovery and site specific effects e.g. the size, shape and tidal range of an estuary or bay change with sea-level[10.5] (Figs 10.4 and 10.5).

The effects of *ice sheet loading* on the earth's crust include (Fig. 10.6):
- A *depressed zone*, with the maximum depth at the centre of the ice sheet and a gradual rise towards the margins.
- A *forebulge zone* (i.e. an uplifted zone), developed to compensate for the crustal depression beneath the ice sheet.

Isostatic recovery has involved:
- *Restrained rebound*: the adjustment that occurs at a site *during* deglaciation. The retreat of a decaying ice results in the migration of the limits of the depressed zone and the forebulge location i.e. the isostatic changes at a particular point can be very complicated.
- *Postglacial rebound*: the gradual adjustment that continues *after* deglaciation, as found in Scandinavia today.
- *Hydro-isostacy*: the depression of the ocean floor crust under the weight of the additional water. This accounts for the sea-level fall of c. 2–3 m since 4 000 BP in the tectonically stable regions around Australia[10.6].

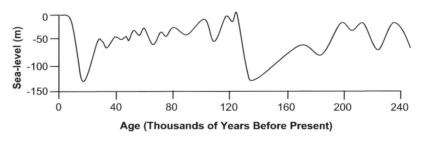

Figure 10.3   Eustatic sea-level curve for the Huon Peninsula, Papua New Guinea (after Aharon 1983)[10.4].

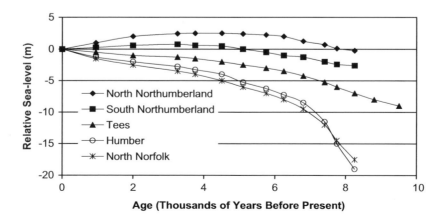

Figure 10.4   Relative sea-level curves for eastern England (after Shennan *et al.* 2000)[10.5].

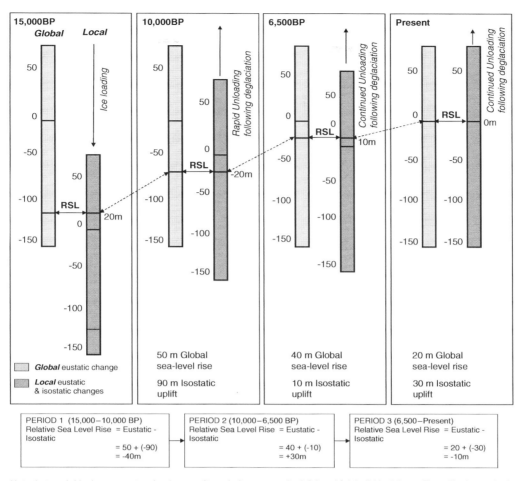

Note that each block represents a land mass. At each time stage, the left hand (global) block is unaffected by isostatic change, whereas the right hand (local) block is affected by both eustatic and isostatic changes.

Figure 10.5 An illustration of the combined effects of eustatic and isostatic sea-level changes.

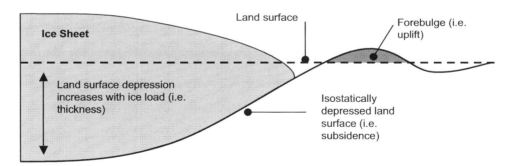

Figure 10.6 Effect of an ice sheet on the land surface (cross section).

Post-glacial sea-level rise involved:

• Rapid rise during the period 15 000 to around 8 000 BP, at around 12.5 mm yr^{-1}. This rise was associated with a combination of the eustatic changes caused by the melting of ice sheets and tectonic/isostatic effects.

• Variable rise since around 8 000 BP largely associated with local tectonic/isostatic changes (including hydro-isostacy) i.e. sea-level change has involved only a relatively minor eustatic component. Variations are shown in Table 10.1 which

reflects different isostatic histories, from 70 m of uplift at Oslo to 20 m subsidence in Delaware, USA[10.7].

Recent Sea-level Changes

Sea-level has risen by between 10–25 cm over the past 100 years, due to the thermal expansion of oceans and the melting of mountain glaciers and the Greenland ice sheet[10.8]. Table 10.2 provides evidence from tide gauges around the UK[10.9]. Tide gauge records from Sydney (1915–1998) and Fremantle, Australia (1897–1998),

Table 10.1   Holocene sea-level history for selected sites, since 8 000 BP (in metres)[10.7].

Year	Delaware, USA	SW England	Caribbean Islands	Panama	Clinton, USA	Jaeren, Norway	Ellesmere, Canada	NE Greenland	Oslo, Norway
1 000 BP	−1	0	−1	−2	−1	0	1	2	4
2 000 BP	−3.5	−0.8	−2	−2	−1.7	1	2	4	7
3 000 BP	−5	−1.3	−2.5	−5	−2.7	1	2	8	10
4 000 BP	−8	−1.9	−3.5	−8	−3.6	3	2	13	16
5 000 BP	−11	−3.4	−5.3	−12	−5	6	3	18	26
6 000 BP	−14.5	−5.6	−7.7	−12	−6.2	3	4	26	32
7 000 BP	−17	−11	−11.5	−13	−7.3	7	12	47	46
8 000 BP	−20	−17.2	−15.5	−15	−8.5	4	34	57	67

Table 10.2   Mean sea-level trends around the UK[10.9].

Tide gauge station	Period	Measured sea-level trend and standard error (mm year⁻¹)	Acceleration in mean sea-level (mm year⁻¹ century⁻¹)	Isostatic component (mm year⁻¹)*
Aberdeen	1901–96	0.57 ± 0.07	0.44 ± 0.20	0.47 ± 0.06
North Shields	1901–96	1.85 ± 0.12		-0.08 ± 0.17
Sheerness	1901–96	1.57 ± 0.09	0.84 ± 0.17	-1.11 ± 0.38
Newlyn	1916–96	1.68 ± 0.12		-1.41 ± 0.10
Liverpool	1916–96	1.23 ± 0.12	0.82 ± 0.36	-0.018 ± 0.04

* Positive implies emergence; negative implies subsidence.

Table 10.3   ENSO events and sea-level in the western Pacific Ocean[10.12].

ENSO condition	Impact on sea-level	Recorded years
Strong El Niño years	Strong sea-level fall	1951, 1958, 1972, 1982 and 1997
Moderate El Niño years	Moderate sea-level fall	1963, 1965, 1969, 1974 and 1987
Strong La Niña years	Strong sea-level rise	1964, 1973, 1975, 1988 and 1998
Moderate La Niña years	Moderate sea-level rise	1956, 1970, 1971, 1984 and 1999

show a rise of 0.86 ± 0.12 mm yr⁻¹ and 1.38 ± 0.18 mm yr⁻¹ respectively[10.10].

Satellite radar altimeter data, e.g. the TOPEX/Poseidon (T/P) mission, have enabled sea levels to be measured on a global basis every 10 days since late 1992. The measurement of sea level variability from the T/P altimeter has indicated a linear trend in mean sea level (MSL) from 1993–1998 of 2–3 mm yr⁻¹ [10.11]. El Niño and La Niña events also have an impact on sea-level, resulting in lower and higher levels respectively in the western Pacific (Table 10.3)[10.12]. Around Papua New Guinea, for example, sea-level may vary from −12 cm to +8 cm relative to the long-term MSL during these events.

The Permanent Service for Mean Sea Level (PSMSL) is the global databank for monthly and annual mean sea levels, containing data from almost 2000 sites (see http://www.pol.ac.uk/home/). The South Pacific Sea Level and Climate Monitoring Project has been established to provide an accurate long term record of sea level in the south Pacific (Table 10.4)[10.13].

Table 10.4 The south Pacific sea level and climate monitoring project: net relative sea-level (RSL) trends (1992–2004)[10.13].

Location	Length of record (Months)	RSL trend (mm year⁻¹)
Cook Is.	133	+ 2.2
Tonga	134	+ 11.2
Fiji	138	+ 3.0
Vanuatu	124	+ 4.8
Samoa	133	+ 3.5
Tuvalu	131	+ 5.2
Kiribati	131	+ 4.7
Nauru	128	+ 7.0
Solomon Is.	113	+ 6.2
PNG	100	+ 7.0
Marshall Is.	126	+ 5.0

Future Sea-level Changes

A global *eustatic* sea-level rise (i.e. change in total volume of water) of 0.09–0.88 m is projected for 1990 to 2100[10.11]. The central value of 0.48 m (4 mm yr⁻¹)

Table 10.5 Changes in the return period of the current 100-year water level with predicted sea-level rise data for various locations in Britain[10.15].

Port	Relative sea-level rise	Approximate return period for the current 100 year water level by 2050s
North Shields	20 cm	20 years
Harwich	31 cm	20 years
Devonport	29 cm	3 years
Fishguard	28 cm	10 years
Heysham	28 cm	40 years

would give an average rate 2–4 times the rate experienced over the 20th century. Although similar rates occurred in the mid-Holocene, this coincided with high sediment availability (many coastal systems are now sediment starved; see Chapters 8 and 29). The actual rate experienced at particular locations (*relative sea-level rise*) will be influenced by land movements (i.e. *isostatic/tectonic* effects).

Many Pacific Small Islands are highly vulnerable to sea-level rise (e.g. Kiribati, Samoa, Fiji); some islands would be inundated by a 1 m rise (e.g. Tuvalu). Coastal cities that have been experiencing subsidence due to groundwater extraction (e.g. Jakarta and Bangkok where the rate of subsidence has been 50 mm yr^{-1} over the 1990s) will also be particularly vulnerable. Elsewhere, sea-level rise will increase the frequency of coastal hazards, accelerate current trends and, in some places, lead to major changes to coastal systems.

Impact on Extreme Events

The frequency of extreme conditions is expected to increase dramatically (Fig. 10.7). The UK Climate Impacts Programme[10.14] (UKCIP02; see http://www.ukcip.org.uk/), for example, has predicted that by the 2050s extreme high water levels will increase in frequency[10.15] (Table 10.5). For example, by 2050 the current 100-year water level at Devonport in South-west England would be the equivalent of a 3-year return period water level. This will increase the wave loadings on coastal defences and increase the chance of failure. Defences that have been built to provide protection against a 1 in 50 year storm sea level will be overtopped far more regularly.

Collapse of the West Antarctic Ice Sheet

The West Antarctic ice sheet (WAIS) is the world's only marine ice sheet that is anchored to bedrock below sea level with floating margins, making it susceptible to collapse. As sea level rises, more of the ice at the edge of the sheet floats, and the forces that hold the ice sheet together are reduced, causing ice to flow more rapidly to the oceans. Collapse of the WAIS would result in a rise in eustatic sea level of 5–6 m. It has been suggested that its collapse might be responsible for the higher sea level (5–6 m above current levels) during the last interglacial period[10.16].

Coastal Response to Sea-level Rise

Sea-level provides a datum for all coastal processes (tidal range, breaking wave height, longshore currents etc). The movement of this datum defines whether the shoreline responds *transgressively* (movement up the shore) or *regressively* (down the shore).

Sea-level rise will result in an increase in wave and tidal energy arriving at the shoreline. The landward end of the shore profile (e.g. the beach face or saltmarsh cliff) would be exposed to a higher energy environment, leading to erosion. The net result could be a landward migration of the shore profile, so that the profile position

Figure 10.7 Venice, a city at risk from an increased frequency of flooding as climate changes, due to the extensive infrastructure built close to or below present sea levels.

and form is maintained relative to sea-level. However, other responses are possible, especially where there is an increase in sediment supply; accelerated erosion in one location will release sediment for deposition elsewhere. For example, the London Clay cliffs at Hadleigh, UK, were abandoned around 6000 BP because of the growth of a broad saltmarsh at the cliff foot[10.17]. This occurred at a time when sea-level was rising at around 4 mm yr^{-1}.

Site-specific changes can only be predicted within the context of the evolution of broad-scale coastal cells (see Chapter 29).

The Bruun Rule

The Bruun Rule[10.18] assumes that an equilibrium profile is maintained as a landform moves inland in response to sea-level rise, by the transfer of eroded material from the upper profile to the lower profile (i.e. in a similar manner to the behaviour of a beach-dune system during a storm; see Chapter 35).

The Bruun Rule can be used to estimate the rate of shore profile migration R (Fig. 10.8):

$$R = \frac{L\ S}{H}$$

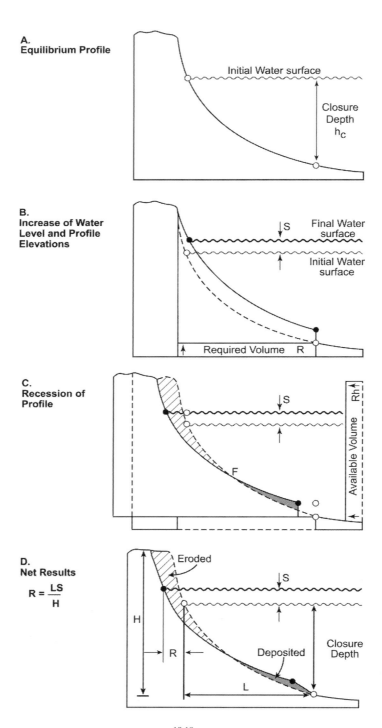

Figure 10.8 The Bruun Rule (see text, after Bruun 1962)[10.18].

Table 10.6 Example of the use of the Bruun Rule to predict future cliff recession rates.

Historical recession rate (m) R_1	Historical rate of sea-level rise (m yr^{-1}) S_1	Future rate of sea-level rise (m yr^{-1}) S_2	S_C	Sea-level rise factor	Future recession rate R_2
6.77	0.00181	0.002	0.00019	0.013	6.783
6.77	0.00181	0.006	0.00419	0.288	7.058
6.77	0.00181	0.008	0.00569	0.426	7.196

Sediment overfill $P = 0.977$; Profile length (m) $L = 1500$; Profile depth (m) $H = 14.8$; Cliff height (m) $B = 7.5$.

where L is the profile length i.e. the offshore distance to the depth of closure; S is the rate of sea-level rise (m yr^{-1}) and H is the profile depth i.e. the depth of closure. The *closure depth* is the boundary of the profile beyond which there is little loss of sediment i.e. the seaward limit of wave-driven sediment transport activity (profile depth).

For example, if sea level rise was 5 mm yr^{-1}, and the depth of closure of 10 m occurs 300 m offshore, the annual predicted profile migration rate would be:

$$R = \frac{L\,S}{H} = \frac{0.005 \times 300}{10} = 0.15 \ \text{m} \ \text{yr}^{-1}$$

The Bruun Rule has been modified to take account of the proportion of eroded sediment that is lost r i.e. not stored elsewhere within the profile:

$$R = \frac{L\,S}{H} \times \left(1 + \frac{r}{100}\right)$$

A modified version of the Bruun Rule can be used to give an indication of the future cliff recession rate (Table 10.6):

$$R_2 = R_1 + \frac{S_c \times L}{P(B + H)}$$

where R_1 = historical recession rate (m yr^{-1}); S_c = change in sea level rise (m yr^{-1}); P = sediment overfill (the proportion of sediment eroded that is sufficiently coarse to remain within the equilibrium profile); B = cliff height (m); H = closure depth (m) and L = length of cliff profile to the closure depth (m).

The Bruun Rule represents complex shoreline behaviour in terms of a simple linear response. It is only a rough guide to future changes. In some circumstances it is potentially misleading; the landward migration of barrier islands involves the landward movement of sediment (roll-over), whereas the Bruun rule assumes a net seaward movement towards the lower profile (see Chapter 34).

References

10.1 Haq, B. U., Hardenbol, J., and Vail, P. R. (1987) Chronology of fluctuating sea-levels since the Triassic. *Science*, **235**, 1156–1166.

10.2 Chappell, J. and Shackleton, N. J. (1986) Oxygen isotopes and sea level. *Nature*, **324**, 137–140.

10.3 Fairbanks, R. G. (1989) A 17,000-year glacio-eustatic sea level record: influence of glacial melting rates on the Younger Dryas event and deep-ocean circulation. *Nature*, **342**, 637–642.

10.4 Aharon, P. (1983) 140,000 year isotope climatic record from raised coastal reefs in New Guinea. *Nature*, **304**, 720.

10.5 Shennan, I., Lambeck, K., Horton, B., Innes, J., Llyod, J., McArthur, J., and Rutherford, M. (2000) Holocene isostacy and relative sea-level changes on the east coast of England. In I. Shennan and J. Andrews (eds.) Holocene Land-Ocean Interaction and Environmental Change around the North Sea. *Geological Society Special Publications*, **166**, 275–298.

10.6 Baker, R. G. V., Haworth, R. J., and Flood, P.G. (2001) Inter-tidal fixed indicators of former Holocene sea-levels in Australia: a summary of sites and a review of methods and models. *Quaternary International*, **83–85**, 257–273.

10.7 Commission on Geosciences, Environment and Resources (1990) *Sea-level Change*. National Academies Press, Washington DC.

10.8 Gornitzs, V., Lebedeff, L. and Hansen, J. (1982) Global sea level trend in the past century. *Science*, **215**, 1611–1614.

10.9 Woodworth, P. L., Tsimplis, M. N., Flather, R. A. and Shennan, I. (1999) A review of the trends observed in British Isles mean sea level data measured by tide gauges. *Geophysics Journal International*, **136**, 651–670.

10.10 Mitchell, W., Chittleborough, J., Ronai, B., and Lennon, G. W. (2000) Sea Level Rise in Australia and the Pacific. *The South Pacific Sea Level and Climate Change Newsletter, Quarterly Newsletter*, **5**, 10–19.

10.11 Inter-Governmental Panel on Climate Change (IPCC) (2001) *The IPCC third assessment report*. Technical Summary of the Working Group 1 Report. Available at http://www.ipcc.ch/index.html

10.12 Chowdhury, R. (2004) Enso and Sea-Level Variability (2): Physical Mechanism (Guam, Marshalls, and American Samoa). Pacific ENSO Update 10, 4. Available at http://www.soest.hawaii.edu/MET/Enso/peu/2004_4th/ special_section.htm

10.13 SPSLCMP (2004) The South Pacific Sea Level & Climate Monitoring Project (Phase III) Sea Level Data Summary Report November 2003–April 2004. Available at http://www.pacificsealevel.org/

10.14 Hulme, M., Jenkins, G. J., Lu, X, Turnpenny, J. R., Mitchell, T. D., Jones, R. G., Lowe, J., Murphy, J. M., Hassell, D., Boorman, P., McDonald, R., and Hill, S. (2002) *Climate change scenarios for the United Kingdom the UKCIP02 Scientific report*. Norwich. Tyndall Centre for Climate Change Research.

10.15 The Hadley Centre for Climate Prediction and Research, Meteorological Office (1998) *Memorandum to the Agriculture Committee*. In the Agriculture Committee Sixth Report: Flood and Coastal Defence, Volume II Minutes of Evidence and Appendices, 328–339.

10.16 Scherer, R. P., Aldahan, A., Tulaczyk, S., Possnert, G., Engelhardt, H., and Kamb, B. (1998) Pleistocene Collapse of the West Antarctic Ice Sheet. *Science*, **28**, 82–85.

10.17 Hutchinson, J. N. and Gostelow, T. P., 1976. The development of an abandoned cliff in London Clay at Hadleigh, Essex. *Philosophical Transactions of the Royal Society*, **A283**, 557–604.

10.18 Bruun, P. (1962) Sea-level rise as a cause of shore erosion. *Journal of the Waterways and Harbours Division* ACSE, **88**, 117–130.

11

The Nature of Change: Rates and Events

Introduction

Over the timescales relevant for most engineering projects, landform change can involve:

- Almost *continuous* or *regular changes* which can be described as an erosion or deposition rate over a particular time period (e.g. m yr^{-1}).
- A *series of events* of different size (*magnitude*) and *frequency*. Events of a particular size are described in terms of a return period or recurrence interval (e.g. a 1 in 100 year event).
- *Discrete events* which can be described in terms of the annual probability of occurrence (e.g. an event with a 0.01 chance of occurring in a given year).

Probabilities are useful for describing the likelihood of future events because of our ignorance of the true frequency of events. For example, it is known that change will occur, but not at what rate or when. There is a chance that the rate could be any value within a particular range. The best estimate will generally coincide with the most likely situation (i.e. highest probability), whereas the worst and best-case scenarios will generally represent the upper and lower boundaries of the range of expected situations. This concept is useful for the *Observational Method* (Chapter 1) where engineering designs have to be prepared to deal with a range of ground conditions that cover the best and worst design scenarios.

Rates of Change

Historical rates of change can be obtained by comparing the position of defined features at different dates (e.g. from topographic maps or aerial photographs) or directly from repeated measurements. Rates are expressed as an average value over a particular time period, usually a year (e.g. Table 11.1):

Average Rate = Total Change/Number of Years

Average (*mean*) values can be misleading because of the variability in the rate. For example, the average annual recession rate of the Holderness coastal cliffs, UK, is estimated to be 1.8 m yr^{-1} but has been over 10 m in a single year on more than one occasion since 1953[11.1]; see Fig. 5.4).

The *standard deviation* provides an indication of the scatter of the observations around the mean. Fig. 11.1

Table 11.1 Sample recession rates for a sandstone cliff prone to occasional rockfalls (average annual recession rates expressed in m year^{-1}).

Location	Period 1 1907–29	Period 2 1929–36	Period 3 1936–62	Period 4 1962–91	Cumulative 1907–91
1	0.09	1.50	1.31	0	0.57
2	2.09	2.00	0.31	0.06	0.83
3	1.63	0.57	0	0.34	0.57
4	0.91	0.57	0.23	0.31	0.48
5	0.64	0.28	0.31	0.28	0.5
6	0	0.28	0.15	0.34	0.19
7	0.09	0	0.54	0.18	0.24
8	1.27	0	0.54	0	0.26
Mean	0.84	0.65	0.42	0.19	0.45
Standard deviation	0.72	0.68	0.38	0.14	0.20

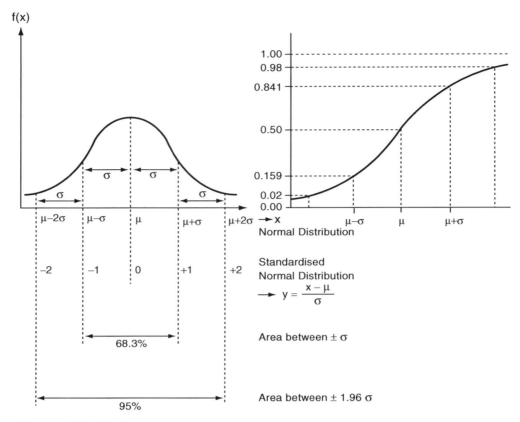

Figure 11.1   The normal distribution.

highlights the relationship between the mean and the standard deviation for a normal distribution. Using the recession data for the period 1907–91 presented in Table 11.1 (mean = 0.45 m yr^{-1}; standard deviation = 0.2):

- 68% of the values lie within 1 standard deviation of the mean (i.e. 0.25–0.65 m yr^{-1})
- 95% are within 2 standard deviations (i.e. 0.05–0.85 m yr^{-1}).

Magnitude-frequency Distributions and Return Periods

Some events can be regarded as part of a random series of events with a distinct magnitude-frequency relationship that can be represented by *distribution laws*. Examples include:

- *Floods*: a variety of probability distributions can be used to model flood frequency/magnitude (e.g. the log-normal distribution, the Gumbel Type I and III distributions, the Pearson Type III distribution[11.2]). The criteria for selection are *goodness of fit* and ease of application.

The Gumbel EV1 distribution is (see Table 11.2):

$$F(Q) = \exp\left\{-\exp\left(\frac{-(x-u)}{\alpha}\right)\right\}$$

where $F(Q)$ is the probability that an annual flood will equal or be less than the discharge Q.

The mean μ and variance v are estimated from the data series. These values are then used to derive the parameters α and u:

$$\alpha = \sqrt{6 \times \left(\frac{v}{3.142^2}\right)}$$

$$u = \mu - 0.5772 \ \alpha$$

The probability of exceedence of the EV1 mean annual flood is:

$$P(\mu) = 1 - F(\mu)$$

- *Earthquakes*: the magnitude/frequency distribution is usually described by a power law, such as the Gutenberg-Richter law[11.3]:

$$\log N(M) = \alpha \ M^{-b}$$

where $N(M)$ is the number of earthquakes with magnitude larger than M, and α and b are constants.

- *Rockfalls*: the magnitude/frequency distribution can be described by a simple power law[11.4]:

$$n(V) = \alpha \ V^{-b}$$

where V is the rockfall volume, $n(V)$ is the number of events per year with a volume equal or greater than V, and α and b are constants.

Table 11.2 Flood frequency data for the River Severn, UK (see Fig. 11.2 for the flood return period plot).

Year	Discharge Q (m³s⁻¹) (mean = 13.8; variance = 20.7)	*P(x)* Probability discharge is equalled/exceeded (Gumbel EV1 distribution)	Rank	Return period (Weibull formula) (years)
1951	10.4	0.766	16	1.31
1952	10.9	0.717	14	1.50
1953	10.6	0.747	15	1.40
1954	10.1	0.794	17	1.24
1955	8.5	0.917	20	1.05
1956	9.6	0.838	18	1.17
1957	23.7	0.033	1	21.00
1958	13.1	0.493	9	2.33
1959	23.1	0.040	2	10.50
1960	17	0.202	5	4.20
1961	11.4	0.666	12	1.75
1962	14.8	0.343	7	3.00
1963	12.3	0.573	11	1.91
1964	20.7	0.076	3	7.00
1965	13.4	0.464	8	2.63
1966	17.5	0.178	4	5.25
1967	12.4	0.562	10	2.10
1968	15.8	0.271	6	3.50
1969	8.8	0.898	19	1.11
1970	11.3	0.676	13	1.62

$P(x) = 1 - \exp\left(-\exp\left(-\left(Q - u\right)/\alpha\right)\right)$; u and α calculated as shown in the text ($u = 11.73$; $\alpha = 3.54$).

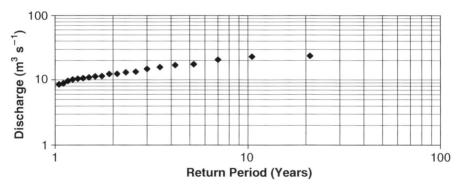

Figure 11.2 Flood frequency data for the River Severn, UK.

The b constant (which controls the shape of the distribution) is independent of geological setting and is around 0.45 (with a range of 0.41–0.46), whereas the constant α varies with the site.

Historical event *frequency* is often used to estimate the likelihood (*probability*) of an event of a particular *magnitude* occurring in the future. For example, the likelihood of a flood of a particular magnitude is generally expressed by the return period or recurrence interval, using the Weibull formula (e.g. Fig. 11.2):

$$\text{Return period} = \frac{N+1}{m}$$

where N is the number of records in the time series (20 in Table 11.2) and m is the order or rank number within the series of the event in question (i.e. Column 4 in Table 11.2).

The relationship between probability (expressed as a percentage chance), return period and the length of period under consideration is shown in Table 11.3. An event with a return period of 100 years (the 100 year event) *will not* have a 100% probability of occurring in a period of 100 years – the true figure is 65%. 100-year events do not occur 100 years apart or once every 100 years. There can be more than one occurrence in any 100-year period and even occurrences in successive years.

Discrete Events

The probability of an event occurring ranges from 0 (impossible) to 1.0 (certain). It is a function of the number of favourable outcomes, such as throwing a 6 with a single die, compared with the total number of *possible* outcomes (there are 6 sides to the die):

$$\text{Probability(event)} = \frac{\text{Number of favourable outcomes}}{\text{Total number of possible outcomes}}$$

$$= \frac{1}{6} = 0.1667$$

An alternative way of expressing probability is in terms of the frequency that a particular outcome occurs during a particular number of trials or experiments:

$$\text{Probability (event)} = \frac{\text{Number of favourable outcomes}}{\text{Total number of trials}}$$

For many systems there are two possible outcomes: change or no change. These two outcomes are *mutually exclusive* so the probability of an event together with the probability of no event must equal unity:

$$\text{Probability (event) + Probability (no event)} = 1$$

It is common practice to use discrete units of time, such as years, as individual trials. The chance of an event occurring in a particular year is, therefore, expressed in terms of an *annual probability*. Thus the probability of a landslide can be calculated as:

$$\text{Annual probability (landslide)} = \frac{\text{Number of recorded landslides}}{\text{Length of record period in years}}$$

An annual probability of 0.01 (1 in 100, or 1%) does not mean that an event will occur once in a 100 years, but rather that if a particular year were repeated 100 times (e.g. as in the film *Groundhog Day*) with each year having randomly sampled energy input parameters there would be 1 with a failure for every 99 with no failures.

Over time the probability of an event accumulates in a predictable manner (i.e. *cumulative probability*: Fig. 11.3). If the annual probability of an event is 0.01, then the

Table 11.3 Percentage probability of the T-Year event occurring in a particular period.

Number of years in period	T = Average Return Period in years							
	5	10	20	50	100	200	500	1000
1	20	10	5	2	1	0.5	0.2	0.1
10	89	65	40	18	10	5	2	1
100	–	99.9	99.4	87	65	39	18	9
1000	–	–	–	–	–	99.3	87	64

Where no figure is inserted the percentage probability > 99.9.

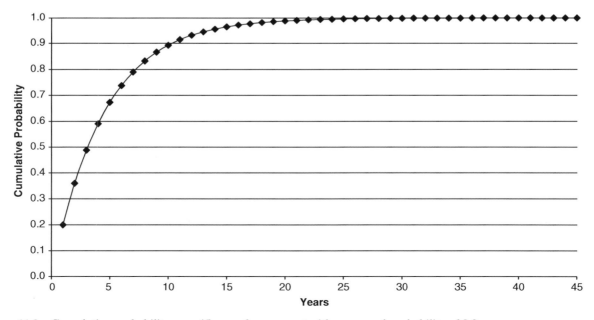

Figure 11.3 Cumulative probability over 45 years for an event with an annual probability of 0.2.

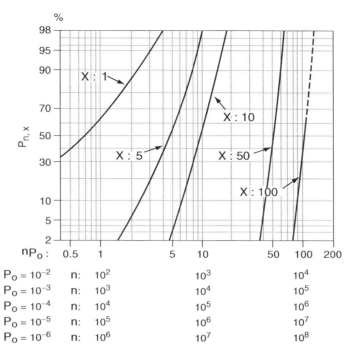

Figure 11.4 Probability of a rare event (P_n, x) occurring at least x times in n years (after Gretener 1967)[11.5].

probability for the event to occur in n years, taking $n = 10$ years as an example, is:

Probability (event, n years)
$$= 1 - (1 - \text{Annual probability (event)})^n$$
$$= 1 - (1 - 0.01)^{10}$$
$$= 1 - (0.99)^{10}$$
$$= 0.10 \text{ or } 10\%$$

This equation can be expanded into a *binomial series* to predict the number of favourable outcomes x (e.g. landslides or debris flows) in n trials (e.g. a given time period) in which the probability p remains constant in each trial:

$$(n\ x)\ p^x\ (1 - p)^{n-x}$$

For given values of n and p the set of probabilities of this form for $x = 0, 1, 2...n-1$ is called a binomial probability distribution. Figure 11.4 presents a chart that can be used to predict the probability of an event occurring at least x times in n years[11.5]. For example, there is a 95% chance that an event (x:1) with an estimated annual probability (Po) of 0.01 (1/100) will occur at least once in 300 years $(nPo = 3)$, at least 5 times (x:5) in 900 years $(nPo = 9)$ and at least 10 times (x:10) in 1600 years $(nPo = 16)$.

Non-stationary Data Series

The use of historical data sets to estimate future event probabilities is based on the assumption that the magnitude-frequency distribution has remained constant over the period of the historical record (i.e. the data series is *stationary*). If the frequency of events has changed because of climate or land use change or progressive weathering of the slope, the data series is *non-stationary*.

Future environmental conditions and system responses could be significantly different from those in the past because of the combined effects of human activity and predicted climate change (see Chapter 9). The historical record cannot be *guaranteed* to provide a reliable estimate of the probability of future events.

References

11.1 Pethick, J. (1996) Coastal slope development: temporal and spatial periodicity in the Holderness Cliff Recession. In M. G. Anderson and S. M. Brooks (eds.) *Advances in Hillslope Processes*, 2, 897–917.

11.2 Stedinger, J. R., Vogel, R. M. and Foufoula-Georgiou, E. (1993) Frequency analysis in extreme events. In D. Maidment (ed.) *Handbook of Hydrology*, McGraw-Hill, New York.

11.3 Gutenberg, B. and Richter, F. (1949) *Seismicity of the earth and associated phenomena*, Princeton University Press, Princeton, New Jersey.

11.4 Dussauge-Peisser, C., Helmstetter, A., Grasso, J-R., Hantz, D., Desvarreux, P., Jeannin, M., and Giraud A. (2002) Probabilistic approach to rock fall hazard assessment: potential of historical data analysis. *Natural Hazards and Earth System Sciences*, 2: 15–26.

11.5 Gretener, P. E. (1967) The significance of the rare event in geology. *The American Association of Petroleum Geologists Bulletin*, **51**(11), 2197–2206.

12

The Implications of Change: Hazards and Risks

Introduction

Landform changes are *natural phenomena*, but they can present hazards and risks to engineering projects:

- *Hazards* are events (i.e. geomorphological processes) that have the potential to adversely affect humans or the things that humans value.
- *Risk* is expressed as the product of the likelihood of a hazard and its adverse consequences[12.1]:

Risk = Probability (Event) × Adverse Consequences

A 50% chance (probability of 0.5) of landslide movement causing $1000 worth of footpath damage has the same risk as a 0.1% chance (probability of 0.001) of a flood causing losses of $0.5M i.e. both have mathematical expectation values of $500.

The *risk assessment process* involves a number of stages from risk analysis to risk evaluation and management (Fig. 12.1; see Table 12.1 for definitions). These stages mark a progression from technical-based activity (risk analysis) to the broader-based judgements and policies involved in evaluation and management. The key questions that need to be addressed are:

- What can go wrong? i.e. hazard assessment
- How likely is it? i.e. estimating the probability of the hazards
- What will be the losses and damages? i.e. consequence assessment
- Does it matter? i.e. risk evaluation
- What should be done about the risks? i.e. risk management.

Risk Analysis: Geomorphological Hazards

For civil engineers, the most important geomorphological hazards include:

- surface erosion, gullying and badland erosion on hillslopes
- landsliding, including first-time failures and reactivation of pre-existing landslides
- ground collapse in karst, underground mines and other terrain
- wind erosion, dust storms and dune migration in and around desert dune fields and sand seas
- aggressive evaporate soils, especially in hot desert environments

- flash floods in steep catchments and dryland channels (e.g. wadis)
- bank erosion, bed scour, channel instability and avulsion (channel switching) along rivers and deltas
- floods along lowland rivers and estuaries
- sedimentation in navigable rivers and estuaries
- erosion and flooding of low-lying coastal areas (including tsunamis)
- erosion of coastal cliffs, including coastal landslide activity.

These hazards are the agents that can cause loss and contribute to the generation of risk. Although the magnitude-frequency characteristics of these hazards (e.g. the 100-year flood) are important in helping to determine risk, they are not risk, merely measures of *hazardousness*.

Hazard models can be used to classify the different types of hazard and quantify their future frequency and magnitude, focussing on (e.g. Fig. 12.2):

- What could happen? i.e. the nature and scale of the events that might occur in the foreseeable future. Hazards develop from incubation, via the occurence of a triggering or initiating event to the system response and possible outcomes Fig.12.3).
- Where could it happen? i.e. a spatial framework for describing variations in hazard across a site or area.
- Why such events might happen? i.e. the circumstances associated with particular events.
- When events might happen? i.e. the timescales within which particular events can be expected to occur.

Consideration must be given to hazards originating off-site or beyond the boundaries of the area under consideration, as well as on the immediate site/area.

The quality of a risk assessment is related to the extent to which the hazards are recognised, understood and explained – this is not necessarily related to the extent to which they are quantified. The temptation for increasing mathematical precision in the risk assessment process needs to be tempered by a degree of pragmatism that reflects the reality of the situation and the limitations of available information[12.2].

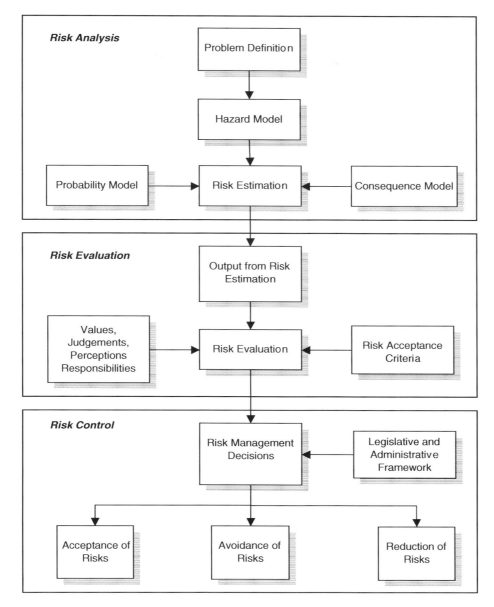

Figure 12.1 The risk assessment process.

Risk Analysis: Event Probability

Common approaches to estimating probability include:

1. The *historical frequency* of hazard events in an area can provide an indication of the future probability of such events. The approach relies on the assumption that the proportion of times any particular event has occurred in a large number of trials (i.e. its *relative frequency*) converges to a limit as the number of repetitions increases.

2. Using *process-response relationships*: many events are associated with a particular triggering event such as a heavy rainstorm or earthquake. An indication of the event probability can be obtained through establishing initiating thresholds between parameters such as earthquakes and landslide activity (see Chapter 19). The probability of an event is the product of the annual probability of the trigger and the *conditional probabilities* of the subsequent system response. For example, suppose a triggering event (E)

has a probability $P(E)$. *Given that this event occurs*, the failure outcome O has the probability $P(O|E)$ (e.g. given that a heavy rainstorm has occurred, then there is a 50% chance of a debris flow.) The probability of this sequence of events occurring is:

$$\text{Probability} = P(E) \times P(O|E)$$

(Note: the symbol | denotes 'given', as in 'given that the previous event has occurred'.)

3. *Expert judgement* (see Chapter 43): historical records allow 'objective' estimates to be made about event probabilities. There are many situations where the lack of available 'hard' information does not allow this type of approach. An alternative strategy involves making judgements about the expected event probability, based on available knowledge plus experience gained from other projects and sites. This is known as 'subjective' probability assessment and relies on the practitioner's

Table 12.1   Risk assessment definitions.

Hazard assessment: analysis of the probability and characteristics of potential hazard events.

Probability: the mathematical expression of chance (e.g. 0.2, 20% or a one in five chance) where possible but in many cases it can be no more than a prospect that can be expressed only qualitatively. The expression applies to the occurrence of a particular event in a given time or as one among a number of possible events.

Adverse consequences: the adverse effects or harm as a result of realising a hazard which causes the quality of human health or the environment to be impaired in the short or longer term.

Risk: a combination of the probability, or frequency, of occurrence of a defined hazard and the magnitude of the adverse consequences of the occurrence.

Risk analysis: the use of available information to estimate the risk to individuals or populations, property or the environment, from hazards.

Risk assessment: the process of risk analysis and risk evaluation.

Risk estimation: the process used to produce a measure of the level of health, property, or environmental risks being analysed. Risk estimation contains the following steps: hazard assessment, consequence analysis, and their integration.

Risk evaluation: the stage at which values and judgements enter the decision-making process, explicitly or implicitly, by including consideration of the importance of the estimated risks, and the associated social, environmental and economic consequences.

Risk perception: the overall view of risk held by a person or group and includes both feeling and judgement.

Risk Management: the complete process of risk assessment and risk control.

Vulnerability: the level of potential damage, or degree of loss, of a particular asset (expressed on a scale of 0 to 1) subjected to a damaging event of a given intensity.

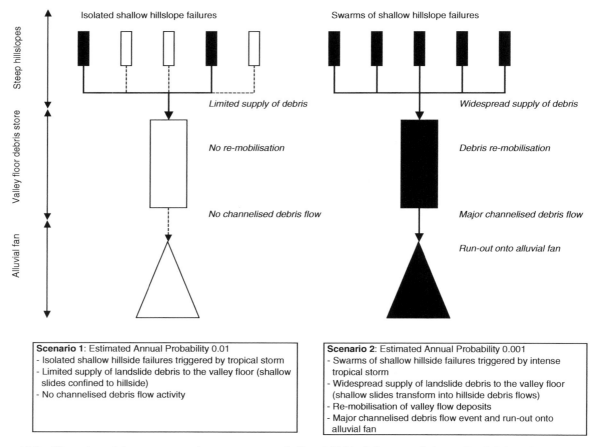

Figure 12.2   Hazard models: a steep catchment prone to shallow hillside failures and channelised debris flows.

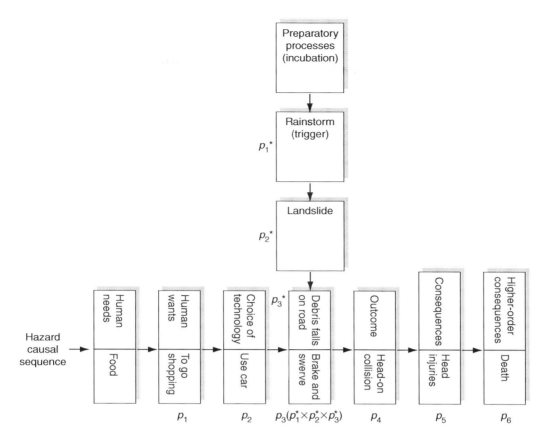

Probability of death $= p_1 \times p_2 \times p_3 \, (p_1^* \times p_2^* \times p_3^*) \times p_4 \times p_5 \times p_6$

Figure 12.3 Hazard model sequence of events (after Lee and Jones 2004)[12.2].

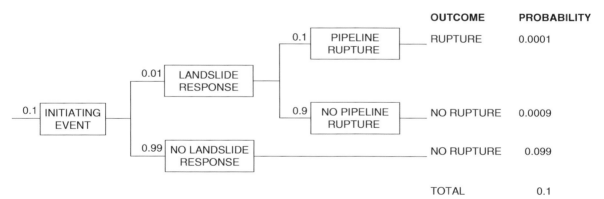

Figure 12.4 Simple event tree model of a pipeline rupture sequence; the numbers at each branch are estimated probabilities (after Lee and Jones 2004)[12.2].

'degree of belief' about the likelihood of a particular outcome.

Event trees are useful for tracing the progression of the various combinations of scenario components to identify a range of possible outcomes (Fig. 12.4). Event scenarios can be broken down into sequences of:

- an *initiating event* (i.e. triggering events)
- the system *response*, controlled by the nature and state of the system components (i.e. the antecedent conditions)
- subsequent *outcomes* determined by the style and intensity of system response and the assets at risk.

The expert judgement process should involve an open-minded but sceptical group of experts from a range of backgrounds, rather than single individuals; this facilitates the pooling of knowledge and experience, as well as limiting bias (see Chapter 43). A useful approach is an *open forum*[12.3].

Open forum discussions should question the available data and the understanding of system behaviour; what do we really know? How much have we assumed? Are the assumptions reasonable? Have we ignored or dismissed particular evidence? How confident are we in our understanding of the relationships between different environmental controls and the surface processes?

None of these methods can be *guaranteed* to provide a reliable estimate. Statements as to the probability of hazard events can only be *estimates*. Estimates will generally need to be 'fit for purpose'[12.2].

Risk Analysis: Consequence Assessment

Adverse consequences include the *direct impact* on people (i.e. loss of life or injury), *direct and indirect economic losses* and *intangible losses* (e.g. effects on the environment, amenity, people's attitudes and behaviour). *Direct effects* are the first-order consequences that are intimately associated with an event, or arise as a direct consequence of it (e.g. destruction), while *indirect effects* emerge later, such as mental illness, longer-term economic problems or relocation costs.

The impact of an event is controlled by its intensity (e.g. spatial extent, water depth, mass and velocity or rate of change) and duration, as well as the assets at risk, the exposure of the assets and their vulnerability to damage. Combining these factors enables *consequence models* to be developed that reflect the damage associated with particular event scenarios, in which:

- *Exposure* is the proportion of time that the assets are in the danger zone.
- *Vulnerability* is the chance of a particular level of damage (e.g. total destruction or death) given a particular event intensity.

In a consequence model, both the *exposure* and *vulnerability* factors range between 0 and 1 and can be used to establish the predicted damages compared with a *total loss event* in which the assets at risk would be completely lost. Thus, for an event of a particular intensity:

Risk = Probability (Event) × Adverse Consequence
= Probability (Event) × (Total loss
× Exposure × Vulnerability)

In most cases it will be necessary to carry out this exercise for each of the individual elements at risk.

Risk Estimation

The main approaches to risk estimation are:

1. *Qualitative risk estimations*: both likelihood and adverse consequences are expressed in qualitative terms (e.g. risk scores). A measure of risk is obtained by combining the scores to produce a matrix (Fig. 12.5).
2. *Semi-quantitative risk estimations*: combinations of qualitative and quantitative measures of likelihood and consequence.
3. *Quantitative risk estimations*: values of consequence are combined with probabilities of occurrence.

Qualitative methods are useful where the available resources or data are limited.

Economic and Environmental Risk: Example

An oil pipeline has been built through dissected hills in the tropics, buried within the colluvium and residual soils that mantle the slopes. The hillslopes are susceptible to shallow (< 3 m deep) debris slides or debris avalanches. A consequence model was developed:

- Total loss: the consequences of a pipe rupture include the environmental damage and clean-up costs associated with an oil spill, estimated to be, on average, \$5M per event. In addition, supply interruptions lasting from days to weeks can be expected. As the pipe carries 40 000 barrels of oil per day (estimated value of \$1 M), a disruption of operation for 10 days would lead to a loss of \$10 M in business.
- Exposure: a factor of 1.0 was used, as the pipeline is permanently in the danger zone.
- Vulnerability; the pipeline is only vulnerable to a proportion of all the landslides that affect the route. It was assumed that only slides > 20 m wide *could* damage the pipe. Available statistics of past landslide events indicate that 15% of all recorded slides are greater than 20 m wide. As these > 20 m wide slides could have a range of intensities and impacts on the pipeline, it was assumed that only 25% *would* lead to rupture of the pipe.

Pipeline Vulnerability Factor = 0.15 × 0.25 = 0.0375

The risk to the pipeline from a landslide event causing a 10 day disruption of operation (Probability = 0.01) is:

Risk = Probability (Event) × Adverse Consequences
= Probability (Event)
× (Total Loss × Exposure Factor
× Vulnerability Factor)
= Probability (Event) × (\$15 M × 1 × 0.0375)
= 0.01 × \$0.56 M
= \$5600

Risk to People: Example

The rockfall risk to a tourist walking along a promenade at the foot of a cliff can be modelled:

Risk = Probability (Landslide)
× Probablity (Hit | Landslide)
× Probability (Fatality | Hit)

Probability (Landslide) = a measure of the expected likelihood of a particular landslide event (Hazards 1 to *h* representing rockfalls of different sizes) occurring in a particular time period.

		Consequence			
		Severe	Moderate	Mild	Negligible
Probability	High	High	High	Medium/low	Near Zero
	Medium	High	Medium	Low	Near Zero
	Low	High/medium	Medium/low	Low	Near Zero
	Negligible	High/med/low	Medium/low	Low	Near Zero

Figure 12.5 A risk evaluation matrix (after Lee and Jones 2004)[12.2].

Probability (Hit | Landslide) = a measure of whether a person is in the wrong place (*where* the landslide happens) at the wrong time (*when* the landslide happens) i.e. the exposure.

Probability (Fatality | Hit) = a measure of the chance that a person would be killed if he/she were hit by the landslide (Hazard *h*) i.e. vulnerability.

The overall risk is estimated by carrying out the following exercise for each of the Hazards (1 to *h*) and summing the results.

The *Location Specific Individual Risk* (LSIR) is a measure of the risk for an individual who is present at a particular location for the entire period under consideration, i.e. 24 hours per day, 365 days per year (i.e. exposure = 1). It is a property of the location in question, rather than behaviour patterns of the population[12.4].

It is calculated for a particular hazard (e.g. rockfall of a particular size class) in a specific unit as the product of the probability of landsliding (Hazard *h*) and the vulnerability to this hazard (Hazard *h*):

$$\text{LSIR (Hazard } h) = \text{Probability (Hazard } h)}$$
$$\times \text{Vulnerability (Hazard } h)$$

The *Individual Specific Individual Risk* (ISIR) is a measure of the risk for an individual who is exposed to the landslide hazards below the cliff for a particular length of time.

$$\text{ISIR} = \text{Probability (Hazard } h) \times \text{Probability}$$
$$\text{(Wrong Place)}$$
$$\times \text{Probability (Wrong Time)}$$
$$\times \text{Vulnerability (Hazard } h)$$

Probability (Wrong Place) = the probability that the path of the landslide intersects the location where the individual could be (i.e. the spatial probability of impact).

Probability (Wrong Time) = the probability that the individual is in the landslide danger zone during the landslide occurrence (i.e. the temporal probability of impact).

The ISIR for a single adult walking along a 1600 m long promenade at 4 km hr^{-1} between 06:00 and 09:00 for 300 days a year is calculated, for a 10 m wide rockfall, as follows:

$$\text{Probability (Hazard } h, \text{ 06:00 and 09:00)}$$
$$= \text{Probability (Hazard } h) \times \text{Proportion of}$$
$$\text{Landslides (06:00 and 09:00)}$$

$$\text{Probability (Wrong Place)}$$
$$= \frac{\text{Size of Landslide Event (m)}}{\text{Length of Promenade (m)}}$$
$$= \frac{10}{1600}$$

$$\text{Exposure (Single Journey)}$$
$$= \frac{\text{Length of Promenade (m)}}{\text{Walking Speed (m hr}^{-1})}$$
$$= \frac{1600}{4000} = 0.4 \text{ hr}$$

The exposure can then be expressed as a proportion of the time (e.g. a year) that a person is potentially in the danger zone (i.e. 3 hours per day on 300 of 365 days per year).

$$\text{Annual Probability (Wrong Time)}$$
$$= \frac{\text{Exposure (Single Journey)}}{3} \times \frac{\text{No. Journeys}}{\text{Year}}$$
$$= \frac{0.4}{3} \times \frac{300}{365}$$
$$= 0.11$$

$$\text{Individual Risk} = \text{Probability (Hazard } h,$$
$$\text{06:00 and 09:00)}$$
$$\times \text{Probability (Wrong Place)}$$
$$\times \text{Probability (Wrong Time)}$$
$$\times \text{Vulnerability (Hazard } h)$$

Societal Risk is a measure of the risk to the entire population that use the promenade at different times throughout the year:

$$\text{Societal Risk} = \text{Probability (Hazard } h)$$
$$\times \text{Probability (Wrong Place)}$$
$$\times \text{Probability (Wrong Time)}$$
$$\times \text{Vulnerability (Hazard } h)$$
$$\times \text{Number of Users}$$

Potential loss of life PLL (equivalent to the *expected value of the number of deaths per year*) is used as a measure of the risk to all individuals exposed to the full range of events that might occur in an area (i.e. the *societal risk*). To calculate PLL it is necessary to estimate, for each event and its possible outcome, the frequency per year *f* and the associated number of fatalities *N*. The PLL is the sum of the outcome of multiplying *f* and *N* for each event:

$$\text{PLL} = \Sigma f_1 N_1 + f_2 N_2 + \cdots + f_n N_n$$

Average Individual Risk (AIR) is a measure of risk to a hypothetical average individual within the overall group of population. It is calculated from the Societal Risk:

$$\text{AIR} = \frac{\text{Societal Risk}}{\text{Exposed Population}}$$

Risk Evaluation and Control

Risk evaluation involves judgements on how significant the estimated risks are and how to establish the best course of future action (*risk management*). A key concept is that there is a degree of risk that is acceptable/tolerable (see http://www.hse.gov.uk/). Above a certain threshold, the risks might be considered *intolerable* or *unacceptable*. Below another threshold, the risk may be considered to be so small that it is *acceptable*. Between these two conditions, the level of risk should be reduced to a level which is *as low as reasonably practicable* (the ALARP principle[12.5–6]; Fig. 12.6). This is often a societal or political decision, rather than one made solely by project engineers. The management principles are:

- If the risk is unacceptable it must be avoided or reduced, irrespective of the benefits, except in extraordinary circumstances

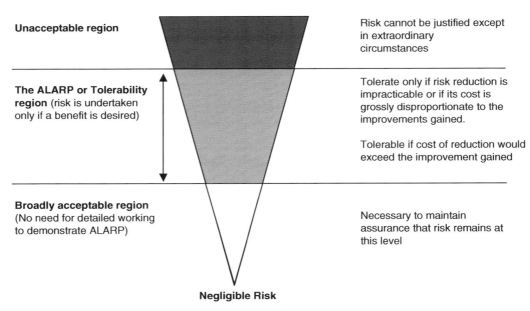

Figure 12.6 The ALARP principle (The Royal Society 1992)[12.1].

• If the risk falls within the ALARP or tolerability region then cost may be taken into account when determining how far to pursue the goal of minimising risk or achieving safety. Beyond a certain point, investment in risk reduction may be an inefficient use of resources. The benefits to be gained from a reduction in risk are normally expected to exceed the costs of achieving such a reduction (i.e. Cost-Benefit Analysis); this approach forms the basis for evaluating flood and coastal defence schemes in the UK[12.7].

References

12.1 The Royal Society (1992) *Risk: Analysis, Perception and Management*. Report of a Royal Society Study Group. The Royal Society, London.

12.2 Lee, E. M. and Jones, D. K. C. (2004) *Landslide Risk Assessment*. Thomas Telford.

12.3 Roberds, W. L. (1990) *Methods for developing defensible subjective probability assessments*. Transportation Research Record 1288, 183–190.

12.4 Institution of Chemical Engineers (IChemE) (1992) *Nomenclature for hazard and risk assessment in the process industries*. Rugby, IChem.

12.5 Health and Safety Executive (HSE) (1988) *The Tolerability of Risk from Nuclear Power Stations*. HMSO.

12.6 Health and Safety Executive (HSE) (1992) *The Tolerability of Risk from Nuclear Power Stations* (revised). HMSO.

12.7 Ministry of Agriculture, Fisheries and Food (MAFF) (1999) FCDPAG3 *Flood and Coastal Defence Project Appraisal Guidance: Economic Appraisal*. MAFF Publications.

13

The Implications of Change: Construction Resources

Introduction

Many deposits are associated with past landform change; an understanding of geomorphological environments and landscape history (e.g. Quaternary changes) can help identify suitable locations for construction resources[13.1–3] (Fig. 13.1; Table 13.1). These materials have been weathered, detached from the parent rocks, transported and sorted prior to or during deposition (e.g. by rivers or wind). Engineering geomorphologists can assist in the search for these deposits.

Geological maps, remote sensing and aerial photography can be useful in identifying potential locations where more detailed field surveys could be undertaken.

Borrow materials are those won from the ground for civil engineering construction purposes and include rock or soil suitable for fills, embankments, breakwaters, roads and concrete.

Aggregates are used for concrete and for road pavements. Sources include superficial deposits (e.g. glacial deposits, river sands and gravels) and strong rock (requires processing e.g. crushing).

Aggregate suitability is usually defined by local Standards or Codes; project specific codes can be developed using experience from similar environments. Careful testing and quality control is essential for successful resource development[13.4].

Sand and gravel properties depend on the rocks from which they are derived and their subsequent history. During transport, weathered or otherwise weakened fragments tend to be selectively worn away i.e. the resulting aggregate material is usually stronger than the crushed weathered parent rock and the particles may be somewhat rounded. Where transport distances are short (e.g. in fluvio-glacial environments, or on alluvial fans),

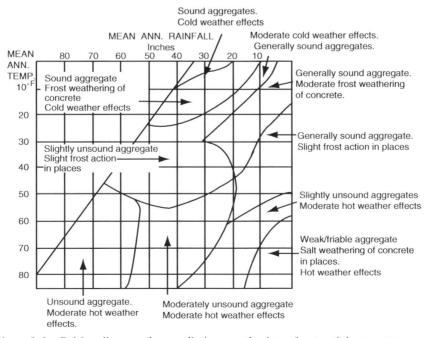

Figure 13.1 Extension of the Peltier diagram for predicting weathering of potential concrete aggregates (after Fookes 1980)[13.3].

Table 13.1 Some geomorphological environments and aggregate sources. Typical processing is crushing (and where necessary, washing) and screening.

Environment	Comment	Landforms and materials
Glacial	Extremely variable, often unsorted. Often needs processing.	*Till*: unsorted material, unlikely to be potential aggregate source unless matrix is sandy. Possible fill.
		Hummocky drift: unsorted, low aggregate potential.
		Drumlins: may contain an upstanding core of solid rock.
		Glacial outwash plains: meltwater deposits, relatively good aggregate sources.
		Kames and eskers: ice contact deposits, relatively good aggregate sources[13.5] (e.g. Fig. 13.2)
Periglacial	Periglacial conditions produce little sand and gravel, but may provide cobbles and boulders for crushing.	*Screes and talus cones* provide a ready source of coarse aggregate. Often poor quality due to weathering.
		Rock fields and boulder streams: may provide rock suitable for crushing to aggregate.
Fluvial	Alluvium often a source of large volumes of dredged sand and gravel. Relict Quaternary sand and gravel sources may lie beneath contemporary floodplains. Generally needs processing.	*River terraces*[13.6] (Fig. 13.3): useful sources of aggregate, but highly variable due to former channel migration and avulsion.
		Alluvial fans[13.7] (Fig. 13.4): coarser sand and gravel deposits at the proximal (upstream) end. Often poorly sorted and interbedded with silts and clays.
Hot drylands (Fig. 13.5)	Dominant upward leaching of salts causes aggressive ground conditions and contamination of aggregates[13.8]. Larger particles may be weathered.	*Coastal sabkhas, salinas*: surface crusts (e.g. calcium and calcium magnesium carbonates, calcium sulphate) severely contaminate sands and gravels. High quality control essential.
		Sand sheets and dunes: rounded single-sized particles, can be used as concreting sands and for embankments with careful mix design. Possible contamination by wind blown salts.
		Alluvial fans: source of poorly sorted sands and gravels around margins of uplands. May be salt free due to downward leaching during periodic floods.
		Ephemeral rivers: limited amounts of sands and gravels, especially around margins of uplands.
		Coastal deposits: carbonate sands may be acceptable fine aggregate; generally require processing (e.g. crushing, screening and washing).
Savannah (Fig. 13.6)	River and alluvial fan deposits, plus extensive duricrusts[13.9] (e.g. laterite).	*Laterite*: strong case hardened material often suitable on crushing for roadstone. May be used for concrete if clean and not friable.
		Laterite and latosols (lateritic soils) may make excellent haul roads on wetting and rolling, and may stabilise fills.
Hot wetlands	Deep tropical weathering results in limited coarse material. Bedrock quarries often required for crushed aggregates in carefully excavated fresh rock.	Ancient strong duricrusts (e.g. *ferricrete, silcretes*) may be useful borrow materials which require crushing.
		Volcanic deposits may produce alluvium with suitable sands and gravels, although fragments may contain macro voids.

mud flakes or fragments of coal or chalk may remain and reduce the quality of a prospective source.

Concrete

Concrete typically comprises coarse aggregate (i.e. gravel, crushed rock, c.50% of unit weight), fine aggregates (i.e. sand, c.25%) together with cement, water and additives to improve workability (c.25%). Several principal cement types are in use to give, for example, quick-setting, sulphate deterioration resistance, low heat rise (during setting), as required. Cement replacements (e.g. silica

fume) may also be used. The 28-day (i.e. the age when tested) unconfined crushing strengths typically range from 20–80 MPa depending on the requirement of the construction.

Reinforced concrete, defined as concrete containing steel bar reinforcement, is used for structural concrete.

Good quality aggregate for concrete has the following properties:

- The unconfined compression strength of the aggregate is equal to or greater than that of the cement matrix (i.e. the bonding material).

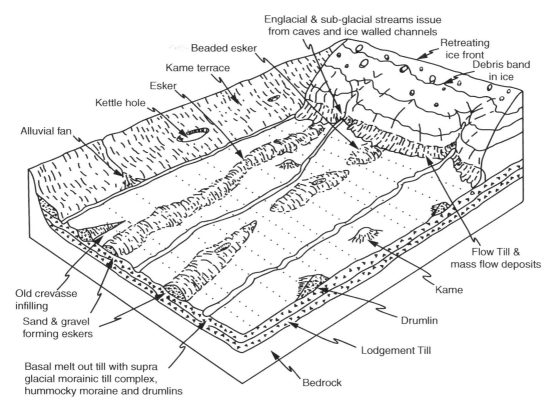

Figure 13.2 Examples of depositional landforms of a deglaciated landscape (after Fookes *et al.* 1975)[13.5].

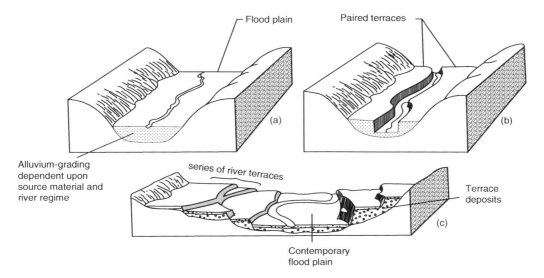

Figure 13.3 The formation of simple river terraces (based on Schumm 1977)[13.6]: (a) deposition of valley-floor deposits; (b) erosion of deposits leaving remnants of old floodplain as paired terraces and (c) unpaired terraces may result from the effects of downward and lateral erosion.

- It is free from deleterious (*unsound*) materials i.e. materials prone to volume change or chemical reaction (e.g. evaporite salts can initiate rusting of reinforcing bars or react with the cement to cause volume change).

- It is clean and free from organic impurities, particles should not be coated by clay or dust.

- Low absorption, good shape to esta blish good interlock between particles, and ideally not too flaky or

elongated which introduces difficulties in workability and strength.

Specific requirements are defined in local Standards (see http://www.concrete.org.uk/). The current Codes of Practice in use in Britain are Euro Standard BS EN206:1, BS 8500:1 and BS 8500:2.

Aggregate for Road Construction

Roads for moderately to heavily trafficked conditions, with flexible pavements particularly in the California

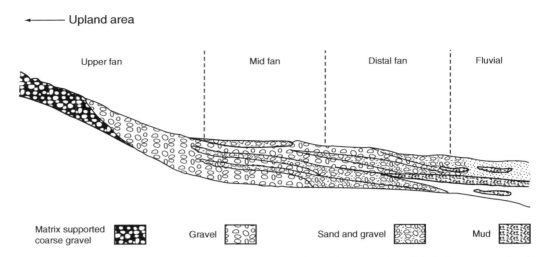

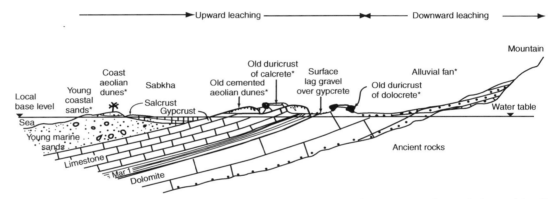

Figure 13.4   Section through an alluvial fan spread on a plain by a river carrying debris from an upland area (after Reineck and Singh 1980)[13.7].

Figure 13.5   Idealised section from hot desert coast to mountain interior showing typical associations of landforms and superficial deposits (after Fookes and Higginbottom 1980)[13.8]. (* Potential borrow materials in otherwise difficult terrain for identifying construction materials).

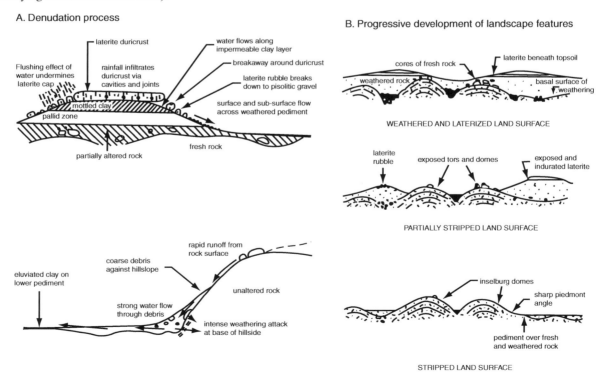

Figure 13.6   Possible aggregate sources in a Savannah landscape (after Thomas 1971)[13.9].

Figure 13.7 Plant used to place and compact a road pavement sub-base on natural ground.

Figure 13.8 A mobile grading rig for winning and processing sand and gravel deposits for use as construction materials.

Bearing Ratio (CBR) design method, commonly comprise four layers (Figs 13.7 and 13.8):

- *Sub-base:* the unbound drainage layer on the bottom of the road. It derives its strength from interlock and compaction of well-graded aggregates (i.e. not single sized) and is placed on the 'formation' (grade in the US) level which is the prepared top of the natural ground or embankment.
- *Base course:* the main load-bearing layer for the road. Aggregates need to have a good crushing strength, should not be porous (to prevent changing on wetting and drying) and be of reasonable shape and grading. Some road bases are bound by cement, lime or bitumen, others by good mechanical interlock and compaction.
- *Road base:* distributes the traffic load from the wearing course onto the base course and provides the highly specified surface on which the wearing course is laid. It is usually bitumen bound and has very good

quality aggregate with high crushing strengths and a close tolerance on grading.

- *Wearing course:* the surface layer bound with bitumen. Highest quality aggregates required with high crushing strengths and the ability to resist polishing and abrasion as well as the ability to bond with bitumen. Alternatively, wearing courses can be of concrete.

All aggregates must be strong and clean with limits on soluble salts and unsound material. Material specifications are increasingly onerous from the sub-base upwards e.g. on the particle shape and grading curve. The shape requirements may mean crushing becomes necessary, increasing construction costs; most roadstone in the upper layers therefore comes from crushed rock. Specifications for road materials in Britain are given in the Highways Agency Design Manual for Roads and Bridges (2005) and various testing procedures are contained in BS 882: 1992.

References

13.1 Cooke, R. U., Brunsden, D., Doornkamp, J. C. and Jones, D. K. C. (1982) *Urban Geomorphology in Drylands*, Oxford University Press.

13.2 Fookes, P. G., Baynes, F. J. and Hutchinson, J. N. (2000) Total geological history: a model approach to the anticipation, observation and understanding of site conditions. GeoEng 2000, an International Conference on *Geotechnical and Geological Engineering*, **1**, 370–460.

13.3 Fookes, P. G., 1980. An introduction to the influence of natural aggregates on the performance and durability of concrete. *Quarterly Journal of Engineering Geology*, **13**, 207–229.

13.4 Smith, M. R. and Collis, L. (eds.) (2001) *Aggregates: Sand, gravel and crushed rock for construction purposes.* 3rd edn. Revised by P. G. Fookes, J. Lay, I. Sims, M. R. Smith, and G. West, Geological Society Engineering Geology Special Publication No. 17. The Geological Society, London.

13.5 Fookes, P. G., Gordon, P. L., and Higginbottom, I. E. (1975) Glacial landforms and engineering characteristics. In *Engineering Behaviour of Glacial Materials.* Proceedings of the Symposium held at Birmingham, 21–23 April, GEO Abstracts, Norwich, 18–51.

13.6 Schumm, S. A. (1977) *The Fluvial System*. Wiley, New York.

13.7 Reineck, H.-E., and Singh, I. B. (1980) *Depositional Sedimentary Environments, with Reference to Terrigenous Clastics.* 2nd Edition, Springer-Verlag, New York.

13.8 Fookes, P. G. and Higginbottom, I. E. (1980) Some problems of construction aggregates in desert areas with particular reference to the Arabian Peninsula; 1. Occurrence and special characteristics. *Proceedings of the Institution of Civil Engineers*, **2**(68), 39–67.

13.9 Thomas, M. (1971) Savanna lands between desert and forest. *Geographical Magazine*, **44**, 185–189.

14

The Implications of Change: Environmental Impacts

The Impact of Engineering on the Landscape

Civil engineering can have significant impacts on surface processes (Fig. 14.1). Potential impacts need to be evaluated in relation to the site-specific conditions. Although some impacts may appear localised, they can result in indirect consequences, affecting the operation of surface processes throughout an earth surface system:

1. *Changes in erosion rates*: accelerated erosion, especially in recently deforested upland areas and hillslopes under arable crops or intensive pasture. For example, the 1930s 'dust bowl' in southwest USA, caused by intensification of agricultural practice (see http://www.pbs.org/wgbh/amex/dustbowl/)[14.2]. In British Columbia, Canada, forest logging activities have resulted in a substantial increase in the number of shallow landslides[14.3].

2. *Changes in slope stability*: development can have a significant effect on slope stability e.g. the artificial recharge of the groundwater table and increased porewater pressures. For example, leakage from septic tanks and water supply pipes was identified as a cause of the 1987–88 ground movements at Luccombe, Isle of Wight, UK[14.4].

   Construction can affect slope stability, especially mining operations, excavations to create level plots for buildings or through road building (e.g. a 94-times increase in landsliding in the Nahwitti watershed, Vancouver Island, associated with access road construction; Table 14.1)[14.3].

3. *Increases in run-off within urban areas*: development is often accompanied by an increase in the amount

Figure 14.1 Tailings lagoon (formerly known as a 'slimes' lake) from iron mining operations in an area of primary rain forest near Bong Town, Liberia.

of impermeable surfaces within a catchment, increasing and accelerating runoff. At Stevenage and Skelmersdale, UK, the mean annual flood increased 2.5 times after urbanisation[14.5].

4. *Changes in river discharge and flood behaviour*: urbanisation can lead to channel modifications downstream following an increase in the magnitude of peak flows and a decrease in sediment supply due to bank protection works. Common downstream changes include bed and bank erosion and undermining of structures. Expansion of development onto floodplains reduces the floodwater storage capacity. This storage dampens the flood wave as it travels down a reach, reducing the peak of the

Table 14.1 Landslides associated with logging roads, Vancouver Island[14.3].

Watershed	Road related landslides	Area of roads (km^2)	Natural landslides	Area unlogged (km^2)	Road density ratio*
Macktush	50	1.5	17	14	27
Artlish	31	3.5	74	99	12
Nahwitti	69	4.3	30	176	94

* Road density ratio is the number of road-related landslides per square kilometre, divided by the number of natural landslides per square kilometre.

flood but extending its duration (i.e. the flood is spread out over a longer period and hence is less severe)[14.6].

Regular inundation by floodwaters can be important in leaching salts from floodplain soils. For example, construction of the Aswan Dams, Egypt (first dam completed in 1902) reduced the flooding of the Nile floodplain causing a build up of ground salts; this has led to problems for agriculture and contributed to the deterioration of the unique ancient monuments downstream[14.7].

Dramatic changes in river discharges can be brought about by the development of extensive irrigation schemes. The most controversial of these has been the diversion of the Amu Darya and Syr Darya rivers, which fed the Aral Sea in Central Asia, to create huge cotton plantations in this semi-arid area. The result is that between 1960 and 1995 the Aral Sea received 1 000 km^3 less water and was reduced in volume by 75% (see http://www.dfd. dlr.de/app/ land/aralsee/). In addition to the resulting ecological disaster, the land that has been exposed is extremely saline and high in pollutants, making it unsuitable for further development[14.8].

5. *Changes in sediment transport along rivers*: on many rivers the transport of coarse sediment has been severely restricted by man-made structures (e.g. weirs, reservoirs, discharge control structures) which act as sand and gravel traps. The removal of coarse material from riverbeds, either as part of flood defence maintenance or for supply of aggregates for construction also reduces the downstream transport. Dams decrease peak flows downstream and impound practically all sediment; channels tend to reduce in size downstream but also incise because of the lack of sediment. Before the Aswan High Dam (1964) the Nile carried an average of 124 million tons of sediment to the sea each year, and deposited another 9.5 million tons on the floodplain; now, 98% of the sediment is retained by the Nasser Reservoir[14.7].

6. *Changes in the delivery of sediment from river channels to floodplains*: the construction of flood embankments along many rivers has reduced the frequency of floodplain inundation and, hence, the deposition of fine sediments. This leads to higher sediment loads within the main channel, increased within-channel siltation, reduced channel capacity and increased flood hazard[14.6].

7. *Changes in the delivery of sediment from rivers to the open coast*: modification of river mouths and estuaries include land reclamation, dredging of navigation channels, and the development of ports and harbours. These activities influence the patterns of deposition and erosion of fine sediment, and the transport of coarse sediment bedload. Many rivers in developed countries now have little or no coarse sediment connection to the coast. Barrages on the Tijuana River, Mexico, for example, retain around 600 000 m^3 of sediment a year; as a result, beaches north of the river mouth have eroded[14.9].

8. *Reduction in coastal sediment inputs from cliff recession*: over the last 100 years or so some 860 km of protection works have been constructed around the English coast to prevent coastal erosion. Sediment inputs have declined by as much as 50% over this period, leading to the degradation of many beaches[14.10]. On the Whitby coast, UK, the cliffs deliver around 3 000 m^{3}yr^{-1} of potential beach-building material to the shoreline (the West Pier to Upgang Ravine section is protected by seawalls; the Upgang Ravine to Raithwaite Gill section is unprotected, see Table 14.2). The current sediment inputs are less than 40% of the pre-coastal defence inputs. This has led to a net decline in beach volume; overall beach losses over the last century are estimated in the order of 500 000 m^3. The impact has been a marked retreat (up to 60–70 m) of the LWM along much of the frontage.

Table 14.2 Contemporary decline in sediment inputs from cliff erosion on the Whitby coast, North Yorkshire, UK[14.10].

Cliff Section	Estimated coarse sediment proportion	Potential coarse sediment yield (m^3 yr^{-1})	Without coast protection — Estimated historical coarse sediment yield (m^3 yr^{-1})	With coast protection — Estimated current coarse sediment yield (m^3 yr^{-1})
West Pier – The Spa (W)	0.25	47	47	0
The Spa (W) – Upgang Ravine	0.25	4920	4920	0
Upgang Ravine – Raithwaite Gill	0.25	2945	2945	2945
		TOTAL	7912	2945
		Decline in coarse sediment yield (m^3 yr^{-1})		4966
		Ratio of current: historical coarse sediment yield		0.37

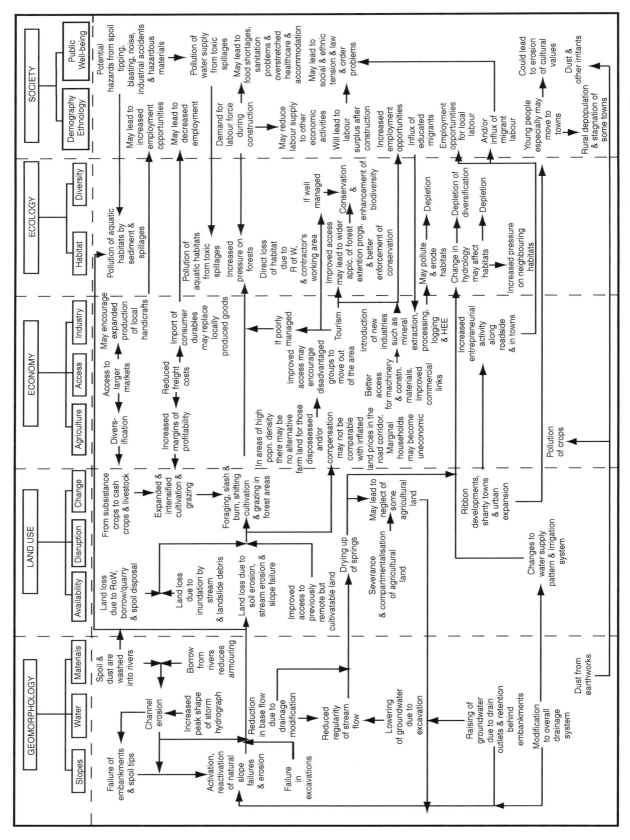

Figure 14.2　Cause-effect interrelationships between environmental impacts along mountain roads serving rural communities (after TRL 1997)[14.1].

9. *Disruption of longshore sediment transport*: groynes, harbour breakwaters and other shoreline structures have had a significant impact on longshore sediment transport processes. For example, construction of breakwaters in 1912 at the mouth of Lagos Lagoon, Nigeria, interrupted sediment transport and resulted in the erosion of the downdrift Victoria Beach. By 1975 the beach had retreated by 1300 m[14.11].

10. *Disruption of cross-shore sediment transport*: beaches and sand dunes form part of an integrated sediment exchange system (see Chapter 29). In many areas the response to foredune erosion has been to construct seawalls or rock revetments, severing the linkage between the dune and beach, causing the erosion of both[14.12].

11. *Decline in the volume of shoreline sediment stores*: beach mining for sand and gravel has had a significant impact on many beach and dune areas. At Hallsands, UK, around 400 000 m^3 of shingle was removed from the seaward side of the beach between 1897 and 1900 for construction in the Royal Navy dockyards at Devonport. The beach levels declined rapidly, falling 3–4 m by 1903, and wave attack led to the abandonment of the village (see http://www.hallsands. org.uk/).

Environmental Impact Assessment

Most large engineering projects will need to be supported by an environmental impact assessment (EIA). EIA is a tool used to identify the environmental, social and economic impacts of a project prior to decision-making. Environmental impact regulations and requirements will vary from country to country (for EU regulations go to http://europa.eu.int/comm/ environment/eia). The key elements of an EIA are:

- *Scoping*: identification of key issues and concerns of interested parties
- *Screening*: deciding whether a formal EIA is required based on information collected
- *Identifying and evaluating alternatives*: listing alternative project sites and construction techniques and the impacts of each; identifying environmentally unacceptable options; production of an *Environmental Action Plan*

- *Mitigating measures*: review of proposed actions to prevent or minimise the potential adverse effects of the project
- *Reporting*: preparation of environmental statements.

An example of an analysis of the impacts along mountain roads is provided in Fig. 14.2[14.1] . The impacts on the landscape and the affected population can be complex, and may take a number of years to become apparent. The earth surface systems approach (see Chapter 4) can provide a framework for identifying potential impacts on water, the soil and slopes.

References

14.1 TRL (1997) *Principles of low cost road engineering in mountainous regions, with special reference to the Nepal Himalaya*. Overseas Road Note 16, TRL, Crowthorne, 149 pages.

14.2 Bonnifield, P. (1979) *The Dust Bowl: Men, Dirt, and Depression*. University of New Mexico Press, Albuquerque.

14.3 Guthrie, R. (2002) The effects of logging on frequency and distribution of landslides in three watersheds on Vancouver Island, British Columbia. *Geomorphology*, **43**, 273–292.

14.4 Lee, E. M. and Moore, R. (1989) *Landsliding in and around Luccombe Village*. HMSO.

14.5 Knight, C. R. (1979) Urbanisation and natural stream channel morphology: the case of two English new towns. In G. E. Hollis (ed.) *Man's impact of the hydrological cycle in the United Kingdom*, 181–198. Geobooks.

14.6 Lee, E. M. (1995) *The Occurrence and Significance of Erosion, Deposition and Flooding in Great Britain*. HMSO.

14.7 Kassas, M. (1973) Impact of river control schemes on the shoreline of the Nile Delta. In M. T. Farvar and J. P. Milton (eds.) *The Careless Technology: ecology and international development*, 179–188, Stacey, London.

14.8 Regional report to the Central Asian States (2000) *State of Environment of the Aral Sea Basin*. Available from http://enrin.grida.no/aral/aralsea/ index.htm.

14.9 Bird, E. (2000) *Coastal Geomorphology: an Introduction*. Wiley and Sons, Chichester.

14.10 Brunsden, D. and Lee, E. M. (2004) Behaviour of coastal landslide systems: an interdisciplinary view. *Zeitschrift fur Geomorphologie*, **134**, 1–112.

14.11 Ibe, A. C. (1988) Nigeria. In H. J. Walker (ed.) *Artificial Structures and Shorelines*. Kluwer, Dordrecht, 287–294.

14.12 Pethick, J. S. and Burd, F. (1994) *Coastal Defence and the Environment*. MAFF Publications, London.

15

Slopes: The Supply of Water and Sediment

Introduction

A significant proportion of all construction takes place on hillslopes. They are systems that transfer water and sediment towards river channels. A slope can be sub-divided into segments[15.1]; each segment has its own inputs, outputs and storage of water and sediments (Fig. 15.1). The output from one segment forms the input to the next.

Slope form can be described as a two-dimensional *profile* (e.g. straight, concave, convex) from crest to toe. However, the three-dimensional form is also important (straight, concave, convex), as contour curvature is a prime control of the water and sediment transport pathways.

The slope form reflects a balance between the *imposed stresses* (e.g. gravity, seismic loads, raindrop impacts, surface water flows and wind), the surface protection provided by *vegetation* and the strength of the *materials* (see Chapter 7). There is a narrow range of slope angles for a given set of conditions (e.g. lithology, geological structure, geomorphological history, soils, vegetation, and climate): the greater the relative relief and overall drainage density, the steeper the hillslopes.

Water Flows: Infiltration and Run-off

On a *bare slope*, only a proportion of the rainfall enters the soil (through *infiltration*) and recharges the ground-water table; the remainder either contributes to *runoff* or is 'lost' by *evaporation* (Fig. 15.2).

On a *vegetated slope*, an even lower proportion of the rainfall contributes to groundwater recharge, *interception* by the tree canopy or other foliage (causing absorption or evaporation losses), root extraction of soil moisture and groundwater, and the subsequent loss to the atmosphere by transpiration (*evapotranspiration* is evaporation plus transpiration of moisture through foliage).

Surface water flow (*runoff*) occurs when the rainfall intensity during a storm or water released by snowmelt exceeds the rate of infiltration into the soil (*infiltration capacity*) or when the soil is saturated. Runoff occurs as shallow flows across a slope (*sheet flow* or *overland flow*) or as channelised flows in rills, gullies and streams. Saturated throughflow can also be important

in generating concentrated discharge adjacent to streams or gully heads (Fig. 15.3).

The amount of runoff from a catchment can be estimated in a number of ways, depending on the data availability and the degree of precision required. For example, the *US Soil Conservation Service Method* was developed in the USA to estimate the runoff from agricultural soil slopes[15.2], but is applicable in other settings:

- Determine the design storm event, e.g. 100 mm hr^{-1}.
- Determine the run-off curve number (Table 15.1), e.g. for Soil group C and paved roads the run-off curve is 90.
- Determine the run-off during the design storm (Fig. 15.4), e.g. for 100 mm rainfall and run-off curve 90, the depth of run-off is 72.5 mm.
- Determine the time of concentration (the longest time taken for water to travel by surface flow from any point in the catchment to the outlet; Fig. 15.5), e.g. for a steep (> 8°) 80 ha catchment, run-off curve 90 (line J on Fig. 15.5), the time of concentration is 0.15 hours.
- Determine the unit peak discharge expected from the design storm (Table 15.2), e.g. for a time of concentration of 0.1 hours, the unit discharge is 0.337 m^3 s^{-1}.
- Convert the unit peak discharge to the actual discharge:

    Peak discharge = Unit discharge
    $\times$ catchment area (km^2)
    $\times$ run-off depth
    = 0.337 $\times$ 0.8 $\times$ 72.5
    = 19.5 m^3 s^{-1}

Sediment Supply

Sediment is detached and removed from hillslopes in a number of ways (Fig. 15.6):

- surface erosion by water or wind, including sub-surface piping (see Chapters 17 and 18)
- mass movement, including soil creep, falls, slides and flows (see Chapter 19)
- in solution, including the removal of soluble rocks in karst terrain (see Chapter 21).

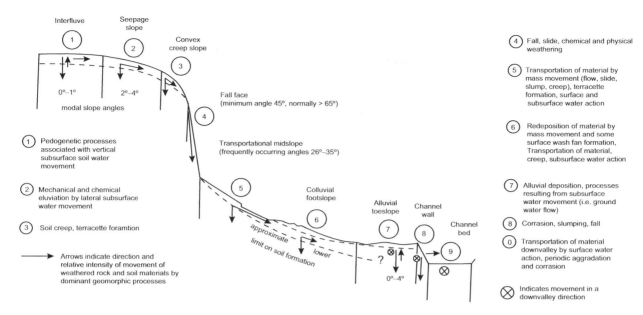

Figure 15.1 The hypothetical nine unit slope model and associated geomorphological processes (after Dalrymple *et al.* 1968)[15.1].

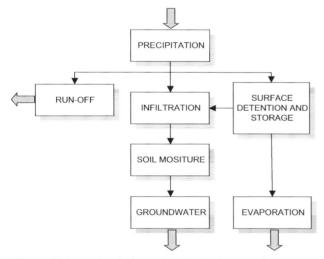

Figure 15.2 A simple bare-slope hydrology model.

High rates of sediment supply (i.e. high erosion or landslide susceptibilities) tend to be associated with areas of high relief mantled by (see Chapter 7):

• deeply weathered residual soils, e.g. in tropical regions (Fig. 15.7)
• glacial deposits, e.g. tills and fluvio-glacial deposits
• spreads of wind-blown silt (loess), e.g. the Yellow River catchment in China.

Climate is an important control on sediment yield[15.3] (Fig. 15.8). The mean annual sediment yield Q_s (tonnes km^{-2} yr^{-1}) from a catchment can be estimated from[15.4]:

$$\log Q_s = 2.65 \log \frac{p^2}{P} + 0.46 \log H \tan S - 1.56$$

where H = mean altitude (m) and S = mean slope (degrees).

The p^2/P index is a measure of the concentration of precipitation in a single month and, hence, is a measure of rainfall intensity (p = the highest mean monthly precipitation, P = the mean annual precipitation, in mm):

p^2/P	$\log p^2/P$	$\log H$	$\tan S$	$\log Q_s$	Q_s tonnes km^{-2} yr^{-1}
101.25	2.005	3.176	0.364	4.29	19322.3

$p = 450$ mm; $P = 2000$ mm; $H = 1500$ m; $S = 20°$; $Q = 10^{\log Q}$

Maximum values are predicted for the hot wetlands and semi-arid areas.

Transfer of sediment from hillslope to river channels involves a series of steps. For example:

• Weathered bedrock or soils may be stripped off a hillslope by gullying or shallow landslides, only to be stored as colluvium or taluvium on the lower slopes.
• This material might be gradually removed by stream erosion at the base of the slope or remain in storage in a debris fan for thousands of years.

The sediment currently being supplied to a river channel might have been detached from the hillslope by processes operating under past environmental conditions e.g. the periglacial solifluction sheets that mantle the lower slopes of many upland catchments in the UK.

Fig. 15.9 provides an indication of the various sediment stores that can be found in mountainous terrain[15.5].

Vegetation and Hillslopes

Vegetation and land use have a significant influence on sediment yield. In central USA, average sediment yields from cultivated slopes are 10–100 times those from forested areas. Vegetation cover can reduce surface erosion by:

• Providing a protective layer against rain-splash (by absorbing raindrop energy), runoff (by increased

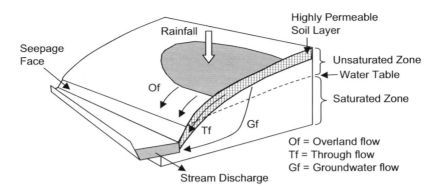

Figure 15.3 Slope-stream hydrology model.

Table 15.1 Runoff curve numbers for use in the Soil Conservation Service Method for estimating run-off [15.2].

Land Use	Hydrological Condition	Runoff Curves for Soil Groups (see below)			
		A	B	C	D
Range or pasture	Poor	68	79	86	89
	Fair	49	69	79	84
	Good	39	61	74	80
Meadow	Good	30	58	71	78
Woodland	Poor	45	66	77	83
	Fair	36	60	73	79
	Good	25	55	70	77
Dirt roads		72	84	87	89
Paved roads		74	84	90	92

Hydrological conditions as defined by the US Soil Conservation Service:

Poor: heavily grazed or plant cover < 50% or regularly burned.
Fair: moderately grazed or plant cover 50–75% or not regularly burned.
Good: lightly grazed, plant cover > 75%.

Soil Groups as defined for the US Soil Conservation Service run-off estimation model:

A: High infiltration capacity; sands, gravels, deeply weathered rocks; well-drained.
B: Moderate infiltration capacity; moderately to deeply weathered rocks; moderately to well-drained, moderately fine to moderately coarse pedological soil texture.
C: Low infiltration capacity; moderately fine to fine pedological soil texture, usually with a horizon that impedes drainage.
D: Very low infiltration capacity; swelling clays, soils with permanent high water tables, soils with clay lenses, shallow soils over impervious materials.

Table 15.2 Unit peak discharges [15.6].

Time of concentration (hours)	Unit peak discharge (m³ s⁻¹)	Time of concentration (hours)	Unit peak discharge (m³ s⁻¹)	Time of concentration (hours)	Unit peak discharge (m³ s⁻¹)
0.1	0.337	1	0.158	4.0	0.063
0.2	0.300	1.5	0.120	5.0	0.054
0.3	0.271	2.0	0.100	10.0	0.034
0.4	0.246	2.5	0.086	20.0	0.021
0.5	0.226	3.0	0.076	24.0	0.019

Discharge rate in m^3 s^{-1} from a run-off depth of 1 mm and a discharge area of 1 km^2.

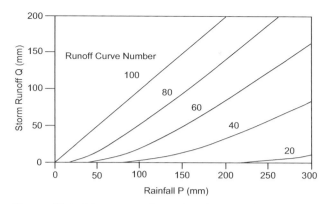

Figure 15.4 Nomogram for estimating runoff from rainfall (after the US Soil Conservation Service 1972)[15.2].

Figure 15.6 Supply of rockfall debris to a mountain stream channel, Aqaba mountains, Egypt.

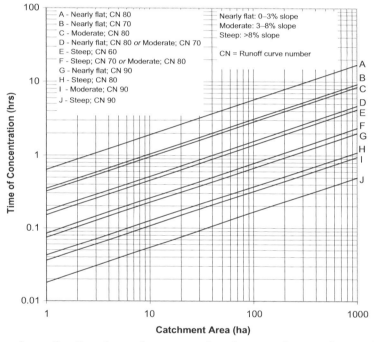

Figure 15.5 Nomogram for estimating time of concentration from catchment characteristics (after the US Soil Conservation Service 1972)[15.2].

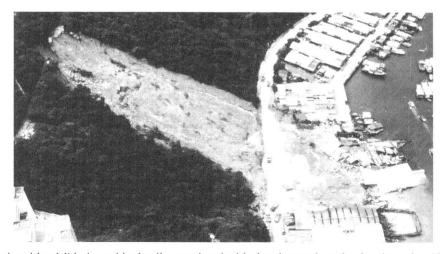

Figure 15.7 Translational landslide in residual soils associated with deeply weathered volcanic rocks, Aberdeen Harbour, Hong Kong.

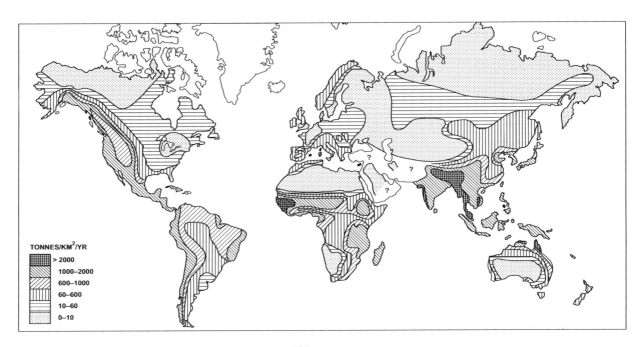

Figure 15.8   Global erosion rates (after Stoddart 1969)[15.3].

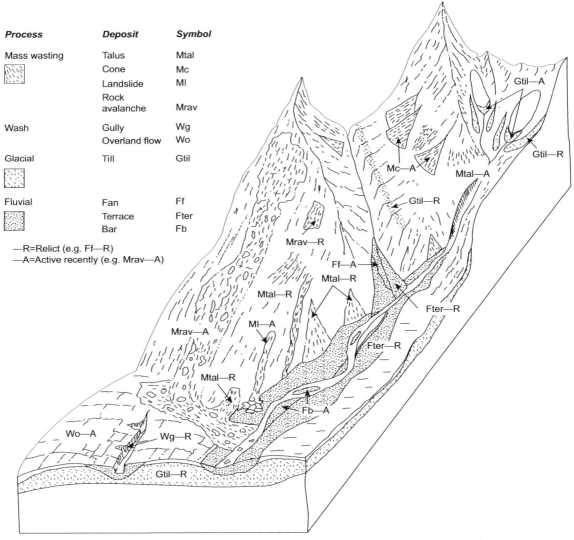

Figure 15.9   Typical sediment stores in mountain terrain (after Selby 1993)[15.5].

surface roughness) or wind erosion (by reducing the shear velocity of wind).

- Increasing the soil strength e.g. it decreases pore water pressure and increases soil suction because of its own water requirement.
- Improving soil structure and porosity through enrichment with organic material and protecting the soil from trampling by animals.

Vegetation can influence the occurrence of shallow landslide activity in a number of ways:

- Modifying soil moisture content and groundwater levels and, hence, the pore water pressures operating along landslide shear surfaces.
- Reinforcing the soil or anchoring along landslide shear surfaces i.e. increasing the shear strength of the slope materials by providing artificial cohesion ΔC (this can be between 1–12 kPa) and increasing the frictional strength component e_r. The Coulomb equation (Chapter 3) can be rewritten as:

$$s = (c' + \Delta C) + (\sigma - u + e_r) \tan \phi'$$

where s = shear strength; c' = effective cohesion and $(\sigma - u) = \sigma'$ = effective stress.

- Roots can have very high tensile strengths; some tree roots can have tensile strengths of up to 90 MPa (i.e. 30% of the tensile strength of mild steel).
- Taproots or sinker roots of large trees can act as slope stabilising piles; if the trees are close together

arching may develop. Decay of root systems removes reinforcement and may promote surface erosion and instability.

- By wedging blocks of rock apart, increasing the potential for detachment and boulder falls.
- By imposing a surcharge (i.e. increased weight, for example, 70 kPa for 60 m-high Douglas fir trees to 5 kPa for 6 m-high Sitka spruce) or dynamic wind-shear force onto the slope materials. Close to the edge of a wood a wind of 90 km hr^{-1} could produce a transmitted shear stress of around 1 kPa.

References

15.1 Dalrymple, J. B., Blong, R. J. and Conacher, A. J. (1968) A hypothetical nine unit land surface model. *Zeitschrift fur Geomorphologie*, **12**, 60–76.

15.2 US Soil Conservation Service (1972) *National Engineering Handbook*; Section 4 Hydrology. US Department of Agriculture, Washington DC.

15.3 Stoddart, D. R., (1969) World erosion and sedimentation. In R. J. Chorley (ed.) *Water, Earth and Man*. Methuen, London, 43–64.

15.4 Fournier, F. (1960) *Climate et érosion: la relation entre l'érosion du sol par l'eau et les précipitation atmosphériques*. Paris, Presses Universitaires de France.

15.5 Selby, M. J. (1993) *Hillslope Materials and Processes*. 2nd Edition Oxford University Press, Oxford.

15.6 FAO (1976) *Hydrological Techniques for Upstream Conservation*. FAO Conservation Guide 2. FAO Rome.

16

Slopes: The Role of Water

Introduction

Water is usually the most important factor in landform change, although wind can be as important in deserts and coastal dunes (Chapters 18 and 35). There are close links between the absolute amounts of precipitation (rainfall, snow, hail and dew), the nature and intensity of surface processes and the generation of hazards. In many regions it may be the extremes, such as high intensity rainfall events or prolonged wet periods (e.g. NAO or El Niño events; see Chapter 9) that trigger soil erosion and landsliding.

Table 16.1 provides an indication of the relative efficiency of rainfall, runoff (overland flow) and channelised flow in eroding and transporting material. Channelised flow (i.e. rills, gullies and streams) is the most effective mechanism and, hence, river channels tend to be the most dynamic part of a landscape (see Chapters 22–25).

Rainfall Intensity and Soil Erosion

Raindrop impact on the ground surface can detach soil particles (i.e. individual grains of material) and lead to the onset of surface erosion (see Chapter 17). Only around 0.2% of raindrop energy is available to initiate erosion, the remainder is dissipated in friction. The soil erosion rate increases with rainfall intensity (Table 16.2).

Table 16.1 Efficiency of forms of water erosion[16.1].

Form	Mass[a]	Typical velocity (m/s)	Kinetic energy[b]	Energy for erosion[c]
Raindrops	R	9	40.5R	0.081R
Overland flow	0.5R	0.01	$2.5 \times 10^{-5}R$	$7.5 \times 10^{-7}R$
Channel flow	0.5R	35[d]	306R	9.2R

[a] Assumes rainfall of mass R of which 50% contributes to runoff.
[b] Based on $0.5mv^2$.
[c] Assumes 0.2% kinetic energy of raindrops and 3% of kinetic energy of runoff is utilised for erosion.
[d] Based on Manning's equation. Assumes a channel 20 m wide, 2.5 m deep, on a slope of 11° and bed roughness coefficient of 0.0197 (see Chapter 23).

Table 16.2 Rainfall intensity and soil erosion rate: Ohio, US 1934–1942[16.2].

Maximum 5-minute intensity (mm hr^{-1})	Number of falls of rain	Average erosion rate (t ha^{-1})
0–25.4	40	3.7
25.5–50.8	61	6.0
50.9–76.2	40	11.8
76.3–101.6	19	11.4
101.7–127.0	13	34.2
127.1–152.4	4	36.3
152.5–177.8	5	38.7
177.9–254.0	1	47.9

A general relationship between rainfall intensity I (mm hr^{-1}) and kinetic energy (J m^{-2} mm^{-1} i.e. the *erosivity* of rainfall) is[16.3]:

$$KE = 11.87 + 8.73 \log I$$

For tropical rainfall[16.4]:

$$KE = 29.8 - \frac{127.5}{I}$$

At rainfall intensities greater than 75 mm hr^{-1}, kinetic energy levels off at around 29 J m^{-2} mm^{-1}.

A number of indices have been developed to express the relationship between kinetic energy and soil loss, including EI_{30} (the product of the kinetic energy E of a storm and the maximum 30-minute rainfall intensity[16.3]; Table 16.3). EI_{30} is used to estimate the rainfall erosivity factor R in the Universal Soil Loss Equation (see Chapter 17):

$$R = \frac{EI_{30}}{1000}$$

where E is measured in J m^{-2} and I_{30} in mm hr^{-1}. Measures of mean annual erosivity E_{va} (J m^{-2}) include[16.1]:

$$E_{va} = 0.865\ P$$
$$E_{va} = 9.28\ P - 8838.15$$

where P is the mean annual precipitation (mm).

Raindrop impact can also lead to soil compaction and the development of a thin surface crust. This will limit the rate of soil detachment but promote greater surface runoff.

Run-off and Surface Erosion

Flow along rills and gullies must attain a threshold velocity before entrainment of particles begins (the velocity of fully turbulent flow can be determined using *Manning's equation*; see Chapter 23).

Once this critical velocity is reached (it varies with particle size; Fig. 3.2), the entrainment rate is dependent on the shear velocity of the flow and the discharge. The transport capacity of the flow T_f is related to the discharge Q (m^3 s^{-1}) and the slope gradient S[16.5]:

$$T_f = 0.0085\ Q^{1.75}\ S^{1.625}\ D_{84}^{-1.11}$$

where D_{84} = the sediment particle size (mm) for which 84% of the sediment mixture is finer.

Rainfall Intensity and Landsliding

Landslide activity may be associated with:
- critical pore water pressure threshold being exceeded;
- increased weight of the saturated soil.

High intensity rainfall can trigger shallow landslides and debris flows within minutes or hours of the event (see Chapter 19). For example, in October 1985 the tropical storm Isabel dropped over 560 mm of rainfall within a 24 hour period on the island of Puerto Rico. This triggered the Mameyes rock slide which destroyed 120 homes and killed 129 people[16.6].

Equations have been developed to define the minimum rainfall intensity required to trigger debris flows and shallow landslides during storms of varying durations[16.7–8]:

$$i_r = 14.82\ D^{-0.39}$$
$$I_r = 4.94\ D^{0.50}$$

where i_r = rainfall intensity (mm hr^{-1}); D = rainfall duration (hours) and I_r = total rainfall (mm).

Threshold rainfall intensities may vary with climate and can be expressed as a percentage of the mean annual precipitation P[16.9]:

$$P\ (\%) = 1.1\ D^{0.56}$$

In Table 16.4, the critical threshold during a 3-hour storm will be 40 mm and 14 mm for areas with mean annual precipitation of 2 000 mm and 700 mm, respectively.

Almost all shallow debris slides in Hong Kong are triggered by heavy rainfall. Rainfall events that trigger landsliding at *medium densities* (1–10 landslides per km^{-2}) occur every 2 years at the local scale i.e. any given site, and 5 times per year for the region i.e. Hong Kong as a whole (see Table 16.5)[16.10].

Groundwater and Landsliding

Landslide activity is often associated with periods of heavy rainfall due to the change in pore water pressures

Table 16.3 Calculation of rainfall Erosivity index EI_{30}.

Time from start (min)	Rainfall (mm)	Intensity I (mm h^{-1})	Kinetic energy (J m^{-2} mm^{-1})	Total kinetic energy E (J m^{-2})	EI_{30} (J mm^{-1} m^{-2} hr^{-1})
0–30	15.74	31.48	25.75	405.3	
30–60	40.38	80.76	28.22	1139.6	
60–90	39.88	79.76	28.20	1124.7	
TOTAL				2669.6	215592.9

I = 30 minute intensity × 2 (i.e. mm hr^{-1}).
Kinetic energy = 29.8 – (127.5 / I).
Total kinetic energy = Rainfall × Kinetic energy.
EI_{30} = Maximum 30-minute intensity (80.76 mm) × Σ(Total kinetic energy).

(see the effect of varying water table on slope stability in Fig. 19.5). For example, 4 periods of landsliding occurred around Lisbon, Portugal between the 1960s and 1990s. Each period was characterised by different rainfall patterns and landslide types[16.11]:

- *Moderate intensity rainfall episodes* (5–10 day episodes with 98–165 mm, 2–5 year return period): river bank failures and shallow translational slides on man-made slopes.
- *High intensity rainfall episodes* (12 hour episodes with short periods of very high intensity e.g. 14 mm in 5 minutes; 70 year return period): shallow translational slides on steep slopes, repeated earth slides and earthfalls.
- *Long lasting rainfall periods* (wet winters e.g. 75 days with a total of 694 mm; 25 year return period): major, deep-seated, translational, rotational and complex slides.

In general, the deeper the slide the longer the period of antecedent heavy rainfall required to initiate failure (i.e. the time required to raise the groundwater table and pore water pressures to a critical level). The period may vary from several days to many months; the antecedent rainfall over a 4-month period was found to be important in controlling landslide reactivation events in the Isle of Wight Undercliff, UK[16.12]. In many areas, it may be that pattern of wet years that controls the occurrence of landslides (e.g. in the Republic of Georgia[16.13]; Fig. 16.1).

Water pressures are important in rock slope stability in the following ways:

- Water pressure in discontinuities (*cleft pressures*) apply a force that may trigger instability. Rock mass strength may be reduced by 30% where clefts are filled with water to the ground surface.

- *Freeze-thaw action*: freezing of groundwater restricts water flows, increasing cleft pressures. For example, prolonged periods of air frost are associated with increased rock fall activity on Chalk cliffs in the UK[16.14]. There was a marked increase in rock fall activity in Norway during the Little Ice Age[16.15] (see Chapter 9).

Groundwater and Solution Weathering

Limestone and other carbonate rock types (dolomite, chalk) are soluble in natural waters that contain carbon dioxide (CO_2). Other non-carbonate rock types such as gypsum, anhydrite (calcium sulphate) and halite (sodium chloride) are also soluble in water. After removal in solution, these rocks may leave almost no residue.

Solution weathering leads to the development of *karst* terrain with underground drainage and distinctive landforms (e.g. caves, closed depressions and sinkholes; see Chapter 21). It tends to widen discontinuities, allowing water to penetrate and widen the fractures further.

The rate of solution of carbonate depends on:

- CO_2 concentration of the water: the chemical reaction of calcium carbonate which leads to weathering is usually simplified as:

$$CaCO_3 + CO_2 + H_2O \rightarrow Ca(HCO_3)_2$$

Calcium bicarbonate ($Ca(HCO_3)_2$) is removed in solution. This is a reversible processes and re-precipitation of calcium carbonate can occur as *tufa* or the more resistant crystalline form *travertine*.

- The amount of water in contact with the rock (i.e. the discharge) and the time of contact (i.e. velocity of flow).

Table 16.4 Rainfall thresholds for initiating debris flows/slides in areas with different mean annual precipitation P[16.9].

Duration D (hours)	P% (= $1.1\ D^{0.56}$)	Threshold rainfall intensity (mm)	
		For climate where P = 2000 mm year^{-1}	For climate where P = 700 mm year^{-1}
3	2	40	14
12	4.4	88	30.8

Threshold rainfall intensity = Mean annual precipitation × P% / 100 = 2000 × 2 / 100 = 40 mm

Table 16.5 Rainfall thresholds for landslide activity in Hong Kong[16.10].

	Threshold I – start of landsliding		Threshold II – start of landsliding at medium densities (1–10 Slides km^{-2})		Threshold III – start of landsliding at high densities (>10 Slides km^{-2})	
	Rainfall	Return period	Rainfall	Return period	Rainfall	Return period
Local	60–70 mm	2.5 times per year	180–220 mm	Every 2 years	380–450 mm	Every 20 years
Regional		25 times per year		5 times per year		Every 2 years

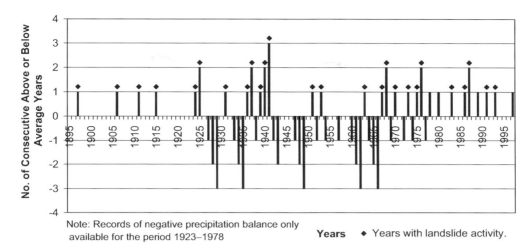

Figure 16.1   Landsliding in Georgia: wet (above average) and dry (below average) year sequences.

- The temperature of the water: higher temperatures give lower CO_2 solubility (at 20°C solubility is 50% of that at 0°C) i.e. potential for limestone solution is higher in colder climates and during glaciations than in dry hot areas.
- The presence of natural organic acids.

Worldwide solution rates are low in arid and semi-arid regions, relatively high in seasonally wet regions (e.g. the Mediterranean) and highest in wet tropical regions due to the presence of high levels of biogenic CO_2 (from vegetation) in the groundwater e.g. Papua New Guinea today.

References

16.1 Morgan, R. P. C. (2004) *Soil Erosion and Conservation*. 3rd edition, Blackwell, Oxford.

16.2 Fournier, F. (1972) *Soil Conservation*. Nature and Environmental Series, Council of Europe.

16.3 Wischmeier, W. H. and Smith, D. D. (1958) Rainfall energy and its relationship to soil loss. *Transactions of the American Geophysical Union*, **39**, 285–291.

16.4 Hudson, N. W. (1995) *Soil Conservation*. B T Batsford, London.

16.5 Carson, M. A. and Kirkby, M. J. (1972) *Hillslope Form and Process*. Cambridge University Press.

16.6 Jibson, R. W. (1992) The Mameyes, Puerto Rico, landslide disaster of October 7, 1985. In J. A. Johnson and J. E. Slosson (eds.) Landslide/Landslide Mitigation. Geological Society of America, *Reviews in Engineering Geology*, **9**, 37–54.

16.7 Caine, N. (1980) The rainfall intensity-duration control of shallow landslides and debris flows. *Geografiska Annala*, **62**, 23–27.

16.8 Innes, J. L. (1983) Debris flows. *Progress in Physical Geography*, **7**, 469–501.

16.9 Sandersen, F. (1997) The influence of meteorological factors on the initiation of debris flows in Norway. In J. A. Matthews, D. Brunsden, B. Frenzel, B. Glaser and M. M. Weiss (eds.) *Rapid mass movement as a source of climatic evidence for the Holocene*, 320–330, Gustav Fischer Verlag.

16.10 Evans, N. C., Huang, S. W. and King, J. P. (1997) *The Natural Terrain Landslide Study: Phase III*. GEO Special Project Report SPR 5/97.

16.11 Zêzere, J. L., Ferreira, A. B., and Rodrigues, M. L. (1999) The role of conditioning and triggering factors in the occurrence of landslides: a case study in the area north of Lisbon (Portugal). *Geomorphology*, **30**, 133–146.

16.12 Lee, E. M., Moore, R., and McInnes, R. G. (1998) Assessment of the probability of landslide reactivation: Isle of Wight Undercliff, UK. In D. Moore and O. Hungr (eds.) *Engineering Geology: The View from the Pacific Rim*, 1315–1321. Balkema.

16.13 Varazashvili, L., Rogava, D. and Tsereteli, E. (1998) Landslide hazards in Georgia and the problems of stable development of population. In D. Moore and O. Hungr (eds.) *Engineering Geology: The View from the Pacific Rim*. 4107–4111. Balkema.

16.14 Hutchinson, J. N. (1972) Field laboratory studies of a fall in Upper Chalk Cliffs at Joss bay, Isle of Thanet. In *Proceedings of the Roscoe Memorial Symposium*, Cambridge Session, **6**, 692–706.

16.15 Grove, J. M. (1972) The incidence of landslides, avalanches and floods in western Norway during the Little Ice Age. *Arctic and Alpine Research*, **4**, 131–138.

17

Slopes: Soil Erosion by Water

Introduction

Soil erosion involves the removal of near-surface materials from hillslopes (Fig. 17.1). It is a three-stage process: *detachment* of material, its *transport* by water or wind and *deposition* when sufficient energy is no longer available to transport the material. Erosion is a function of the power of water or wind (*erosivity*; see Chapter 16) and the resistance/strength of the material (*erodibility*; see Chapter 7). The severity of erosion depends on the quantity of material supplied by detachment and the ability of the running water or wind to carry it (see Chapter 18 for wind erosion).

The highest rates of surface water erosion are associated with deforestation or construction[17.1]

Figure 17.1 Soil erosion removing all near-surface materials and exposing bedrock in the Ethiopian Highlands.

(Table 17.1). For example, during construction in Maryland, USA, sediment yields reached 55 000 t km^{-2} yr^{-1}; neighbouring undisturbed areas yielded <400 t km^{-2} yr^{-1} [17.2].

Forms of Soil Erosion

- *Overland flow*: rarely occurs as a single 'sheet' (despite its common name of sheetflow) but more often as a collection of unstable *anastomosing* very shallow water courses with no defined channel. Individual soil particle entrainment and transport is related to the velocity of flow (see Chapter 16).
- *Rill erosion*: initiated downslope at the point where overland flow becomes channelised. Rills form when the ratio between the shear stress exerted by the flow and the shear strength of the soil > 0.0001–0.0005. Once formed, rills can migrate upslope by headward retreat.
- *Gully erosion*: more permanent steep-sided channels that become integrated with the stream network. Gullies are most common in weak materials (e.g. loess, alluvium, colluvium, gravels, deeply-weathered soils). Initiation is often accompanied by rapid incision, channel widening and gully head retreat. Over time the channel gradient is reduced and the gully may stabilise as the channel infills and the flanks degrade and re-vegetate. The next phase of activity may be triggered by climate, a change in slope gradient (e.g. undercutting by a stream or pipe collapse) or land use change.
- *Piping*: sub-surface tunnels or pipes can develop where throughflow is concentrated along erodible soil layers. Common settings include a permeable soil horizon below an impermeable horizon or the occurrence of a weak clay horizon. Piping can be widespread in semi-arid regions where the soils contain swelling clays that crack on drying. Pipes often collapse and initiate gullying; they are often associated with landslide backscars.
- *Badlands*: terrain comprising steep slopes dissected by rills and gullies, narrow elongate ridgelines, high drainage density, thin bare soils, rapid erosion and shallow landsliding. Badlands are often associated with weak, impermeable, smectite-rich mudrocks in

Table 17.1 Representative rates of erosion in the USA[17.1].

Land use	Sediment yield (t km^{-2} yr^{-1})	Relative to forestry = 1
Forest	8.5	1
Grassland	85	Increase 10 ×
Abandoned surface mines	850	Increase 100 ×
Cropland	1700	Increase 200 ×
Construction sites	17 000	Increase 2000 ×

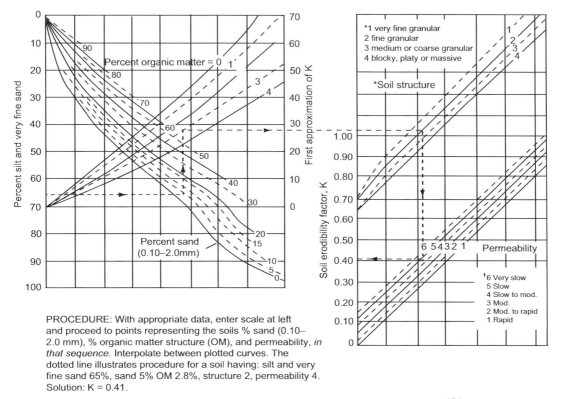

PROCEDURE: With appropriate data, enter scale at left and proceed to points representing the soils % sand (0.10–2.0 mm), % organic matter structure (OM), and permeability, *in that sequence*. Interpolate between plotted curves. The dotted line illustrates procedure for a soil having: silt and very fine sand 65%, sand 5% OM 2.8%, structure 2, permeability 4. Solution: K = 0.41.

Figure 17.2 Nomogram for estimating the erodibility factor K (after Wischmeier *et al*. 1971)[17.4].

semi-arid upland areas. Other weakly cemented sediments such as loess may also develop badlands; the sediment yield from the Loess Plateau of China is around 38,000 t km^{-2} yr^{-1} [17.3].

Soil Erodibility

Erodibility is the resistance of the slope materials to surface erosion (i.e. particle detachment and transport). Silts and fine sands are most susceptible to erosion (see Fig. 3.2). Other soils that are particularly susceptible to erosion include: dry organic soils, some residual soils, and lightweight man-made fill materials e.g. pulverised fuel ash (PFA).

The most widely used *erodibility index* is the K value that represents the soil loss per unit of EI_{30} (as measured on a standard bare soil plot, 22 m long and on a 5° slope; see Chapter 16). K values can be estimated from nomographs[17.4] (Fig. 17.2).

Hazards

- *On-site* problems, including the encroachment of gully-heads onto a route alignment (roads, railways and pipelines) or site, loss of land, or exposure of services.
- *Off-site* impacts, including blocked road or railway culverts and drainage ditches, and damage to property. In the UK, over 30 separate mud floods were reported in the South Downs between 1976 and 1990, affecting around 200 houses[17.5].
- Loss of water storage capacity and power head as eroded soil is deposited in reservoirs (Fig. 17.3). For example, the capacity of the San Gabriel Dam, Los Angeles, declined from 65 Mm^3 in 1938 to 55 Mm^3 by 1980[17.6]. During this period some 22 Mm^3 of sediment had been removed at a cost of $20M (1981 prices).
- Decline in water quality, involving the transfer of agricultural chemicals to watercourses, discoloration and

increased cost of treatment. For example, between 1960 and 1978 there was a 90% decline in sea trout catches in the River Fleet, Scotland, believed to be due to increased sediment loads following the expansion of forestry in the catchment area[17.7].

- Post-construction reinstatement difficulties; this can be a major issue for projects such as roads, railways or pipelines.
- Sedimentation in watercourses, leading to reduced channel capacity and, hence, increased flood risk. High sediment loads may also trigger significant alluvial river channel changes (see Chapter 25).
- Surface erosion can severely limit the long-term sustainable use of agricultural land (i.e. *soil degradation*, see http://www.unesco.org/ or http://www.fao.org/). The Ethiopian Highlands, for example, have lost around 1 Mtonnes of topsoil per year as a result of overuse. Enhanced soil erosion is associated with farming on steep slopes, forest and natural vegetation clearance, overgrazing, overcropping and bad farming practices (e.g. leaving fields bare during the rainy season, ploughing downslope).

Erosion Hazard Assessment

The most widely used equation for predicting general soil loss from rain-splash and runoff is the Universal Soil Loss Equation (USLE, see http://www.ars.usda.gov/Research/)[17.8–9]. It needs to be used with caution, because it was derived from measurements of soil loss from agricultural plots in the USA (Table 17.2):

$$E = R \times K \times LS \times C \times P$$

E = average annual soil loss (t ha–1); R 5 rainfall erosivity factor, based on the mean annual EI30 (see Chapter 16; for R to be in metric units, R 5 EI30/ 1000); K 5 soil erodibility index (see Fig. 17.2); C 5 crop management factor and represents the ratio of soil loss under a given crop to that from bare soil, and P 5 erosion control factor (P 5 1.0 for no erosion control).

Figure 17.3   Reservoir behind a dam infilled with sediment prior to its intended completion in 1973 (southeast Spain).

LS is the combined slope length L (m) and slope steepness factor S (%) derived from:

$$LS = \left(\frac{L}{22.13}\right)^{n}(0.065 + 0.045S + 0.0065S^{2})$$

Table 17.2   Universal soil loss equation: worked example.

Factor	Source	Value
R	Mean annual erosivity = 0.865 P (Chapter 16) P (mean annual precipitation) = 3000 mm	2595

K	Using Fig. 17.1 assuming:
	     proportion of sand and silt = 65%
	     proportion of sand = 5%
	     amount of organic matter = 2.8%
	     structural class = 2
	     permeability class = 4
		0.41

LS	$LS = (L/22.13)^{0.5}(0.065 + 0.045\ S + 0.0065\ S^{2})$ $L = 50$ m $S = 12\%$ (i.e. 0.045×12, etc.)	2.31
C	Bare soil	1.0
P	No erosion practice	1.0
Mean annual soil loss $E = R \times K \times LS \times C \times P$ (t ha^{-1})		= 2464 t ha^{-1}

The exponent n varies with slope steepness; for 3° slopes n = 0.4; for 2° slopes n = 0.2 and for < 1° slopes n = 0.1. For slopes > 6° n = 0.6.

Soil erosion classes provide a useful basis for any investigation (Table 17.3). Erosion rates greater than 10 t ha^{-1} yr^{-1} generally exceed the capacity of the soil to naturally replace itself; remedial action will be necessary[17.10]. For engineering projects, acceptable levels are dependent on the risk to the scheme. For example, in a reservoir design the soil erosion rates may be one of the critical factors in defining the design life[17.11]. On a road or railway project, if erosion rates are known to be high, culverts can be designed to be self-cleaning.

Table 17.3 Soil erosion classes: erosion by wind or water.

Units	None/slight	Moderate	High	Very high
t ha^{-1} yr^{-1}	< 10	10–50	50–200	> 200
mm yr^{-1}	< 0.6	0.6–3.3	3.3–13.3	> 13.3

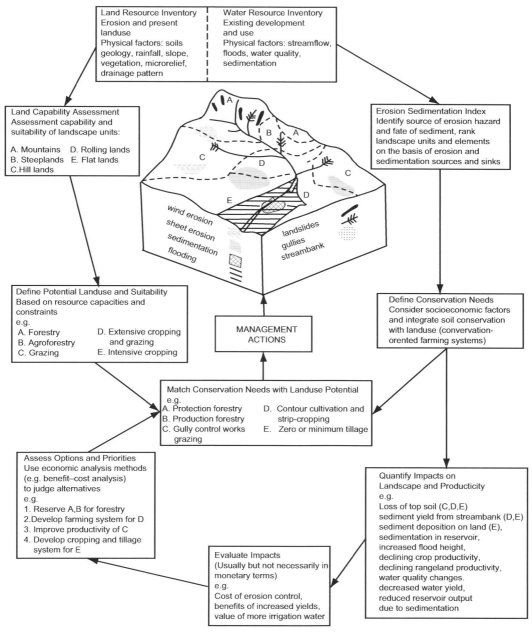

*A, B, C, D, E refer to landscape units shown in the diagram.

Figure 17.4   Development of a soil erosion strategy (after Perrens and Trustrum 1984)[17.12].

Figure 17.5 Agronomic measures employed to reduce soil erosion in the Ethiopian Highlands (trash lines i.e. piles of vegetation).

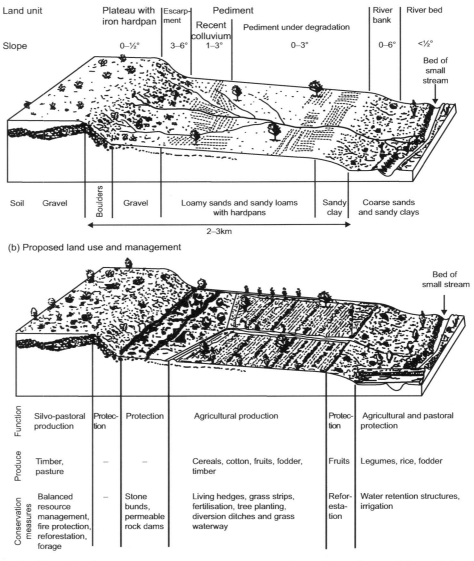

Figure 17.6 Developing a land use management strategy to restrict soil erosion problems (after Hijkoop *et al.* 1991)[17.13].

Management and Soil Conservation

Soil conservation strategies may comprise (e.g. Fig. 17.4; see http://www.nrcs.usda.gov/ or http://www.fao.org/):

- *Agronomic measures*, such as mulching, high-density planting, crop rotation, multiple, strip and cover cropping, agroforestry (Fig. 17.5).
- *Farming management techniques*: contour tillage, ridging, controlled or limited tillage; controlled grazing, barn-raised livestock, increased training of farmers.
- *Mechanical methods*: trash racks or lines, bunds, terraces, built channels for controlled removal of water (waterways), erosion control structures in gullies (check dams), armouring over buried structures.

It is necessary to understand both the causes of accelerated erosion and to measure or estimate the rate at which soil is being lost. Remedial measures must be part of an overall land use and management strategy[17.13] (e.g. Fig. 17.6).

References

17.1 US Environmental Protection Agency (1973) *Methods for Identifying and Evaluating the Nature and Extent of Nonpoint Sources of Pollutants*. Washington DC.

17.2 Wolman, M. G. and Schick, A. P. (1967) Effects of construction on fluvial sediment, urban and suburban areas of Maryland. *Water Resources Research*, **3**, 451–464.

17.3 Chen, Y. (1983) A preliminary analysis of the processes of sediment production in a small catchment on the Loess Plateau. *Geographical Research* (China), **2**, 35–47.

17.4 Wischmeier, W. H., Johnson, C. B. and Cross, B. V. (1971) A soil erodibility nomograph for farmland and construction sites. *Journal of Soil and Water Conservation*, **26**, 189–193.

17.5 Boardman, J. (1990) Soil erosion on the South Downs: a review. In, J. Boardman, I. D. L. Foster and J. A. Dearing (eds.), *Soil Erosion on Agricultural Land*, 87–105. John Wiley and Sons, Chichester.

17.6 Bruington, A. E. (1982) Fire-loosened sediment menaces the city. In Proceedings of the Symposium on *Dynamics and Management of Mediterranean-type ecosystems*, June 22-26, 1981, San Diego, California. USDAFS General Technical Report PSW-58, 40–422.

17.7 Lee, E. M. (1994) *Occurrence and Significance of Erosion, Deposition and Flooding in Great Britain*. HMSO.

17.8 Wischmeier, W. H. and Smith, D. D. (1978) *Predicting rainfall erosion losses*. USDA Agricultural Research Service Handbook 537.

17.9 Renard, K. G., Foster, G. R., Weesies, G. A. and Porter, J. P. (1991) RUSLE: revised universal soil loss equation. *Journal of Soil and Water Conservation*, **46**, 30–33.

17.10 Food and Agriculture Organisation (FAO) (1996) *Methods and materials in soil conservation*. FAO, Rome. Available from: http://ftp.fao.org/ agl/agll/docs/mmsoilc.pdf

17.11 Griffiths, J. S. and Richards, K. S. (1989) Application of a low-cost database to soil erosion and soil conservation studies in the Awash Basin, Ethiopia. *Land Degradation & Rehabilitation*, **1**, 241–262.

17.12 Perrens, S. J. and Trustrum, N. A. (1984) *Assessment and evaluation for soil conservation policy*. East-West Environment Policy Institute, Workshop Report, Honolulu, Hawaii.

17.13 Hijkoop, J., van der Poel, P. and Kaya, B. (1991) *Une lutte de longue haleine*. Aménagements anti-érosifs et gestion de terroir. IER Bamako/KIT, Amsterdam.

18

Slopes: Wind Erosion and Deposition

Introduction

Air velocity and surface sediment grain size are the main factors controlling the erosion and deposition of sediment by wind (*aeolian* processes). Wind speeds are lowest nearest the ground due to surface roughness (i.e. obstacles such as stones and vegetation causes the air flow to become turbulent). Winds generate shear forces on the ground surface. This force is expressed in terms of a *drag* or *shear velocity*; a measure of the capability of the wind to *entrain* (i.e. move) individual soil particles. The drag velocity u^* (m s^{-1}) at a particular height above the ground surface increases with wind speed[18.1] (Table 18.1):

$$u^* = \frac{u_z}{5.75 \log \dfrac{z}{z_0}}$$

where u_z = velocity at the height of measurement z and z_0 is the surface roughness length, approximately equal to $d/30$ where d is the mean particle diameter (m).

Wind erosion commences when air pressure on a loose, dry soil surface overcomes the force of gravity acting on the particles. Particles (generally < 0.2 mm, i.e. fine sand and smaller) are moved through the air by *saltation* (bouncing), *surface creep* or *suspension* within the flow. The most erodible particles are 0.10–0.15 mm in size, although particles between 0.05–0.5 mm may be removed by wind. The *saltation curtain* is generally less than 0.5 m high. Dust is carried in suspension, often at great heights and for considerable distances.

Threshold shear velocities u^*_t are related to particle size[18.1] (Fig. 3.2 and Table 18.2):

$$u^*_t = A \sqrt{\frac{\rho_p - \rho_a}{\rho_a} g d}$$

where ρ_a = density of air; ρ_p = density of the particle; g is the gravitational acceleration; d = particle diameter and A is a constant:

- The *fluid threshold* is the velocity needed to initiate saltation (A = 0.1).
- The *impact threshold* is that needed to initiate entrainment by the impact of bouncing particles (A = 0.08).

Table 18.1 Change in shear velocity with height.

Wind velocity (m s^{-1})	Height z (m)	Surface roughness z_0	Drag or shear velocity u^* (m s^{-1})
1	0.1	0.00008	0.056
2	0.1	0.00008	0.112
4	0.1	0.00008	0.225
6	0.1	0.00008	0.337

Table 18.2 Threshold shear velocity for loose particles.

Particle size d (cm)	Fluid threshold velocity u^*_t (cm s^{-1})	Impact threshold velocity u^*_t (cm s^{-1})
0.1	45.52	36.42
0.15	55.75	44.6
0.2	64.38	51.5

$\rho_a = 0.00125$ g cm^{-3}; g = 982 cm s^{-2}; $\rho_p = 2.65$ g cm^{-3}; A = 0.1 (fluid threshold) or 0.08 (impact threshold).

Fine-grained soil surfaces comprise aggregates (silt/clay particles held in a single structure such as a clod) or crusts that limit the potential for wind erosion. Wind erosion can selectively remove silt and fine sand particles; the remaining coarser material (the *lag deposit*) can armour the ground surface, protecting it from further erosion unless it is disrupted.

Meteorological station measurements of wind velocity are often made at 10 m (i.e. z = 10 m). The rate of sand movement q (g m^{-1} s^{-1}) can be estimated from[18.1–3] (Fig. 18.1 and Table 18.3):

$$q = \alpha C \sqrt{\frac{d}{D}} \frac{\rho_a}{g} (u_{10} - u_t)^3$$

where C is a constant related to grain size (1.8 for naturally graded sand, 2.8 for poorly sorted sand, 3.5 for pebbly surfaces); d = particle diameter (mm); D = standard grain diameter (0.25 mm); ρ_a = density of air; g = gravitational acceleration; u_{10} = wind velocity at 10 m and u_t = fluid threshold velocity.

Wind Erosion Forms

Wind erosion involves:

1. *Deflation*: entrainment of loose sediments by the wind, causing progressive surface lowering. Deflation is usually limited by sediment supply.

Estimated surface lowering rates are in the range 0.5–5 mm yr^{-1} [18.4]. Deflation forms include:

- *Hollows and depressions* in the rock or cohesive soil (e.g. the Qattara depression in the Western Desert, Egypt). Salt pans (*playas*) often occur in depressions, as deflation lowers the surface down to the capillary fringe, allowing dissolved salts to precipitate. Construction conditions may be extremely difficult.
- *Lag gravels and stone pavements*: a surface pavement of closely packed stones (*reg* or *sarir* in the Sahara, *gobi* in Asia and *gibber plain* in Australia) can develop as the finer materials have been removed by deflation. These deposits are a source of aggregates; they are usually gap-graded as the sand and silt particles have been removed by wind. Surface sealing is complete when the stones are touching each other.

2. *Abrasion*: materials are worn down by sand-laden winds (*sandblasting*) and *vorticity* within wind flows around obstacles (*aerodynamic erosion*). Abrasion rates can be in the range 1–6 mm yr^{-1}, depending on rock type and wind speed[18.4].

Yardangs are aerodynamically shaped landforms resulting from wind abrasion of rock surfaces, typically a series of parallel ridges and troughs. The mega-yardangs in the Sahara cover an area of 650000 km^2 and reach

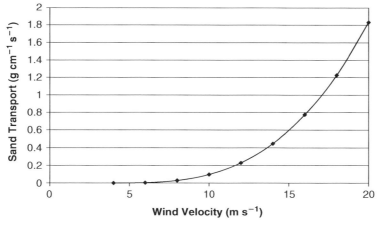

Figure 18.1 Sand transport rates and wind velocity (for 0.25 mm soil particles).

Table 18.3 Rates of sand movement for an area with 0.25 mm soil particles.

Wind velocity u_{10} (m s^{-1})	Threshold velocity u_t (m s^{-1})	Rate of sand movement q (tonnes m^{-1} s^{-1})	Rate of sand movement q (g cm^{-1} s^{-1})
4	4	0	0
6	4	3.57×10^{-7}	0.004
8	4	2.86×10^{-6}	0.029
10	4	9.65×10^{-6}	0.097
20	4	1.83×10^{-4}	1.831

d = 0.25 mm; C = 1.8; ρ_a = 0.00125 g cm^{-3}; α = (0.174 / (log (z / k'))^3; z = 10 m; k' = 0.01 m; d / D = 1.

200 metres in height with a spacing of 0.5–2 km. Most are associated with unidirectional wind regimes and are relict features from the Pleistocene. Yardangs indicate bedrock at the surface.

Depositional Forms: Dunes

Typical dune features are shown in Fig. 18.2 (see http://pubs.usgs.gov/gip/deserts/eolian/)[18.5]. Many desert plains contain *sand sheets*, *dune fields* (a collection of dunes in an area less than 30 000 km^2) or *sand seas* (ergs, areas of dunes of varying forms and sizes extending over 30 000 km^2). Dunes accumulate where concentrated sand flows converge or where the wind energy declines, as in topographic lows away from upland areas.

There are three broad classes of dunes: *stabilised dunes* (immobilised by cementation or vegetation after their formation) found both in and adjacent to arid areas, *anchored dunes* (fixed by topographic obstacles or plants), and *mobile dunes*[18.6] (Figs. 18.3 and 18.4).

Mobile dune types appear to be related to the volume of sand and the wind direction variability (Fig. 18.5). Barchans occur where there is little sand-moving wind variability and limited sand supply[18.7] (Fig. 18.6). Linear dunes (*seifs*) are associated with restricted sand supply, but variable winds from two predominant directions. Star dunes occur where sand is abundant and the wind direction is highly variable.

Wind Erosion Hazards

- *'Dust bowl' erosion*: the disastrous wind erosion of the 1930s in southern USA, damaging 9 Mha of agricultural land and generating large dust storms[18.8]. Santon Downham in Suffolk, UK, was engulfed by blown sand around 1630[18.9].
- *Dust storms and visibility problems*: creating problems for airports (e.g. Sharjah and Bahrain) and road

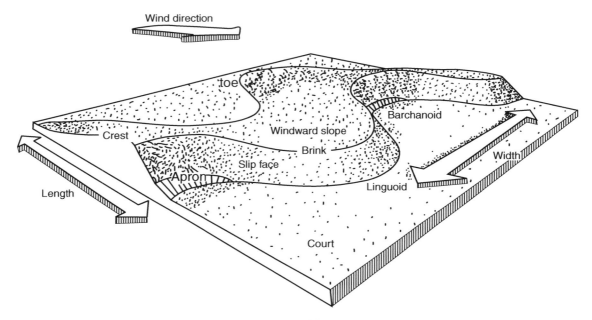

Figure 18.2   Features of a sand dune (after Cooke *et al.* 1993)[18.4].

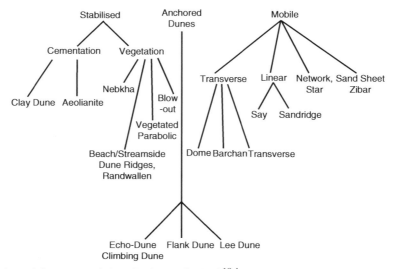

Figure 18.3   Classification of dune types (after Cooke *et al.* 1993)[18.4].

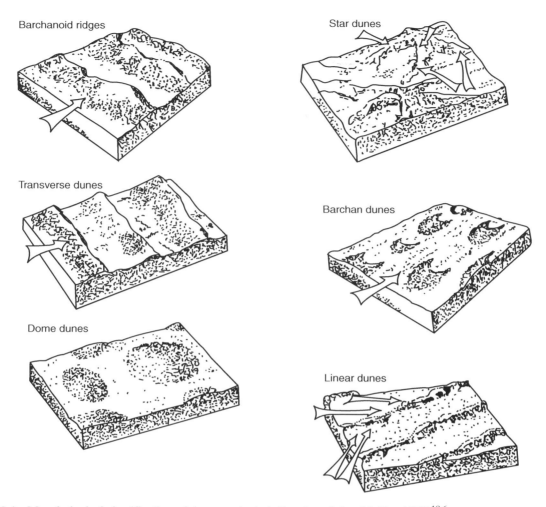

Figure 18.4 Morphological classification of dunes and wind directions (after McKee 1979)[18.6].

Figure 18.5 Mobile dunes close to a built-up area in Gran Canaria.

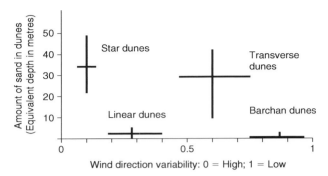

Figure 18.6 Dune types and wind direction variability (after Wasson and Hyde 1983)[18.7].

traffic; public health problems. In deserts, stone pavements provide surface protection against wind erosion, but they are easily destroyed by vehicle traffic or during construction, triggering accelerated dust problems.

- *Abrasion* by moving sand of exposed surfaces, generally within 0.25 m of the ground surface e.g. telegraph poles (wooden or concrete) and fences become sand-blasted up to 0.5 m above their bases.
- *Dune mobility*: mobile dunes may blow over and block roads (e.g. the Trans Saharan Highway in Algeria) or accumulate against buildings and structures. Pipelines through dune fields can be vulnerable to wind scour (creating unsupported spans and causing pipe failure) and loading by encroaching dunes (e.g. in the Turkmen desert where there has been an estimated 2 pipe ruptures per 1000 km yr^{-1})[18.10].

Wind Erosion Hazard Assessment

A similar technique to the USLE (see Chapter 17) can be used for wind erosion prediction[18.11]:

$$WE = \text{fn}\ (I, C, K, L, V)$$

where WE = soil loss t ha^{-1} yr^{-1}; I = mean annual soil erodibility (t ha^{-1}) determined by wind tunnel experiments, although it may be estimated from the percentage of dry stable aggregates larger than 0.4 mm in the soil; C = wind energy expressed by a climatic factor that takes account of wind velocity and soil moisture content (dimensionless); K = surface roughness (dimensionless); L = length of open wind blow (m); and V = quantity of vegetative cover expressed as a small grain equivalent (kg ha^{-1}).

The equation is more complex than the USLE and is best-solved using nomographs and tables or computer software; a downloadable version of the model is available at www.csrl.ars.usda.gov/wewc/rweq.

Dune migration rate is related to dune height[18.12]:

$$\text{Migration Rate} = \frac{TEq(0)}{\gamma H}$$

where TE is the sand trapping efficiency (see Fig. 18.7); $q(0)$ is sand transport rate at the dune crest (kg m^{-1} s^{-1}); γ is the sand bulk density (around 2650 kg m^{-3}) and H is dune height (m).

In general, the bigger the dune, the less mobile it is[18.12–13] (Fig. 18.8; Table 18.4).

Wind Erosion Management

Erosion control measures include[18.2] (see http://pubs.usgs.gov/gip/deserts/desertification/ or http://www.unccd.int/):

- *Vegetation stabilisation*: the only effective method of permanent dune stabilisation, not suited to hyper-arid regions. Suitable vegetation is planted in a grid with each plant surrounded by temporary plant frond or board windbreaks to provide protection in the early stages of growth. This helps create turbulent air flow which cannot move sand grains as efficiently as laminar flow.
- *Excavation scheduling*: sand in-filling open excavations during construction can be minimised by careful scheduling and only having short lengths of trenches or foundations open at any one time.

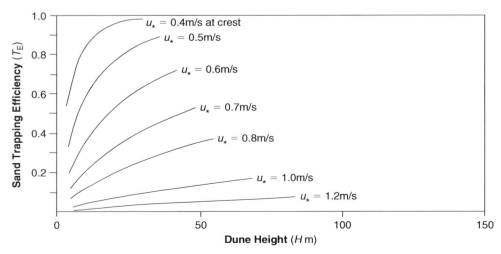

Figure 18.7 Sand trapping efficiency and dune height for different wind shear velocities (after Momiji and Warren 2000)[18.12].

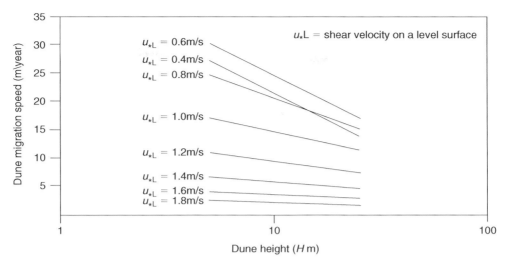

Figure 18.8 Dune migration speed related to dune height (after Momiji and Warren 2000)[18.12].

Table 18.4 Examples of dune activity rates [18.13].

Dune type	Location	Dune crest height (m)	Net migration rate (m yr^{-1})
Barchan	Pampa de la Joya, Peru	1–7	9–32
	Algodones, California	6	20
	Abu Moharic, Egypt	6–14	5–8
	El-Arish, Sinai	1.9–4	6.2–13.1
	Mauritania	3–17	18–63
	Jafurah Sand Sea, Saudi Arabia	<30	15
Transverse	Namib Sand Sea	2.5	5
	Erg Oriental	35	0.3
	Erg Oriental	240	0.16

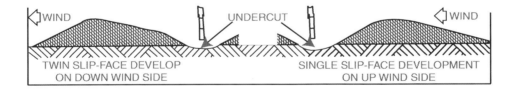

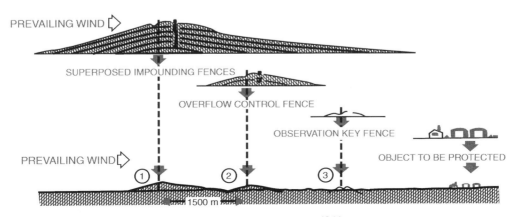

Figure 18.9 The use of fences as sand traps (after Kerr and Nigra 1952)[18.14].

- *Land management*: cover crops, surface mulches, use of vegetation strips and soil conditioning can protect the surface in windy seasons.
- *Surface stabilisation* with gravel, gypsum, saline water or a chemical/oil spray. Oil has been used successfully to stabilise large areas at low cost but is unsightly[18.14]. Permanent stabilisation can only be achieved through the establishment of a vegetation cover.
- *Wind-breaks* e.g. hedges and lines of trees or artificial materials (e.g. geotextiles, stakes or palm fronds). Windbreaks should be set at right angles to the dominant wind. The degree of protection is determined by the barrier spacing and their width, height and porosity.
- *Control of sand movement with fences or palm fronds*: construction of panels or fences to divert or stop sand transport reaching a vulnerable site (e.g. a village). A system of 3 parallel fences provides effective protection to a site (the 'Arab Little Stick' method, Fig. 18.9). Sand accumulates around fence 1; as the effectiveness of this fence declines, sand accumulates increasingly at fence 2, and so on[18.2]. Once the fence is covered it no longer acts as a trap; new sand arriving from up-wind sweeps across its surface and is carried down wind (i.e. no reduction in transport capacity). Increasing the fence height will restore the trap efficiency.
- *Careful pipeline alignment*: risks to pipelines through areas susceptible to wind erosion can be reduced by burying the pipe beneath the limit of wind deflation[18.10, 15]. Major dune cuttings may be needed so that pipelines can be trenched in the underlying ground surface. Minimise route length through mobile dunes i.e. make full use of inter-dune corridors.

References

18.1 Bagnold, R. A. (1941) *The Physics of Blown Sand and Desert Dunes*. Chapman and Hall, London.
18.2 Cooke, R. U., Brunsden, D., Doornkamp, J. C. and Jones, D. K. C. (1982) *Urban Geomorphology in Drylands*. Oxford University Press.
18.3 Lettau, K. and Lettau, H. H. (1978) Experimental and micrometeorological field studies of dune migration. In H. H. Lettau and K. Lettau (eds.) *Exploring the World's Dryest Climates*. Institute of Environmental Science Report 101. Centre for Climatic Research, University of Wisconsin, 110–147.
18.4 Cooke, R. U., Warren, A. and Goudie, A. S. (1993) *Desert geomorphology*. UCL Press.
18.5 Warren, A. (1994) Sand dunes: highly mobile and unstable surfaces. In P. G. Fookes and R. H. G. Parry (eds.) *Engineering Characteristics of Arid Soils*, Balkema; Rotterdam, 47–53.
18.6 McKee, E. D. (ed.) (1979) *A study of global sand seas*. US Geological Survey Professional Paper 1052.
18.7 Wasson, R. J. and Hyde, R. (1983) A test of granulometric control of desert dune geometry. *Earth Surface Processes and Landforms*, **8**, 301–312.
18.8 Worster, D. (1979) *The Dust Bowl*. OUP, New York.
18.9 Lee, E. M. (1994) *The Occurrence and Significance of Erosion, Deposition and Flooding in Great Britain*. HMSO.
18.10 Fookes, P. G., Lee, E. M. and Sweeny, M. (2004) In Salah Gas Project, Algeria – Part 1: Terrain Evaluation for Desert Pipeline Routing. In M. Sweeney (ed.) *Terrain and Geohazard Challenges Facing Onshore Oil and Gas Pipelines*, Thomas Telford, 144–161.
18.11 Skidmore, E. L. and Williams, J. R. (1991) Modified EPIC wind erosion model. In *Modeling Plant and Soil Systems*. ASA-CSSA-SSSA Agronomy Monograph **31**, 457–69.
18.12 Momiji, H. and Warren, A. (2000) Relations of sand trapping efficiency and migration speed of transverse dunes to wind velocity. *Earth Surface Processes and Landforms*, **25**, 1069–1084.
18.13 Thomas, D. S. G. (1992) Desert dune activity: concepts and significance. *Journal of Arid Environments*, **22**, 31–38.
18.14 Kerr, R. C. and Nigra, J. O. (1952) Eolian sand control. *American Association of Petroleum Geologists*, Bulletin **36**, 1541–1573.
18.15 Cherednichenko, V. P. (1973) *Morphology of aeolian topography and pipeline construction in deserts* (in Russian). Turkmenistan SSR Academy of Sciences Red Banner Institute for Deserts. Ylym, Ashkhabad.

19

Slopes: Landslides

Introduction

Landslides involve the movement of rock, debris or earth down a slope, and can involve a variety of *mechanisms* (see http://landslides.usgs.gov/)[19.1–3]:

1. *Falling*: detachment of material from a steep face/cliff, from a surface on which little or no shear displacement occurs. It then descends by falling, rolling or bouncing (Fig. 19.1).

2. *Toppling*: the forward rotation of material out of a slope about a point or axis below the centre of gravity of the displaced mass.

3. *Spreading*: extension of a cohesive soil or rock mass combined with a general subsidence of the fractured mass into softer underlying material. Spreads may result from liquefaction (see below) or the flow and extrusion of the softer underlying material.

4. *Flowing*: turbulent movement of a fluidised mass over a more rigid bed, with either water or air as the pore fluid (e.g. like wet concrete or running dry sand[19.4]). There is a gradation from flows to slides depending on water content and mobility.

5. *Sliding*: downslope movement of a soil or rock mass as a coherent body on surfaces of rupture or on zones of intense shear strain (i.e. a *shear surface* at the contact between the moving mass and the underlying soil or rock; see Fig. 19.2).

Liquefaction is triggered by strong shaking during earthquakes or by working site equipment which temporarily increases pore pressures. It is most likely to occur (see Table 20.3):

- in saturated, relatively uniform, cohesionless, fine sands, silty sands, or coarse silts of low relative density (loose)
- generally at depths < 20 m
- in areas where the water table is within 5 m of the ground surface.

Key classification systems for describing landslides include: landslide type[19.5] (Table 19.1), activity state (Table 19.2) and causes (Table 19.3). The main components of a multiple rotational landslide are shown in Fig. 19.3[19.6].

Figure 19.1   Rockfalls and rolling boulders affecting a mountain road in Gran Canaria.

Figure 19.2   Major debris slide in the Río Agua valley, southeast Spain.

Many landslides are associated with a particular triggering event (see Chapter 3):

- *Heavy rainfall*: in 1996 several hundred landslides were triggered by an exceptional rainstorm in Washington State, USA (over 50 cm in 7 days)[19.7].
- *Snowmelt*: over 4000 landslides were generated in Umbria (Italy) by snowmelt following a sudden change in temperature on 1 January 1997[19.8].
- *Earthquakes*[19.9]: a magnitude 7.6 earthquake off the coast of El Salvador in 2001 triggered numerous landslides throughout the country, including the Las Colinas landslide, which buried more than 400 homes and killed 1000 people.
- *Volcanic eruptions*: around 25 000 people were killed in the town of Armero, Colombia, by debris/mud flows generated by the 1985 eruption of the Nevado del Ruiz volcano. The flows travelled over 40 km from the volcano (see http://volcanoes.usgs.gov/Hazards/What/Lahars/RuizLahars.html)[19.10].

Other events may be associated with prolonged periods of above average rainfall (e.g. reactivation of the large Falli-Holli landslide near Freiberg, Switzerland[19.11]) or the effects of man (e.g. the 1966 Aberfan disaster in south Wales: the colliery spoil tip collapsed and flowed downslope, the primary school was overwhelmed, 144 people died[19.12]).

Shear Strength and Soil Behaviour

- *Effective shear strength*: the maximum shear stress that a material can withstand before failure occurs (see Chapter 3).
- *Mobilised shear strength*: the amount of shear strength used to resist deformation. When failure occurs the shear strength is fully mobilised along a *shear surface*.

Frictional shear strength is reduced by the presence of water in the pore spaces between soil particles (i.e. water pressure reduces the contact between grains). Pore pressure can be negative when suction exists i.e. *apparent cohesion* (e.g. in a child's sand-castle, or an unconfined sample of clay).

Deformation occurs when a load (i.e. *stress*) is applied to a soil. If the stress is small, then the response may be elastic. Once the stress exceeds a threshold value, large irreversible deformation takes place. A number of types of soil behaviour occur:

1. *Coarse granular soils* (gravels, sands and non-plastic silts): shear strength depends on internal friction (*cohesionless*, except where suction occurs). When *dense* sands are sheared they become loosely packed and increase in volume (*dilate*), giving a high strength (peak strength). If shearing continues the strength drops and eventually the sand may be sheared at a constant volume. *Loose* granular soils contract during initial shearing as the packing becomes closer, strength increases with further deformation until it shears at a constant volume.
2. *Fine-grained soils* (soils with significant amounts of clay and/or silt). When clays are sheared in a *drained condition* (sufficient time for dissipation of pore pressures generated during shearing) the behaviour is characterised by changing shear strength (Fig. 19.4):
   - *Peak strength* obtained by rapid application of stress to previously unsheared material.
   - *Fully softened strength* which occurs in weathered clays, or where stress is applied very slowly allowing progressive deformation (strain softening, e.g. due to high lateral stresses developing within the slope).
   - A minimum shear strength achieved at large displacements, concentrated along pre-existing narrow

Table 19.1 Landslide classification[19.3].

Type	Form of initial failure surface	Subsequent deformation
Fall (rock or stiff soil)		
Detachment form: pre-existing discontinuities or tension failure surfaces	a) Planar surface b) Wedge (two or more intersecting joints) c) Stepped surface d) Vertical surface	Free fall, may break up, bounce, slide or flow down slopes. May involve fluidisation, liquefaction, cohesionless grain flow, heat generation or other secondary effects on disintegration when failed rock hits the ground surface.
Topple (rock or stiff soil)		
Detachment form: pre-existing discontinuities or tension failure surfaces	a) Single b) Multiple	As above
Slide		
Rotational movement (failure surface essentially circular; occurs in soils)	a) Single b) Multiple c) Successive	Toe area may deform in a complex way. The ground may bulge, the slide may creep or even flow, possibly overriding existing failures. Failure might be retrogressive or progressive.
Non-rotational compound movement (non-circular failure surface; may be listric or bi-planar; found in soils and rocks):	a) Single b) Progressive c) Multi-storied	Graben often develops at the head of the landslide. It may include a toe failure of a different type.
Translational movement (often associated with discontinuity controlled failures in bedded or foliated rocks)	a) Planar b) Stepped c) Wedge d) Non-rotational	May develop complex run-out forms after disintegrating (see falls and flows).
Spread (soils and weak rock)		
Lateral spreading of ductile or soft material that deforms	a) Soft layer beneath a hard rock b) Weak interstratified layer c) Collapsing structure	Can develop sudden spreading failures in quick clays when the slope opens up in blocks and fissures followed by liquefaction. Might be a slow movement associated with denudational unloading. Can be represented by cambering and valley bulging.
Flow (usually associated with soils but rock flows do occur – see below)		
Debris movement by flow	a) Unconfined b) Channelised	Flow involves complex run-out mechanisms. It may be catastrophic in effect and it may move in sheets or lobes. The form of movement is a function of the rheological properties of the material.
Creep movement	Failure surface rarely clearly defined	Creep may be a superficial gravity movement, seasonal movements or it might represent pre-failure and progressive movements prior to a larger scale failure.
Rock flow (sometimes referred to as sagging or Sackung). Usually associated with mountain terrain or areas of rapid and deep incision.	a) Single-sided b) Double-sided c) Stepped (Failure surface may be rotational, compound, listric, biplanar or intermittent.)	May be slow gravity creep or the early stages of larger scale movements that only show as bulging in the topography without a clearly defined toe deformation. Where controlled by discontinuities it may involve toppling.

(continued)

Table 19.1　(continued)

Type	Form of initial failure surface	Subsequent deformation
Complex		
a) Movements involving two or more of the above mechanisms (referred to as compound when two types of movement occur currently)	Dependent on the form of failure as described above	As described for the various categories above.
b) Rock or debris avalanche	Often initiated as fall/slide of rock and/or debris	Complex long-run-out mechanisms, including fluidisation and cohesionless grain flow.

Table 19.2　Landslide activity states[19.2].

Activity State	Description
Active	Currently moving
Suspended	Moved within the last 12 months but not currently active
Dormant	An inactive landslide that can be reactivated
Abandoned	A landslide which is no longer affected by its original cause and is no longer likely to be reactivated. For example, the toe of the slide has been protected by a build up of material, such as a floodplain or beach
Stabilised	A landslide which has been protected from its original causes by man-made remedial measures
Relict	An inactive landslide developed under climatic or geomorphological conditions different from those at present.

Table 19.3　Classification of landslide causes[19.3].

External process(es)	Causal effects	Description of typical changes	Examples of specific changes on slope
Weathering: physical, chemical and biological	Changes in physical and chemical properties; horizonation; changes in regolith thickness, strength.	Changes in grading; cation exchange; cementation; formation of weak discontinuities or hard bands; increased depth of low strength materials	Changes in density, strength, permeability; stress, pore and cleft water pressure.
Erosion of material from face or base of slope by fluvial, glacial and/or coastal processes	Changes in slope geometry; unloading	Alterations to relief: slope height, length, angle and aspect	Changes in stress, permeability and strength
Ground subsidence	Loss of support; natural and man-made undermining	Mechanical eluviation of fines; solution; loss of cement; leaching seepage erosion; backsapping; piping	Loss of support; consolidation; changes in porewater pressure; loss of strength
Deposition of material to face or top of slope by fluvial, glacial or mass movement processes	Loading; long-term (drained) or short-term (undrained)	Alterations to relief: slope height, length, angle and aspect	Changes in stress, permeability, strength, loading and porewater pressure.
Seismic activity and general shocks and vibrations (the smallest earthquakes likely to cause landsliding are around Magnitude 4.0[19.9])	Rapid and repeated vertical and horizontal displacements	Disturbance to intergranular bonds; transient high porewater pressures; materials subject to transient and repeated periods of compression and tension	Changes in stress; loss of strength; high porewater pressures; potential for liquefaction

(*continued*)

Table 19.3   (continued)

External process(es)	Causal effects	Description of typical changes	Examples of specific changes on slope
Air fall of loess or tephra	Mantling slopes with fines; adding fines to existing soils	New slope created with well-defined discontinuity boundary	Changes in stress; strength; water content and water pressure
Water regime change	Rising or falling ground water; development of perched water tables; saturation of surface; flooding	Piping, floods, lake bursts; 'wet' years; intense precipitation; snow and ice melt; rapid drawdown	Excess porewater pressures; changes in bulk density; reduction in effective shear strength
Complex 'follow-on' or runout processes after initial failure	Liquefaction; remoulding; fluidisation; acoustic grain flow	Long runout landslides; low values for ratio of initial failure volume to total failure volume; low angles of reach; low breadth to length ratios	Changes in effective shear strength, water distribution, bulk density and rheological characteristics
Human Interference	Excavation at toe of slope; top loading of slopes; flooding (e.g. leaking services)	Same as for *natural* erosion Same as for *natural* deposition Same as for *natural* water regime change	Same as for *natural* erosion Same as for *natural* deposition Same as for *natural* water regime change

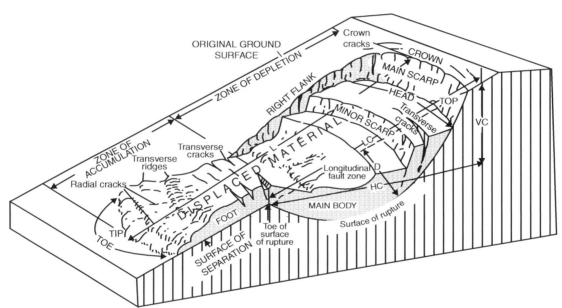

Figure 19.3   Definition of the characteristics of a multiple rotational landslide (based on the UNESCO World Landslide Inventory 1995)[19.6].

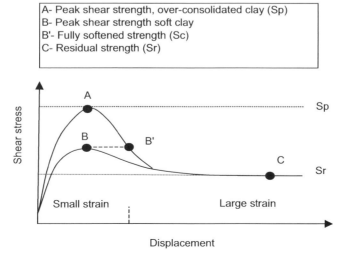

Figure 19.4   Shear behaviour: fine-grained soils.

Infinite slope failure analysis

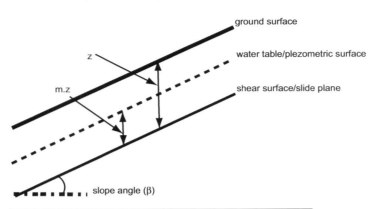

$$F = (c' + (\gamma - m\gamma_w)\, z\, \cos^2\beta\, \tan\Phi')/(\gamma\, z\, \sin\beta\, \cos\beta)$$

F = factor of safety
c' = effective cohesion (kN/m^2)
γ = unit weight of soil at natural moisture content (kN/m^3)
γ_w = unit weight of water (9.81 kN/m^3)
Φ' = angle of friction with respect to effective stresses (degrees)
β = slope angle (degrees)
m = vertical height of the water table above the slide plane as a fraction of the soil thickness above the plane (if the water table is at the ground surface $m = 1$; if the water table is just below the slide plane $m = 0$)
$\cos^2\beta = 0.8009$
z = vertical depth to slide plane (m)

Example calculations

c' = 0.25 kN/m^2
γ = 19 kN/m^3
γ_w = 9.81 kN/m^3
Φ' = 32 degrees
β = 26.5 degrees (1 vertical to 2 horizontal); $\cos^2\beta$ =
z = 3 m

therefore ($\gamma\, z\, \sin\beta\, \cos\beta$) = 22.76

a) for $m = 1$
   $c' + (\gamma - m\gamma_w)\, z\, \cos^2\beta\, \tan\Phi'$ = 14.05
   thus $F = 14.05/22.76 = 0.62$

b) for $m = 0.5$
   $c' + (\gamma - m\gamma_w)\, z\, \cos^2\beta\, \tan\Phi'$ = 21.41
   thus $F = 21.41/22.76 = 0.94$

c) for $m = 1$
   $c' + (\gamma - m\gamma_w)\, z\, \cos^2\beta\, \tan\Phi'$ = 28.78
   thus $F = 28.78/22.76 = 1.26$

Figure 19.5   Stability analysis using the 'infinite slope' model.

shear zones – the *residual strength*. This is produced by the re-orientation of clay particles along the sheared surface.
• *Brittleness* is a measure of the difference between peak and residual strength values.
When saturated clays and silts are confined and sheared rapidly in an *undrained condition* (insufficient time for

the dissipation of pore pressures), they behave as though they are a purely *cohesive* material with no frictional strength (undrained shear strength s_u; see Chapter 7).

Clays and silts also behave as a cohesionless material when shearing occurs slowly and excess pore pressures do not build up.

Landslide Behaviour

- *First-time failures* of ground that was previously intact, i.e. unsheared. Such slides often involve large, rapid displacements, particularly in very brittle materials.
- *First-time failures on pre-existing shear surfaces of non-landslide origin* e.g. flexural shearing during the folding of sequences commonly of hard rocks and clay-rich strata (i.e. 'bed over bed' slip), and periglacial solifluction.
- *Reactivation of pre-existing landslides* where part or all of a landslide mass is involved in new movements, along pre-existing shear surfaces.

Four movement stages can be defined:

1. *Pre-failure movements*: small displacements due to the progressive development of shear surfaces, from isolated shear zones to continuous displacement surfaces. Provides forewarning of failure e.g. development of tension cracks or minor settlement behind a cliff face, and bulging on the slope or at the slope foot.

2. *Failure* occurs when the disturbing forces acting on the slope exceed the forces resisting failure (shear stress > shear strength):
   - *peak strength failures* in which the peak strength is mobilised during first time failure.
   - *progressive failure*: the fully softened strength or the residual strength is mobilised during first-time failure.

3. *Post failure movements*: some of the *potential energy* of the landslide (a function of slope height and geometry, PE = m h g; see Chapter 2) is lost through friction as the material moves along the shear surface. The remainder is dissipated in the break-up and remoulding of the moving material and in accelerating it to a particular velocity (*kinetic energy*, KE = 0.5 m v^2). In brittle materials, the kinetic energy can be very large, giving rise to long run-out landslides. In 1970 an offshore earthquake (M = 7.7) triggered a massive avalanche from the summit of the Huascaràn mountain, Peruvian Andes. A turbulent flow of mud and boulders (estimated at 50–100 Mm^3) descended 2700 m as a 30 m high wave travelling at an average speed of 270–360 km hr^{-1}. The flow destroyed the town of Yungay, 15 km away, killing between 15 000–20 000 people[19.13].

4. *Reactivation*: part or all of previously failed mass is involved in renewed movements (i.e. materials are at residual strength and non-brittle). Failures are generally slow moving with relatively limited displacements, although larger displacements can occur[19.14]. Reactivation is an episodic process with phases of movement, often associated with periods of prolonged heavy rainfall or earthquakes, separated by inactivity.

Potential for Landsliding

Hillslopes can be classified into[19.15]:

- areas where pre-existing landslides are present (the *slid areas*) i.e. potential for reactivation
- areas which have not been affected by landsliding (the *unslid areas*) i.e. first-time, landslides may occur in previously unsheared ground.

Once a landslide has occurred it can be made to move again under conditions that the slope, prior to failure, could have resisted. Reactivations can be triggered much more readily than first-time failures. For example, they may be associated with lower rainfall/groundwater level thresholds than first-time failures in the same materials.

Stability Analysis

Factor of safety *F* is calculated as:

$$F = \frac{\text{Resisting forces}}{\text{Destabilising forces}} = \frac{\text{Shear strength}}{\text{Shear stress}}$$

The factor of safety of a slope at the point of failure is assumed to be 1.0, i.e. *limiting equilibrium*. Specified F values (e.g. F > 1.3) are often required for the design engineered slopes.

Stability analysis is used to determine the factor of safety; it provides an index of slope stability. It is an empirical tool that can support field observations, historical records, and the performance history in evaluating the potential for slope failure[19.16–17].

Both *finite element* and *limit equilibrium* methods are available; most methods are computer-based. The *infinite slope analysis* can be solved by hand calculation and is applicable to a wide range of situations (Fig. 19.5). The soil is assumed to slide on a shear surface parallel to the ground surface, down a slope of infinite extent. The slide depth can be any value below the ground surface, and the soil and groundwater conditions are assumed to be constant throughout the failure mass.

References

19.1 Cruden, D. (1991) A simple definition of a landslide. *Bulletin of the International Association of Engineering Geology*, **43**, 27–29.

19.2 Cruden, D. M. and Varnes, D. J. (1996) Landslide types and processes. In A. K. Turner and R. L. Schuster (eds.) *Landslides: Investigation and Mitigation*, Transportation Research Board, Special Report 247, National Research Council, National Academy Press, Washington DC, 36–75.

19.3 Dikau, R., Brunsden, D., Schrott, L. and Ibsen, M.-L. (1996) *Landslide Recognition*. John Wiley and Sons, Chichester.

19.4 Hungr, O., Evans, S. G., Bovis, M. J. and Hutchinson, J. N. (2001) A review of the classification of landslides of flow type. *Environmental and Engineering Geoscience*, **3**, 221–238.

19.5 Hutchinson, J. N. (1988) General report: Morphological and geotechnical parameters of landslides in relation to geology and hydrogeology. In C. Bonnard (ed.) *Landslides*. Balkema, Rotterdam, 3–35.

19.6 WP/WLI, (International Geotechnical Societies' UNESCO Working Party for World Landslide Inventory) (1995) A suggested method of a landslide summary. *Bulletin of the International Association of Engineering Geology*, **43**, 101–110.

19.7 Harp, E. L. (1997) *Landslides and landslide hazards in Washington State due to February 5-9 1996 storm*. US Geological Survey Administrative Report.

19.8 Cardinali, M., Reichenbach, P., Guzzetti, F., Ardizzone, F., Antonini, G., Galli, M., Cacciano, M., Castellani, M.

and Salvati, P. (2002) A geomorphological approach to the estimation of landslide hazards and risks in Umbria, Central Italy. *Natural Hazards and Earth System Sciences*, **2**, 57–72.

19.9 Keefer, D. K. (1984) Landslides caused by earthquakes. *Geological Society of America Bulletin*, **95**, 406–421.

19.10 Herd, D. G. and the Comite de Estudios Vulcanologies (1986) The 1985 Ruiz volcano disaster. *Eos*, **67**, 457–460.

19.11 Lateltin, O. and Bonnard, C. (1999) Reactivation of the Falli-Holli landslide in the Pre-Alps of Freiburg, Switzerland. In K. Sassa (ed.) *Landslides of the World*. Kyoto University Press, 331–335.

19.12 Miller, J. (1974) *Aberfan – a disaster and its aftermath*. Constable, London.

19.13 Plafker, G. and Ericksen, G. E. (1978) Nevados Huascaràn avalanches, Peru. In B Voight (ed.) *Rockslides and Avalanches – 1 Natural Phenomena*. Elsevier, Amsterdam, 277–314.

19.14 Hutchinson, J. N. (1987) Mechanisms producing large displacements in landslides on pre-existing shears. *Memoir of the Geological Society of China*, **9**, 175–200.

19.15 Hutchinson, J. N. (1992) Landslide hazard assessment. In D. H. Bell (ed.) *Landslides*. Balkema, Rotterdam, 3, 1805–1841.

19.16 Morgenstern, N. (1995) The role of stability analysis in the evaluation of slope stability. In D. H. Bell (ed.) *Landslides*. Balkema, Rotterdam, 2, 1615–1629.

19.17 Duncan, J. M. (1996) Soil slope stability analysis. In A. K. Turner and R. L. Schuster (eds.) *Landslides: Investigation and Mitigation*, Transportation Research Board, Special Report 247, National Research Council, National Academy Press, Washington DC, 337–371.

20

Slopes: Landslide Hazard and Risk

Destructive Significance of Landslides of Different Velocity Classes

Not all landslides result in the total destruction of an area. The hazard posed by an individual landslide is a function of its mass and velocity[20.1] (Table 20.1). Speeds range from the imperceptible up to 400 km hr^{-1}.

Broad categories of landslide intensity are: *sluggish, intermediate* and *catastrophic*. Slides tend to be associated with the sluggish and intermediate categories. Flows are intermediate to catastrophic, as they can have extremely high kinetic energy.

Pre-failure Hazards

The hazard is usually restricted to minor falls, the formation of tension cracks and minimal displacements (Fig. 20.1). This is the time for monitoring, the development of hazard scenarios and the formulation and of emergency actions.

Main Failure Hazards

The key elements are:
1. Loss of *cliff top* or *slope crest* land
2. *Differential ground movement* within the main landslide body. There is, a close association between landslide type and style of ground disturbance (see Fig. 19.3):
   - vertical settlement and contra-tilt occurs at the slide *head*
   - downslope movement of the *main body* generates significant lateral loads
   - differential horizontal and vertical settlement occurs between the individual blocks within the *main body*
   - compression and uplift (heave) occurs in the *toe area*.
3. *Run-out* beyond the source area: the distance the slide mass will travel and its velocity determine the extent to which the landslide will affect property and people downslope, and their ability to escape. The *travel distance* of landslide debris is generally estimated in terms of a travel angle (defined as the slope of a line joining the *tip* of the debris to the crest of the landslide *main scarp*).

4. The *impact* of falling boulders or debris: falling rocks present a public safety issue, especially along road/railway cuttings or on bathing beaches. Falls can cause considerable property damage and disruption to services.

Secondary Hazards

Landslides may produce *secondary hazards* that can be even greater than those posed by the failure itself:
- *Landslide dams*: caused by large landslides (e.g. rock avalanches and rock slides), which block narrow, steep-sided mountain valleys[20.2] (see Table 20.2). Landslide dams can result in *upstream* (backwater) *flooding* as a result of the impoundment of water. In 1983, the town of Thistle, Utah was inundated by a 50–60 m deep lake that formed behind a landslide dam on the Spanish Fork River[20.2]. The total direct costs of this event at the time were c. $200M.
- Failure of the dam can cause *downstream flooding*. For example, in 1840–41 a landslide in the Nanga Parbat Massif, Pakistan completely blocked the River Indus, causing the impoundment of a 60–65 km long lake. The dam breached in June 1841; its lake emptied in 24 hours causing The Great Indus flood during which hundreds of villages and towns were swept away[20.3].
- *Tsunamis*: caused by fast moving landslides entering water (see http://vulcan.wr.usgs.gov/Glossary/Tsunami/). In 1792, a landslide off the flanks of Mount Unzen, Japan, generated a tsunami in Ariake Bay which caused 14 500 deaths around the shoreline. Collapse of the flank of the Cumbre Vieja volcano in the Canary Islands could generate a massive tsunami toward the coasts of Africa, Europe, and America. Economic losses are predicted to be in the multi-trillion US dollar range[20.4].

Reactivation Hazards

Reactivated landslides present only a minor threat to life, as movements usually involve only slow to extremely slow displacements. Typical reactivation hazards are differential ground movement and distortion. The cumulative effects of episodes of slow movement can cause considerable damage to property. Reactivation of the

121

Table 20.1   Landslide impact classification[20.1].

Velocity class	Description	Velocity (mm s^{-1})	Typical velocity	Probable destructive significance
7	Extremely rapid			Catastrophe of major violence; buildings destroyed by impact of displaced material; many deaths; escape unlikely (e.g. Hauscarán debris avalanche in Peru that reached a velocity of over 300 km hr^{-1})
		$> 5 \times 10^3$	5 m s^{-1}	
6	Very rapid			Some lives lost; velocity too great to allow all persons to escape
		5×10^1	3 m/min	
5	Rapid			Escape evacuation possible; structures, possessions and equipment destroyed (e.g. the slide that destroyed the site of the Ok Ma tailings dam in Papua New Guinea moved at a maximum rate of 5.2 m hr^{-1})
		5×10^{-1}	1.8 m/hr	
4	Moderate			Some temporary and insensitive structures can be temporarily maintained
		5×10^{-3}	13 m/month	
3	Slow			Remedial construction can be undertaken during movement; insensitive structures can be maintained with frequent maintenance work if total movement is not too large during a particular acceleration phase
		5×10^{-5}	1.6 m/yr	
2	Very slow			Some permanent structures undamaged by movement
		5×10^{-7}	16 mm/yr	
1	Extremely slow			Imperceptible without instruments; construction possible with precautions (e.g. the town of Ventnor on the Isle of Wight, UK, is built on a slowly moving landslide)

Table 20.2   A selection of historic landslide dams[20.2].

Landslide	Year	Dammed river	Landslide volume m^3	Lake Volume m^3
Deixi landslide, China	1933	Min River	150×10^6	400×10^6
Cerro Codor Sencca rockslide, Peru	1945	Mantaro River	5.5×10^6	300×10^6
Tanggudong debris slide, China	1967	Yalong River	68×10^6	680×10^6
Mayunmarca rock slide, Peru	1974	Mantaro River	1.6×10^9	670×10^6
Thistle earth slide, USA	1983	Spanish Fork River	22×10^6	78×10^6
La Josefina rockslide, Ecuador	1993	Paute River	20–44×10^6	177×10^6

Figure 20.1 Tension cracks developed 20 m behind the 25-metre high backscar of a large landslide, southeast Spain.

Portuguese Bend landslide in the Palos Verdes Hills, California, resulted in $45M of damage (equivalent to 1996 prices) to roads, homes and other structures between 1956–59[20.5].

Liquefaction

Liquefaction damage during earthquakes is usually related to lateral spreads or settlement in flat, low-lying areas (Table 20.3). Most liquefaction occurs in areas of poorly engineered hydraulic fills and in geologically young fine-grained deposits (often fluvial, lacustrine or estuarine), commonly less than 1500 years old. The 1964 M_w 9.2 Great Alaska Earthquake triggered large liquefaction induced flow failures that demolished port facilities in Valdez, Seward and Whittier and carried large parts of those towns into the sea (see http:// neic.usgs.gov/ neis/eq_depot/usa/).

Landslide Hazard Assessment

- *landslide susceptibility* is the potential for landsliding to occur
- *landslide hazard* is the potential for landsliding to cause adverse consequences.

The underlying assumptions in landslide hazard assessments are[20.6]:

1. The location of future slope failures will largely be determined by the distribution of existing or past landslides i.e. known landslide locations will continue to be a source of hazard.
2. Future landslides will occur under similar conditions (e.g. conditioning or controlling factors) to those occurring at the sites of existing or past landslides. Mapping these factors provides a reasonable indication of the relative tendency for slopes to fail.
3. The distribution of existing and future landslides can be approximated by reference to conditioning factors alone (e.g. rock type or slope angle).

Landslide susceptibility mapping identifies the areas that either have failed or are likely to fail; it forms the basis for landslide hazard assessments.

Landslide hazard mapping needs to include a measure of frequency of occurrence and the likelihood/ potential to cause damage. This could use an ordinal scaling (e.g. very low to very high) or in terms of probability of occurrence.

There are two basic methodologies[20.7]:

- *Direct mapping approaches*: hazard is defined in terms of the distribution of past landslide events e.g. landslide inventories; geomorphological mapping of landslide features and qualitative map combinations (the identification of key factors that appear to control the pattern of landsliding).

Table 20.3   Liquefaction failure modes and their characteristic ground damage effects.

Liquefaction failure mode	Typical ground damage and effects
Lateral spreads	Small to large lateral displacements of surficial blocks of sediments, on gentle slopes (< 3°). Movements, commonly of several metres to tens of metres, are usually towards a free face, particularly in incised stream channels, canals, or open cuts. Lateral spreads are particularly damaging to pipelines, bridges, and structures with shallow foundations, especially on floodplains adjacent to river channels.
Flow failures	Flow failures, commonly the most catastrophic mode of liquefaction failure, are usually developed on slopes greater than 3°, with movements ranging from tens of metres to several kilometres, at very rapid velocities. Such flows involve great volumes of material, and are highly damaging to structures located on them, or in their path.
Ground oscillation	Oscillation occurs when liquefaction is triggered at depth, or within confined liquefied layers. May produce visible ground oscillation waves, ground settlements, opening and closing of fissures, and ejections of sand and water from cracks and fissures (sand "boils"). Subsurface structures (pipes, tanks, etc.) may be damaged from this phenomenon, but damage is typically less than from lateral spread or flow failures.
Loss of bearing strength	Strength loss caused by liquefaction or densification can cause ground collapse and settlements. Structures may settle and topple, and buried structures (pipelines, septic tanks, etc.) may float to the surface. Spreading and collapse of embankment fills may occur from liquefaction of foundation soils.

- *Indirect mapping approaches*, involving the mapping and statistical analysis of large numbers of relevant parameters to derive a predictive relationship between the terrain conditions and the occurrence of landsliding.

Landslide hazard assessment increasingly involves the use of GIS and databases (see Chapter 38).

Figure 20.2 provides an example of landslide susceptibility mapping leading to a hazard zonation scheme for a river basin in Nepal[20.8].

Probability of Landsliding

Approaches to estimating the probability of landsliding include[20.9] (see Chapter 12):

1. *The historical frequency*: e.g. the Scarborough cliffs, UK, can be divided into 8 separate sections which are either large, pre-existing landslides (the oldest occurred in 1737/38) or intervening intact (i.e. unfailed) steep slopes. From historical research, the frequency of failure of the intact steep slopes was estimated to be 4 events in 256 years (i.e. 1 in 64 years)[20.10]. The annual probability of failure of any one of the eight original intact slopes was estimated to be:

$$\text{Annual Probability Failure (slope } s) = 4/(8 \times 256) = 0.00195 \text{ (1 in 512)}$$

2. *Using process-response relationships*: e.g. the probability of an earthquake triggered landslide at a site in California, USA, was estimated from the return period of the *maximum-magnitude earthquake* that could occur on a nearby fault (a M_W 8.5 event, with a return period of 1 in 498 i.e. 0.0020). The probability of this earthquake occurring over the 30-year design life of the plant was calculated using the *Poisson distribution*:

$$\text{Probability (Earthquake)} = 1 - \exp\left(-\frac{30}{498}\right) = 0.06$$

An estimate was made of the probability that the maximum-size earthquake would actually trigger a landslide event, based on a geotechnical appraisal of the site conditions and the level of ground acceleration and shaking that would occur at the site during the earthquake. The probability was judged to be 0.16. The conditional probability of landsliding given a maximum-size earthquake during the 30-year design life of the power plant was:

$$\text{Probability (Landslide)} = \text{Probability (Earthquake)} \times \text{Probability (Landslide | Earthquake)}$$
$$\text{Probability (Landslide)} = 0.06 \times 0.16 = 0.01$$

3. *Expert judgement*: e.g. an existing pipeline passes through a number of landslide-prone terrains in the Caucasus. The likelihood of pipeline rupture due to landsliding was estimated from the consensus expert judgement of a team of specialists (see Chapter 43), taking into account:
   - The frequency of landslide triggering events (earthquakes and heavy rainfall, annual probability of 0.1).
   - The likelihood of a landslide event, *given* the occurrence of a landslide triggering event i.e.

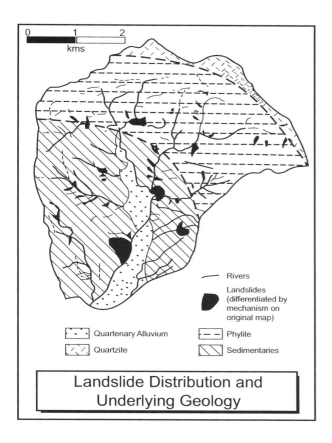

Landslide Distribution and Underlying Geology

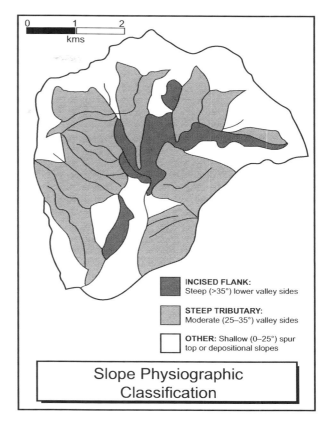

Slope Physiographic Classification

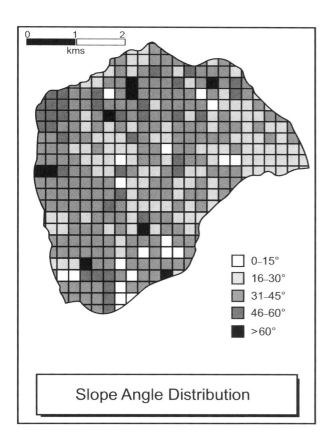

Slope Angle Distribution

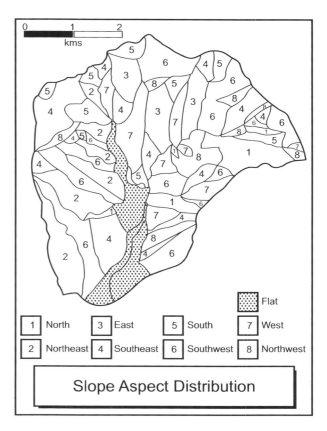

Slope Aspect Distribution

Figure 20.2 Landslide hazard mapping for route alignment through an unstable river basin in east Nepal (after TRL 1997)[20.8].

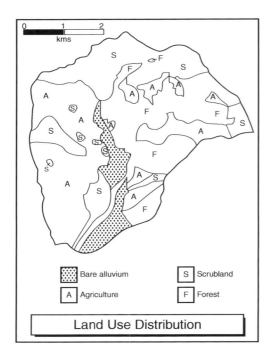

Land Use Distribution

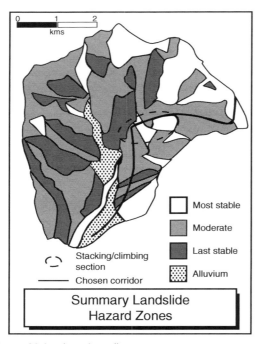

Summary Landslide
Hazard Zones

Figure 20.2 (continued).

FACTORS	CATEGORY	O/E	HAZARD RANK
ROCK TYPE	Sedimentaries	1.2	3
	Phyllite	0.8	2
	Quartzite	0.3	1
SLOPE ASPECT	North	0.7	1
	South	2.0	2.5
	East	2.2	3
	West	0.7	1
	Northwest	0.6	0
	Northeast	0.9	2
	Southeast	0.9	2
	Southwest	0.5	0
PHYSIO-GRAPHY	Incised flank	2.9	3
	Steep tributary	1.1	2
	Other	0.9	1
LAND USE	Scrub	1.0	N
	Agriculture	1.1	O
	Forest	0.9	T
SLOPE ANGLE	0–15°	0.7	
	16–30°	1.0	S
	31–45°	1.3	I
	46–60°	0.8	G
	>60°	0.4	N
CHANNEL PROXIMITY	Stream rank:		I
	First Order	0.9	F
	Second Order	1.1	I
	Third Order	1.5	C
	Fourth Order	0.9	A
	Fifth Order	1.0	N
			T

Build up of hazard rank for the illustrated catchment. Hazard ranks for each factor category are summed for every terrain unit and assigned to one of three hazard classes. 3 is the most unstable condition.

N.B. The expected number of landslides (E) is calculated on the basis of the percentage study area coverage of each factor category multiplied by the total number of observed landslides (O) in the study area. Thus, higher O/E ratios indicate a greater occurrence of instability than would be expected from a random distribution.

not all triggering events will result in significant landslide activity i.e. for Site A, Probability (Landslide) = 0.01.

- The probability of ground movement of sufficient intensity to rupture the pipeline, *given* the occurrence of a landslide i.e. Probability (Movement | Landslide) = 0.1.

The conditional annual probability of a pipeline rupture at Site A was calculated as follows:

Probability (Rupture) = Probability (Initiating
Event) × Probability
(Response)
× Probability
(Outcome)
= 0.1 × 0.01 × 0.1
= 0.0001

Landslide Risk Assessment

Risk assessment needs to take account of the elements at risk from landslides, the scale of the hazard and the adverse consequences should a landslide occur (see Chapter 12).

Event trees provide a framework for modeling landslide hazards. Landslide reactivation scenarios were developed for the unstable coastal slopes at Lyme Regis, UK[20.11]:

1. *Identification and characterisation of landslide systems*: detailed geomorphological mapping (Chapter 42) of the coastal slopes defined a series of discrete landslide units within broader landslide systems.

2. *Identification of landslide reactivation scenarios*: combining an understanding of the landslide systems with monitoring data and analysis of past events and building damage. Each scenario was developed from an *initiating event* (i.e. seawall failure, wet years sequences). Subsequent *responses* and *outcomes*, identified as the effects of the initiating event, were transmitted through the adjacent landslide units.

3. *Development of event trees*: each sequence of initiating event-response-outcome was simplified to a series of simple event trees (Fig. 20.3), with responses to a previous event either occurring or not occurring (i.e. yes/no options).

4. *Estimation of the annual probability of initiating events*: based on structural inspections of the seawalls and analysis of historical rainfall records.

5. *Estimation of the probability of responses*, through use of an open forum (see Chapter 43).

6. *Calculation of conditional probabilities for each scenario*: for year 1 (the initiating event and response occur in year 1) the conditional probability

associated with each branch of an event tree was calculated as follows:

$$\text{Scenario Probability} = P(\text{Initiating Event})$$
$$\times \; P(\text{Response 1})$$
$$\times \; P(\text{Response 2})$$
$$\times \; P(\text{Response n})$$

Landslide Risk Evaluation

The ALARP (As Low As Reasonably Practicable) principle can provide the basis for establishing *landslide risk acceptability criteria* (see Chapter 12). For example, the Hong Kong Government has published interim risk guidelines for natural terrain landslide hazards for trial use[20.12] (Fig. 20.4).

Figure 20.5 presents a matrix relating the quantitative measures of landslide likelihood and threat for an oil pipeline route[20.13]. Three zones were defined within the matrix on the basis of the landslide risk levels along the pipeline route:

- Zone A: the risks are *unacceptable*, defined as the annual probability of pipeline rupture being greater than 1 in 10 000.
- Zone B: the risks should be reduced to a level which is *as low as reasonably practicable* (the ALARP zone; the annual probability of pipeline rupture is between 1 in 10 000 and 1 in 100 000).
- Zone C: the risks are *acceptable*, the annual probability of pipeline rupture is less than 1 in 100 000.

S1 Localised damage to property, seawalls, services etc.
S2 Extensive loss of amenity gardens, sea front property, seawalls, services etc.
S3 Loss of up to 20m of cliff top land, including gardens, tennis courts, access lane to gardens, property

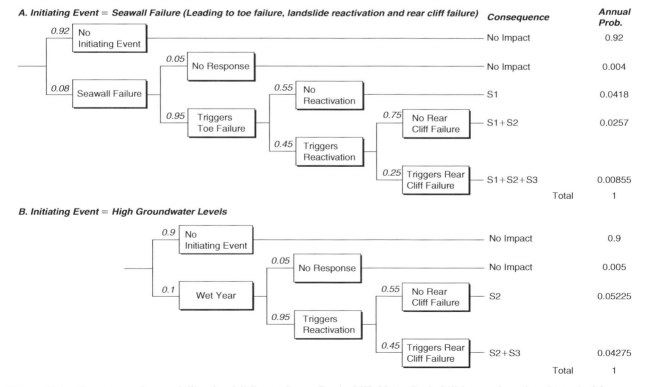

Figure 20.3 Event trees for modelling landsliding at Lyme Regis, UK. Note: Probabilities are based on historical frequencies and expert judgement (after Lee *et al.* 2000)[20.11].

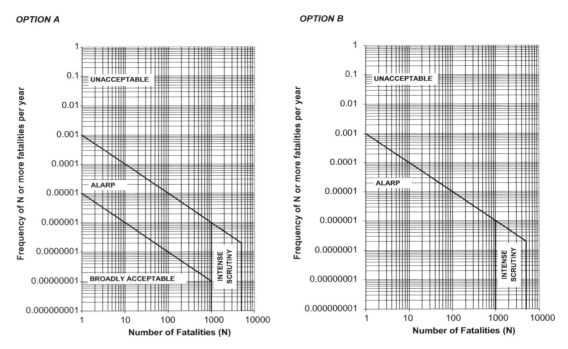

Figure 20.4   Proposed landslide risk acceptability criteria, Hong Kong (after Reeves *et al.* 1999)[20.12].

		Threat to the Pipelines				
		Negligible	Very Low	Low	Medium	High
Likelihood Class		0.000	0.033	0.066	0.330	0.495
1 Almost Certain	0.1	0.0000	0.0033000	0.0066000	0.0330000	0.0495000
2 Likely	0.01	0.0000	0.0003300	0.0006600	0.0033000	0.0049500
3 Possible	0.001	0.0000	0.0000330	0.0000660	0.0003300	0.0004950
4 Unlikely	0.0001	0.0000	0.0000033	0.0000066	0.0000330	0.0000495
5 Rare	0.00001	0.0000	0.0000003	0.0000007	0.0000033	0.0000050
6 Not Credible	0.000001	0.0000	0.0000000	0.0000001	0.0000003	0.0000005

Unacceptable >1 in 10,000 Annual Probability of Rupture
ALARP Zone (1 in 10,000 to 1 in 100,000 Annual Probability of Rupture)
Acceptable < 1 in 100,000 Annual Probability of Rupture

Figure 20.5   Landslide risk acceptability criteria for a pipeline project (after Lee and Charman 2005)[20.13].

Table 20.4   Landslide stabilisation methods[20.14].

	Best Application	Limitations	Remarks
A) Reduce the driving forces			
1. Change exact location of site, alignment or grade	During preliminary design phase of project	For highways it will affect alignment adjacent to land-slide area; pipeline – may increase total length; for all projects it may increase land-take requirements	The most reliable
2. Drain surface	Appropriate for any scheme both as part of design and in remedial measures	Will only correct surface infiltration or seepage to surface infiltration	Slope vegetation should be considered in all cases
3. Drain subsurface	On any slope where lower-ing of groundwater table will increase slope stability	Not effective when sliding mass is impervious	Stability analyses should include consideration of seepage forces

(continued)

Table 20.4   (continued)

	Best Application	Limitations	Remarks
4. Reduce weight	At any existing or potential landslide	May require use of light-weight materials that may be costly or unavailable; excavation waste may create problems; may require access to land beyond site area	Very careful stability analyses required to ensure excavation or placement of lightweight fill will improve stability
B) Increase resisting forces by applying an external force			
1. Use buttress and counter-weight fills; toe berms	At an existing landslide in combination with other methods	May not be effective on deep-seated landslides; must be placed on firm foundations	Where space is limited reinforced walls and slopes with geotextiles may be a valid alternative
2. Use structural systems	To prevent movement before excavation; where space is limited	Will not withstand large deformations; must be founded well below the sliding surface	Stability analyses must incorporate soil-structure interaction
3. Install anchors (appropriate for both soils and rocks)	Where space is limited	Foundation materials must be able to resist shear forces by anchor tension; can inhibit future development on adjacent land	*In situ* strengths of soils and rocks must be known; economics dependent on anchor capacity, depth and frequency
C) Increase resisting forces by increasing internal strength			
1. Drain subsurface	At any landslide where water table is above the shear surface	Requires expertise to install and ensure long-term effective operation	Can be very effective
2. Use geosynthetic reinforcement in backfill	For embankments and fill slopes on steep sidelong ground; in landslide reconstruction	Geosynthetic reinforcement must be durable in the long-term	Need to consider stresses imposed on reinforcement during construction
3. Install *in situ* reinforcement	As temporary structures in stiff soils; for small scale failures on rock slopes	Requires long-term durability of nails, anchors, micropiles and masonry dentition	Design methods still undergoing development
4. Use biotechnical stabilisation	On soil slopes of modest heights	Climate: may require irrigation in dry seasons; longevity of selected plants	Design is by trial and error linked to local experience
5. Treat chemically	Where sliding surface is well-defined and soil reacts positively to treatment	May be reversible; long-term effectiveness has not been proved; environmental stability unknown	Field installation must evaluate possible long-term environmental effects
6. Use electro-osmosis	To relieve excess porewater pressures and increase shear strength at required construction rate	Only a short-term construction solution as requires constant direct current power supply and maintenance	Used when nothing else works; possible application for emergency stabilisation of landslides
7. Treat thermally	To reduce sensitivity of clay soils to action of water	Requires expensive and carefully designed systems to artificially dry or freeze subsoils	Methods are experimental and costly

Landslide Risk Management

(See http://www.fema.gov/hazard/landslide/index.shtm)

The decision on whether to do nothing, avoid, or mitigate by whatever means will be a function of the costs involved and the benefits derived.

- *Do nothing*: accept the consequences associated with the level of landslide risk.
- *Avoid*: through careful planning and investigation the level of hazard and risk are identified and the hazard avoided.

Figure 20.6   Rock bolts supporting a road cutting through Carboniferous rocks in South Wales.

- *Plan for losses* through insurance and the establishment of reserve funds e.g. the New Zealand Earthquake Commission which provides natural disaster insurance cover to residential property owners (details at: www.eqc.govt.nz).
- *Spread the losses* e.g. emergency aid or disaster relief.
- *Mitigate the losses* through early warning and emergency action plans.
- *Mitigate the loss potential through careful planning.* This approach will only be appropriate if the landslide is not considered to be an immediate threat because of its slow velocity (i.e. velocity classes 1, 2, and possibly 3).
- *Mitigate the problem* by land management or landslide stabilisation (Table 20.4; Fig. 20.6).

References

20.1 WP/WLI (International Geotechnical Societies' UNESCO Working Party for World Landslide Inventory) (1995) A suggested method of a landslide summary. *Bulletin of the International Association of Engineering Geology*, **43**, 101–110.

20.2 Schuster, R. L. (ed.) (1986) *Landslide Dams: Processes, Risk and Mitigation*. Geotechnical Special Publication No. 3, American Society of Civil Engineers.

20.3 Mason, K. (1929) Indus floods and Shyock glaciers: the Himalayan Journal. *Records of the Himalayan Club, Calcutta*, **1**, 10–29.

20.4 McGuire, W. J., Mason, I. and Kilburn, C. (2002) *Natural Hazards and Environmental Change*. Arnold, London.

20.5 Merraim, R. (1960) The Portuguese Bend landslide, Palos Verdes. California. *Journal of Geology*, **68**, 140–153.

20.6 Hearn, G. J. and Griffiths, J. S. (2001) Landslide hazard mapping and risk assessment. In J. S. Griffiths (ed.) *Land Surface Evaluation for Engineering Practice*. Geological Society, London, Engineering Group Special Publication, **18**, 43–52.

20.7 Soeters, R. and van Westen, C. J. (1996) Slope instability recognition, analysis and zonation. In A. K. Turner and R. L. Schuster (eds.) *Landslides: Investigation and Mitigation*, Transportation Research Board, Special Report 247, National Research Council, National Academy Press, Washington DC, 129–177.

20.8 TRL (1997) Principles of low cost road engineering in mountainous regions, with special reference to the Nepal Himalaya. Overseas Road Note 16, TRL, Crowthorne.

20.9 Lee E. M. and Jones D. K. C. (2004) *Landslide Risk Assessment*. Thomas Telford.

20.10 Lee, E. M., Clark, A. R., and Guest, S. (1998) An assessment of coastal landslide risk, Scarborough, UK. In D. Moore and O. Hungr (eds.) *Engineering Geology: The View from the Pacific Rim*, 1787–1794.

20.11 Lee, E. M., Brunsden, D. and Sellwood, M. (2000) Quantitative risk assessment of coastal landslide problems, Lyme Regis, UK. In E. N. Bromhead, N. Dixon and M.-L. Ibsen (eds.) *Landslides: In Research, Theory and Practice*, Thomas Telford, 899–904.

20.12 Reeves, A., Ho, K. K. S. and Lo, D. O. K. (1999) Interim risk criteria for landslides and boulder falls from natural terrain. Proceedings of the Seminar on *Geotechnical Risk Management*, Geotechnical Division, Hong Kong Institution of Engineers, 127–136.

20.13 Lee, E. M. and Charman, J. H. (2005) Geohazards and risk assessment for pipeline route selection. In M. Sweeney (ed.) *Terrain and Geohazard Challenges Facing Onshore Oil and Gas Pipelines*, Thomas Telford, 95–116.

20.14 Holtz, R. D. and Schuster, R. L. (1996) Stabilisation of soil slopes. Chapter 17 in A. K. Turner and R. L. Schuster (eds.) *Landslides: Investigation and Mitigation*, Transportation Research Board, Special Report 247, National Research Council, National Academy Press, Washington DC, 439–473.

21

Slopes: Karst Terrain

Introduction

Karst terrain develops where solution predominates over other weathering and erosion processes (see Chapter 16). The key surface features of karst terrain (Fig. 21.1) include dry valleys and closed depressions (*sinkholes* or *dolines*) of varying sizes, *pinnacled rockhead* through to *rock hills, cones* and *towers*, together with small-scale solution sculpturing of exposed rock. Underground drainage can be channelled into *caves*, *conduits* or *fissures* formed by solution, or as diffuse flow through fractures and smaller voids throughout the whole rock mass.

Karst types include (see http://water.usgs.gov/ogw/karst/):
- *Bare or soil covered karst*: rockhead is < 1 or 2 m below the surface.
- *Mantled karst*: solution affected topography is covered by 10+ m of unconsolidated deposits. This may include insoluble residue of dissolved limestone or the insoluble remains of the strata which once overlaid the limestone.

Mantled karst has a subdued topography with frequent ponds filling collapsed hollows (*suffusion dolines*) especially in lowland areas with a shallow water table.

- *Interstratal karst*: solution of karst rocks takes place below overlying insoluble strata. Collapse of solution cavities may produce boulder filled *pipes* in the overlying strata and *dolines* may form in the ground surface.

Carbonates are the commonest of the soluble rocks; the main karst forming rocks are the limestones (calcium carbonate). The largest caves and the most rugged surface karst landforms are formed in the older, stronger limestones and also the less common but strong dolomites (magnesium carbonates).

Gypsum (calcium sulphate) karst typically has thinner beds, wider development of interstratal karst and a greater number of breccia pipes. It does not develop to cone karst or tower karst. Gypsum (also anhydrite and other sulphates) goes into solution more quickly than carbonates and as a result is associated with arid or semi-arid regions. Gypsum solution is significant over engineering time[21.2].

Pseudokarst occurs in rocks which are prone to piping (see Chapter 16). In weathered granite, for example, the gruss (clayey sand) weathering product gets washed out (it often widens joints down which the weathering has penetrated) and forms open passages.

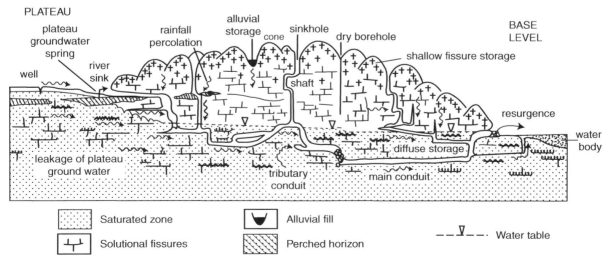

Figure 21.1 Elements of groundwater flow and storage in the limestone of Gunung Sewu, an area of cone karst in Java (after Waltham 2005)[21.1].

Hazards

- *Problem foundation conditions* i.e. irregular, pinnacled rockhead surface. No other terrain has such marked contrast in bearing capacity, between strong rock, soil-filled fissures and caves.
- *Reservoir leakage*: in China there are over 5000 reservoirs impounded on karst in Guangxi, Guizhou and Hunan provinces alone, and about a third of these suffer significant leakage[21.3].
- *Sinkhole collapse* (Fig. 21.2), formed by downwashing of the soil cover into fissures/caves e.g. during rainstorms, as a result of dewatering (i.e. water table decline) or uncontrolled surface drainage. For example, the Dserzhinsk region of Belarus averages >4 new sinkholes km^{-2} yr^{-1} developed over both limestone and gypsum. A machine factory was lost in a sinkhole in 1992[21.1].

Collapses in sandy soils are generally slow. In clays cavitation starts above a rockhead fissure, and grows slowly beneath an arched soil roof; the cavity propagates upward until the surface collapses instantly and without warning i.e. a major engineering hazard in a soil-covered karst.

- Cave collapse: in limestone, collapse is generally confined to situations where the solid cover thickness is less than the cave width[21.1].

Engineering Issues

Figure 21.3 presents an engineering classification highlighting the complexity likely to be encountered by foundation engineers; increasing difficulties are created by more mature karstic erosion, cave and landform evolution (e.g. Table 21.1).

Table 21.1 also presents an outline to ground investigation and foundation measures. Investigations should define[21.4]:

- karst class, including lithology
- amplitude of rockhead relief
- sinkhole density
- typical and maximum cave passage widths
- frequency of new sinkhole collapses.

The distribution of sinkholes and caves can be largely unpredictable. Geophysical surveys can produce useful results in certain situations[21.5]. All geophysical anomalies require verification by drilling.

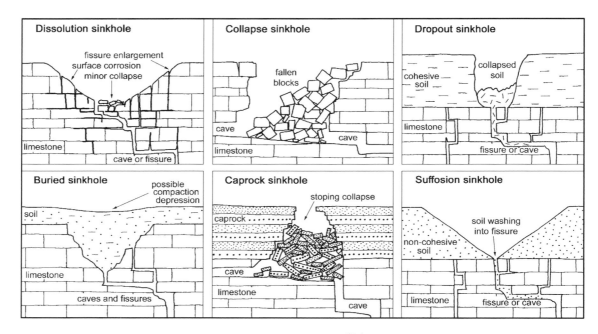

Figure 21.2   Sinkhole classification (after Waltham and Fookes 2003)[21.4].

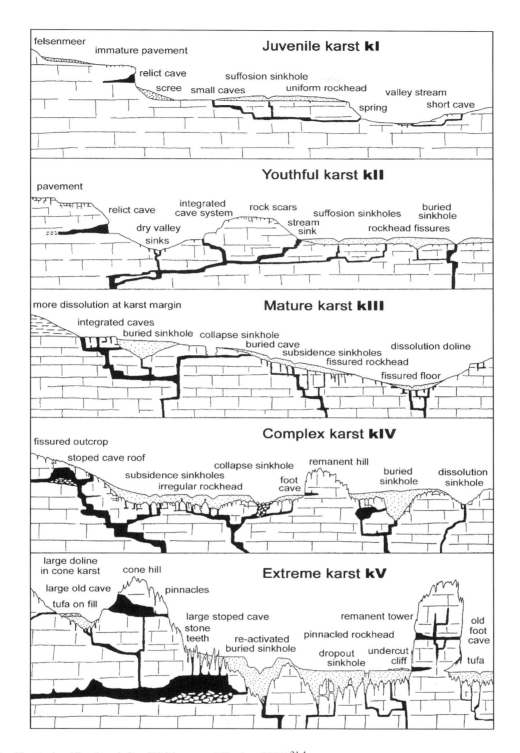

Figure 21.3 Karst classification (after Waltham and Fookes 2003)[21.4].

Table 21.1 An engineering classification of karst.

Karst class	Locations	Sinkholes	Rockhead	Fissuring	Caves	Ground investigation	Foundations
kI *Juvenile*	Only in deserts and periglacial or zones, or on impure carbonates	Rare NSH* <0.001	Almost uniform; minor fissures	Minimal; low secondary permeability	Rare and small; some isolated relict features	Conventional	Conventional with consideration for provision in the contract for the Observational Method (Chapter 1)
kII *Youthful*	The minimum in temperate regions	Small suffusion or dropout sinkholes; open stream sinks; NSH 0.001–0.05	Many small fissures	Widespread in the few metres nearest surface	Many small caves; most <3 m across	Mainly conventional, probe rock to 3 m, check fissures in rockhead	Grout open fissures; control drainage
kIII *Mature*	Common in temperate regions; the minimum in the wet tropics	Many suffusion and dropout sinkholes, large dissolution sinkholes; small collapse and buried sinkholes; NSH 0.05–1.0	Extensive fissuring; relief of <5 m; loose blocks in cover soil	Extensive secondary opening of most fissures	Many <5 m across at multiple levels	Probe to rockhead, probe rock to 4 m, microgravity survey	Rafts or ground beams, consider geogrids, driven piles to rockhead; control drainage
kIV *Complex*	Localized in temperate regions; normal in tropical regions	Many large dissolution sinkholes; numerous subsidence sinkholes; scattered collapse and buried sinkholes; NSH 0.5–2.0	Pinnacled; relief of 5–20 m; loose pillars	Extensive large dissolutional openings, on and away from major fissures	Many >5 m across at multiple levels	Probe to rockhead, probe rock to 5 m with splayed probes, microgravity survey	Bored piles to rockhead, or cap grouting at rockhead, control drainage and abstraction
kV *Extreme*	Only in wet tropics	Very large sinkholes of all types; remnant arches soil compaction in buried sinkholes; NSH > 1	Tall pinnacles; relief of >20 m; loose pillars undercut between deep soil fissures	Abundant and very complex dissolution cavities	Numerous complex 3-D cave systems with galleries and chambers >15 m across	Make individual ground investigation for every pile site	Bear in soils with geogrid, load on, proven pinnacles or on deep bored piles; control all drainage and control abstraction

This table provides outline descriptions of selected parameters; these are not mutually exclusive and give only broad indications of likely ground conditions that can show enormous variation in local detail. It should be viewed in conjunction with Fig. 21.3 which shows some of the typical morphological features. The comments on ground investigation and foundations are only broad guidelines to good practice in the various classes of karst. *NSH = approximate rate of formation of new sinkholes per km^2 per year.

References

21.1 Waltham, A. C. (2005) Karst terrains. In P. G. Fookes, E. M. Lee and G. Milligan (eds) *Geomorphology for Engineers*. Whittles Publishing, 662–687.

21.2 James, A. N. and Lupton, A. R. R. (1978) Gypsum and anhydrite in foundations of hydraulic structures. *Geotechnique*, **28**, 249–72.

21.3 Yuan, D. (ed.) (1991) *Karst of China*. Geological Publishing House, Beijing.

21.4 Waltham, A. C. and Fookes, P. G. (2003) Engineering classification of karst ground conditions. *Quarterly Journal of Engineering Geology and Hydrogeology*, **36**, 101–118.

21.5 Cooper, S. S. and Ballard, L. F. (1988) Geophysical exploration for cavity detection in karst terrain. In Sitarn (ed.) *Geotechnical aspects of karst terrains: exploration, foundation design and performance, and remedial measures*. Geotechnical Special Publication 14. American Society of Civil Engineers. New York 25–39.

22

Rivers: The Drainage Basin

Introduction

River systems are open channels that carry water and sediment through a *drainage basin* or *catchment* (the area over which all precipitation flows downslope to exit at a single point; Fig. 22.1[22.1]). An overall drainage basin system comprises three broad zones, each with a characteristic form and function[22.2] (see Chapter 4; Fig. 4.1):

- Zone 1: the water and sediment source area
- Zone 2: the sediment transport zone
- Zone 3: the sediment store or deposition zone (e.g. *alluvial fan, floodplain, delta*).

This sub-division illustrates how the dominant processes in different parts of the system are interlinked into a water and sediment cascade. However, in reality sediments can be supplied (i.e. eroded), transported and deposited in all of the zones.

The supply of sediment is generally intermittent, with rare floods amongst the most effective events in delivering sediment into the river channel network. Once in the channel, the velocity of flow and sediment size is important in determining how far it is carried before being temporarily stored in features (e.g. point bars, levees, floodplains, or lakes). In short rivers, the suspended load may reach the estuary in a single flood, but coarser sediments may move only tens of metres from an eroding riverbank to a gravel bar within the channel. Sediment storage is generally the prevailing condition for most contemporary river systems.

Drainage Basin Form

The main controls of drainage basin form are geology, climate and hydrology. The boundaries between catchments form *watersheds*. In certain rock types, there may be differences between the surface water catchment and the sub-surface water catchment.

The *drainage density* D_d is:

$$D_d = \frac{\Sigma L}{A_d}$$

where ΣL = total length of all streams within the catchment and A_d = catchment area.

The channel *pattern* tends to vary with rock type and geological structure (Table 22.1); this link is useful in remote sensing interpretation[22.3] (Fig. 22.2).

Floodplains

A floodplain is the low relief area, of variable width, adjacent to a river that is inundated by water during floods (Fig. 22.3). Floodplains represent the main sediment store within a drainage basin (floodplain deposits are called *alluvium*) and are potentially an important source of natural aggregates (Chapter 13). Floodplains are not just confined to the lower reaches of a basin; they can be found wherever the river has deposited alluvium alongside its channel.

Floodplains accumulate by two processes:

- *Vertical accretion* from overbank flow, producing thin layers of fine sediments.
- *Lateral accretion* from the migration of channels across a floodplain. This results in a fining upwards sequence of gravels and sands, capped by finer sediments.

The three-dimensional stratigraphy of the floodplain alluvium reflects past channel migration[22.4–5] (Figs 22.4 to 22.6).

Floodplain accretion rates in temperate rivers vary between fractions of mm yr^{-1} to a few cm yr^{-1}. Faster rates occur along rivers prone to flash floods or debris flows. In Bijou Creek, Kansas, USA, 1 m of sediment was deposited in a flash flood in 1965[22.6].

Floodplain landforms consist of:

- The main channel, which is likely to have a *meandering* or *braided* form (see Chapter 24)
- Natural *levees* alongside the main channel(s): broad long ridges composed of sand and silt deposited by floodwaters as they overtop the channel banks.

Lateral migration by the meanders can lead to *abandoned meanders* being left on the floodplain containing *ox-bow* lakes that slowly infill. *Channel migration* leads to undercutting of the river bank on the outside of the meander bend and deposition of *point bar deposits* on the inside, possibly backed by a low-lying swampy area: the *backswamp*.

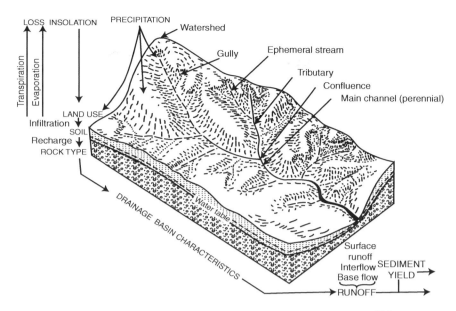

Figure 22.1   Characteristics of the river catchment (based on Gregory and Walling 1971)[22.1].

Table 22.1   Drainage patterns.

Drainage Pattern	Typical Geological Characteristics
Dendritic	Homogenous rock type; density will reflect infiltration characteristics of the materials
Rectangular	Well-developed orthogonal discontinuity sets
Radial	Homogenous dome structure, such as a sub-surface pluton or extrusive volcano
Centripetal	Enclosed synform or internal drainage in basin-and-range areas
Trellis	Cuesta sequence in gently dipping sedimentary rocks of variable strength
Parallel	Dip slope of a relatively homogenous sedimentary sequence
Annular	Antiform developed in variable materials often with radial discontinuity systems
Contorted	Strong structural geology influence resulting from neotectonics and metamorphic activity
Multi-basin	Hummocky ground, typically from glacial deposition, or intermittent drainage on limestone
Deranged	Uneven ground, as typically in sloping outwash plains

Floodplain limits may be defined differently by geomorphologists, engineers and insurers[22.7]. In Fig. 22.7 buildings impinge on the non-regulated floodplain area subject to flooding by a 1 in 10 year flood (Chapter 11); even the floodway fringe can expect to be flooded every 10 to 100 years. For the geomorphologist the floodplain is defined by the up-slope limits of contemporary fluvial landforms and may include low terraces. This is likely to include the area encompassed by at least the 1 in 100 year flood.

Water Management

Catchment uses are varied; management involves engineers, hydrogeologists, hydrologists and environmental managers. Engineering geomorphology can be used to:

• Identify the causes and significance of site specific problems, through a fluvial geomorphology survey[22.8] (Table 22.2).
• Identify the potential impacts (upstream and downstream) of proposed works[22.9] (Table 22.3).
• Identify suitable mitigation measures, including channel restoration operations.

Many problems in water management are related to the way water and sediments are transported from supply areas and held in, or released from, temporary stores such as bars or spreads on the river bed. Figure 22.8 highlights the linkages between different parts of a river system and illustrates the knock-on effects of a large sediment input in the uplands[22.10].

Often rivers are treated on a site-by-site basis and this can lead to frequent and costly maintenance. Site problems are a symptom of broader-scale processes that operate throughout the drainage basin.

The Tennessee Valley Authority, for example, was set up in 1933 (see http://www.tva.gov/). Initially based around the need to reduce the very high soil

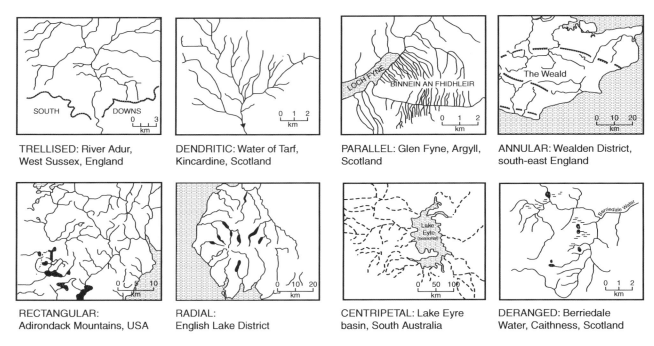

Figure 22.2   Classification of drainage networks with examples (after Smithson *et al.* 2002)[22.3].

Figure 22.3   River floodplain, in southeast Iceland.

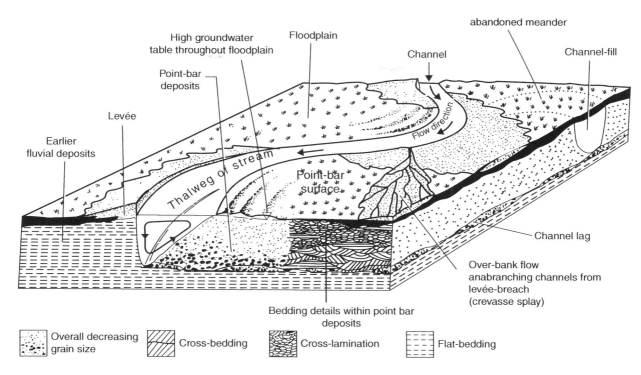

Figure 22.4   Meandering river model: floodplain (based on Collinson 1978)[22.4].

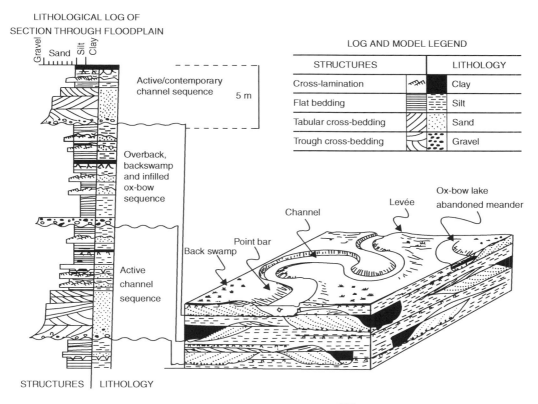

Figure 22.5   Meandering river channel stratigraphy (based on Selley 1976)[22.5].

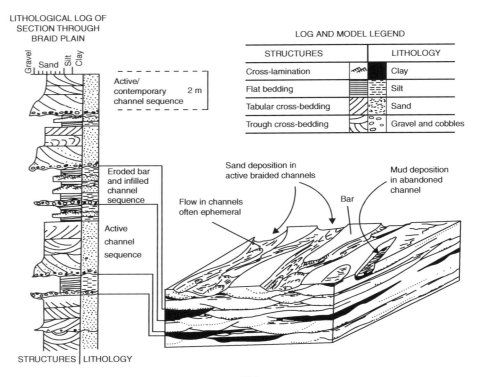

Figure 22.6   Braided river stratigraphy (based on Selley 1976)[22.5].

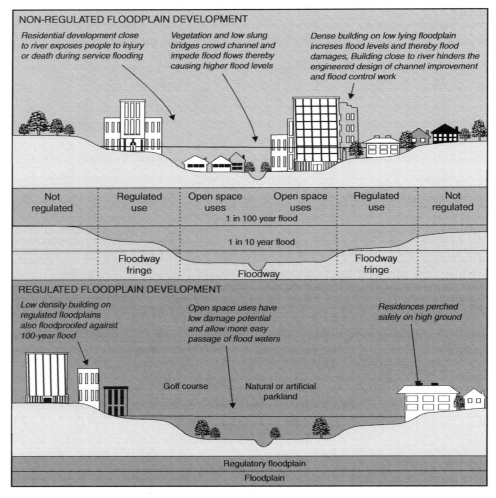

Figure 22.7   Zoning of river floodplains (after Newson 1995)[22.7].

Table 22.2 Carrying out a Fluvial Geomorphology Survey [22.8].

Main study context	Geomorphological topic	Input data required	Output	Practical engineering outcomes
Flow regime, channel geometry and dynamics	1. Determination of dominant discharge and range of effective flows	Flow duration curves, sediment rating curve	Dominant discharge defined by flow magnitude, duration and return period. Range of effective flows for transporting a large percentage of the total sediment load	Reservoir sedimentation rates; flood control measures; reservoir storage capacity; run of the river hydroelectric scheme capacity; river training works; flood hazard
A. Morphological analysis of overall channel features	2. Specific gauge analysis	Current and historical stage-discharge curves	Identification of water level changes through time for a range of specified constant discharges, indicating possible bed level changes and/or changes in bedforms, bars and flow resistance	Flood control measures; bridge pier and abutment foundations; run of the river hydroelectric scheme capacity
B. Morphological analysis of in-channel and sub-channel morphology	3. Hydraulic geometry analysis of primary channel cross-sections	Dominant discharge, sediment rating curves, bed material size distribution, bed material properties, river cross-sections	Predictions of expected hydraulic geometry from existing equations of comparisons with observed data. Development of empirical relationships between driving variables (flow, sediment load, effective rainfall etc) and primary channel dimensions	Waterway capacity of bridges, culverts and spillways; erosion control; flood hazard; dredging for navigation
C. Bank erosion and recent channel changes	4. Analysis of channel geometry and pattern changes with changing stage	River cross-sections, interpretation of remote sensing data, flowlines for dominant and other effective flows	Determination of morphological links between dominant discharge, effective range of sediment transporting flows, and significant channel features such as bar-full, full channel width, and bank-full stages	Location of river crossing sites; river training works
	5. Long-profile analysis of thalweg, bar top, bank top and water surface profiles	Long profiles of present and historic river, flowlines for selected flows	Establish link between selected flows and channel morphology along reach; define planform/profile associations; identification of aggradation and degradation trends	Long term control measures; identification of potential dam sites
	6. River planform analysis	Remote sensing data and range of historical and contemporary maps	Establish link between selected flows and channel morphology along reach; define planform/profile associations; identification of aggradation and degradation trend	Long term river control and training measures; identification of potential dam sites; location of stable river crossing sites; scour estimation around bridging structures
	7. Braiding intensity analysis	River cross-sections and remote sensing data	Establish present historical braiding intensity; determine relationship between braiding pattern and cross-sectional morphology	Prevent uncontrolled sand and gravel extraction
	8. Braiding processes and dynamics	Remote sensing data, bed topography; flow velocity profiles and sediment transport at bars, diffluences and confluences	Understanding of flow processes and sediment transport mechanics involved in the formation, enlargement and movement of braid bars, anabranches and confluence scour holes	Prevent uncontrolled sand and gravel extraction

(continued)

Table 22.2   (continued)

Main study context	Geomorphological topic	Input data required	Output	Practical engineering outcomes
	9. Meandering: bendway scour analysis	River cross-sections and remote sensing data; field measurements of near-bank velocities and scour depths for a range of stage heights	Empirical prediction of maximum scour depth at channel or anabranch bend resulting from curvature effects; testing of analytical bend scour predictors	Needed for assessment of bridge pier locations and bank platform
	10. Meandering: outer bank velocity analysis		Empirical prediction of depth averaged velocity over the toe of the outer bank; testing of analytical bend velocity models	Needed for assessment of bridge pier locations and bank platform
	11. Meandering: migration and bank erosion analysis		Empirical prediction of bank erosion and bend migration rates; calibration of various models	Needed for assessment of bridge pier locations and bank platform
	12. Recent bankline changes and width adjustments	Historical remote sensing and GIS	Determination of recent (e.g. 20 year) bank erosion rates and bankline changes; identification of recent trends of width adjustment and planform evolution	River bank stabilisation
	13. Field observations of bank stability and failure	Fieldwork: observation and recording of bank conditions and morphology	Inventory of current bank conditions, materials, stratigraphy, profiles, erosion processes and bank failure mechanisms; determination of severity, distribution and extent of erosion in study reach	River bank stabilisation measures; river training works
	14. Bank stability analyses	Bank profile surveys, soil properties and stratigraphy, bank stability models	Definition of role of mass instability in contributing to bank retreat; identification of critical conditions for bank collapse; estimation of bank sensitivity to basal scour and/or lateral erosion by the flow	River bank stabilisation; river training works
	15. Sediment mass balance analysis	Historical remote sensing data, GIS processing, river cross-sections, sediment load measurements, sediment load/stage height relationships	Determination of mass balance between bank erosion sediment yield and bar sediment storage; mass balance for sediment flux through study reach	Reservoir sedimentation rates; soil erosion estimation and control; sand and gravel extraction
	16. Comparison of bank and bar sediment size distributions	Bank and bar material samples; sediment transport data	Undertake separate mass balance for sand and silt fractions of perimeter and through study reach	
Physiographic and geological aspects	17. Floodplain topography	Existing topography maps and field survey	Identification of floodplain morphology, drainage pattern and potential pathways for major channel avulsions	River training works; location of stable crossing points
	18. Geological mapping	Existing geological maps, field mapping, remote sensing data	Identification of geological influence on river due to outcrops of erosion-resistant materials, presence of faults, subsidence or other neotectonic phenomena	Geohazard assessment for all structures; location of construction materials

(continued)

Table 22.2 (continued)

Main study context	Geomorphological topic	Input data required	Output	Practical engineering outcomes
	19. Geomorphological mapping	Existing superficial geology maps, field mapping and remote sensing data	Classification of significant geomorphological features and sediments, identification of past river courses and potential pathways for avulsions	Flood hazard; locating suitable crossing or damming sites; scour potential; sedimentation rates; identify potential construction materials
	20. Land use and settlement mapping	Existing maps or socio-economic surveys	Information on flood-plain use, riparian environments and hazards caused by bank erosion, channel migration or avulsions	Erosion control; risk assessment for floods; irrigation and land drainage requirements
Long term channel development	21. Long term planform pattern evolution and channel migration	Historical maps and remote sensing data, long term river surveys	Establishment of a long-term history and chronology of channel development; determination of past channel patterns, widths, and rates of bank erosion; identification of changes in sinuosity or braiding over the long-term; relationship of channel process to overall landscape development	Applicable to the design and construction of all forms of drainage control, river crossing, or river training structures

Table 22.3 River engineering and geomorphological inputs[22.9].

Subject	Engineering application	Geomorphological input		
		Design aspects	Local impact	Regional impact
1. Drainage basin				
Irrigation schemes for agriculture	Pattern, shape size and slope of waterways (e.g. canals)	Use of flow regime equations	Changes to water table, soil erosion, sedimentation, water quality	Changed magnitude and frequency in the river system
Gully control (e.g. in badlands)	General gully protection measures (e.g. check dams)	Knowledge of gully development and erosion patterns	Effects of downstream scour	Changes to scour and erosion loci on main channels
Land drainage for agriculture	Drain design and installation	Efficiency of alternative drain design	Increase discharge downstream; changes in local water balance	Channel adjustment downstream of drained area
Road drainage	Types of road drainage; culverts	Road runoff and culvert flow estimation; increase in peak discharges to stream network	Roadside gully development; increased scour and flood potential from local streams	Channel adjustment resulting from higher peak flows; flooding potential increased
Urban stormwater drainage	Design and installation of stormwater drainage network including increased storage capacity	Urban runoff estimation through sewerage and drainage network	Erosion and deposition at discharge loci.	Channel adjustment resulting from higher or lower peak flows; flooding potential within urban area and downstream

(continued)

Table 22.3   (continued)

Subject	Engineering application	Geomorphological input		
		Design aspects	Local impact	Regional impact
2. Channel reach				
River aggregate extraction	Sand and gravel extraction methods and quantity estimation	Identification of suitable sites; environmental impact; quantities	Changes to sedimentation in channel stretch;	Downstream channel adjustment; changes in the sediment supply
River diversion and artificial cut-offs	Design and construction of diversion channel	Stable channel design for range of flows associated with river; location of suitable diversion route	Channel adjustment to new route and flow regime	Downstream channel adjustment
Channelization	Widening, deepening, straightening and clearing channel; bank strengthening and protection; dyke construction	Understanding of water and sediment regime; bank stability; location of optimum site for bank protection	Channel adjustment to new route and flow regime	Downstream channel adjustment
Building on floodplain	Construction on soft alluvial materials within a flooding zone	Flood magnitude and frequency for height estimation and inundation zonation schemes; floodplain ground model for foundation design; establish channel migration pattern	Increase flood protection measures; insurance assessments	Increase regional flood protection costs; downstream channel adjustments
3. Specific locations				
Dam construction	Design of general civil, electrical and mechanical engineering works associated with the dam; coffer dam and diversion channel and tunnel design	Ground model for foundations, tunnels and excavation stability; flood and drought frequency analysis; reservoir sedimentation rates; downstream scour assessment	Changes in local groundwater levels; reservoir rim stability; increased scour downstream	Upstream and downstream channel adjustment; possible induced local seismic activity; micro-climate adjustments
Bridge construction	Design of structure with suitable waterway area as well as to be able to carry design loads, withstand pier scour, wind loading and impacts from water borne objects including shipping and ice	Flood frequency analysis; bridge pier scour estimation; ground model for bridge foundations and abutments; channel and river bank stability assessment;	Localised increased flows and scour	Upstream and downstream channel adjustments
Dredging for navigation	Depth, frequency and method of sediment removal	Rate of sedimentation	Changes in the local sediment regime	Changes in the downstream sediment supply

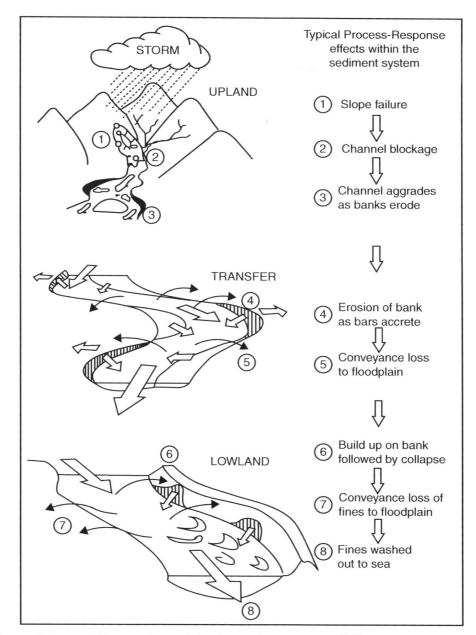

Figure 22.8   Process linkages within a catchment (after Sear and Newson 1992)[22.10].

erosion rates and improve agriculture, the present plan (2004) covers all aspects of river management including power generation from hydroelectricity, nuclear and fossil fuels.

References

22.1   Gregory, K. J. and Walling, D. E. (1971) Field measurements in the drainage basin. *Geography*, **56**, 277–92.

22.2   Schumm, S. A. (1977) *The Fluvial System*. Wiley, New York.

22.3   Smithson, P., Addison, K. and Atkinson, K. (2002) *Fundametnals of the Physical Environment*. 3rd Edition, Routledge, London and New York.

22.4   Collinson, J. D. (1978) Alluvial sediments. In H. G. Reading (ed.) *Sedimentary Environments and Facies*. Elsevier, New York, 15–60.

22.5   Selley, R. C. (1976) *An Introduction to Sedimentology*. Academic Press, London.

22.6   McKee, E. D., Crosby, E. J. and Berryhill, H. L. (1967) Flood deposits of Bijou Creek, Colorado, June 1965. *Journal of Sedimentary Petrology*, **37**, 829–851.

22.7   Newson, M. (1995) *Hydrology and the River Environmment*. Clarendon, Oxford.

22.8   Thorne, C. R. (1993) *Guidelines for the use of stream reconnaissance sheets in the field*. Contract Report HL-93-2, US Army Engineer Waterways Experimental Station, Vicksburg, Mississippi.

22.9   Cooke, R. U. and Doornkamp, J. C. (1990) *Geomorphology in Environmental management*. 2nd Edition. Oxford University Press.

22.10 Sear, D. A. and Newson, M. D. (1992) *Sediment and gravel transportation in rivers including the use of gravel traps*. NRA Project Report 232/1/T.

23

Rivers: Water and Sediment Loads

Introduction

Rivers are moving bodies of water that transport rock/soil particles and dissolved materials (the *sediment load*; Fig. 23.1) downslope in a clearly defined path carved in recent alluvial sediments or bedrock (i.e. the *river channel*). Rivers in bedrock tend to follow a stable course whereas alluvial channels tend to change both their location (*channel change* and *avulsion*) and behaviour over engineering time (see Chapter 25).

Water Flow

The flow Q (*discharge*; see Table 23.1) measured in $m^3 s^{-1}$ is calculated as:

$$Q = w\,d\,v$$

where w = channel width (m), d = average depth (m) and v = velocity of flow (m s^{-1}).

Discharge measurements are made using various gauging structures (e.g weirs, flumes, or rated channel sections). Variations in discharge over time are plotted on a *flood hydrograph* (Fig. 23.2) which shows the response of the river flow to an input of water (e.g. precipitation, snow melt or lake outbursts):

- *Baseflow*: the long-term flow, resulting from drainage out of the groundwater storage. This may be absent in *ephemeral rivers*.
- *Quickflow*: the response to an input of water. This results from surface and channel water flow i.e. *runoff* (see Chapter 15).

The hydrograph has a *rising limb*, typically quite steep in areas where the precipitation arrives as short duration intense downpours (e.g. in drylands), and a shallower *recession limb*.

The time from the peak precipitation to the peak flow is the *lag time*. Urbanisation leads to higher peaks and more rapid flow routing which must be accommodated in the design drainage systems (see Chapter 14).

Water flows downhill under the influence of gravity, but the velocity of flow is resisted by the frictional forces between the water molecules and between the water and the channel boundaries. The resistance (i.e. *roughness*) is controlled by grain size of bed and bank material,

Figure 23.1   Bedload exposed in the ephemeral river Mithiwan Nallbh, northwest Pakistan.

146

Table 23.1 Examples of maximum river discharge values.

River	Drainage basin area (km²)	Discharge (m³ s⁻¹)	Specific discharge (m³ s⁻¹ km⁻²)
Amazon	6 million	200 000	0.033
Yangtse	1.8 million	30 000	0.017
Mississippi	2.98 million	15 500	0.052
Tay (Scotland, Jan 1993)	865	2269*	2.623
Lynmouth, UK (1954)	101.5	511**	5.034

* The highest recorded discharge on a UK river; ** only exceeded twice by the Thames since 1883.

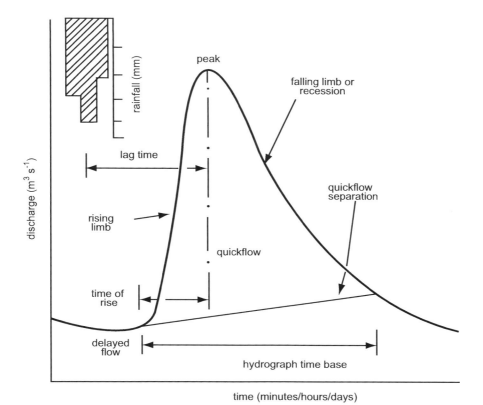

Figure 23.2 Basic properties of a flood hydrograph.

the channel vegetation and the irregularity of the adjacent channel cross-sections (Fig. 23.3). For simple, artificial and sand-bedded channels, *Manning's roughness coefficient n* can be calculated from the median grain diameter D_{50} (mm):

$$n = 0.0151 D_{50}^{0.167}$$

Typical values of n in natural channels (Table 23.2): sand-bedded channels = 0.011–0.035; winding natural streams = 0.035; mountain streams with rocky beds = 0.040–0.050. A full range of values for n can be found in standard texts[23.1].

Velocity V (m s⁻¹) of flow can be estimated using Manning's equation (Table 23.3).

$$V = \frac{R^{0.667} S^{0.5}}{n}$$

where R is the hydraulic radius (A / P; A = cross-sectional area, P = wetted perimeter) and S = channel gradient.

When the ratio of channel depth d to mean particle size diameter D is ≤ 10 in shallow rocky channels, the power of R increases to 0.75. Channel depth can be substituted for R when the channel is wide relative to the depth.

Flow velocity can also be calculated from the Darcy-Weisbach equation:

$$V^2 = \frac{8 \mathrm{g} R S}{f}$$

where g is the gravitational acceleration and f is the Darcy-Weisbach friction factor[23.1] (Table 23.2).

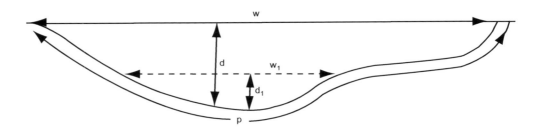

Dimensions of the channel cross-section; w-wetted perimeter;
d-bankfull depth; p-bankfull wetted perimeter; w_1-width at
mean annual discharge; d_1 – depth at mean annual discharge

Figure 23.3   Dimensions of the channel cross-section.

Table 23.2   Channel types and typical roughness coefficient values.

Channel	Channel Slope (%)	Bed material D_{50} (mm)	Darcy-Weisbach f	Manning n
Sand-bed	< 0.01	< 2	0.01–0.25	0.01–0.04
Gravel/cobble bed	0.05–0.5	10–100	0.01–1	0.02–0.07
Boulder-bed	0.5–5	> 100	0.05–5	0.03–0.2

Table 23.3   Calculation of flow velocity using the Darcy-Weisbach and Manning equations (for grain size $D_{50} = 2$ mm).

Width (m)	Depth (m)	Slope gradient	Cross section A (m²)	Wetted perimeter P (m)	Hydraulic Radius R	Manning resistance coefficient n	Darcy-Weisbach friction factor f	Darcy-Weisbach equation		Manning's equation	
								Velocity (m s⁻¹)	Discharge Q (m³ s⁻¹)	Velocity (m s⁻¹)	Discharge Q (m³ s⁻¹)
10	1.00	0.090	10	12	0.83	0.017	0.024	17.2	171.7	15.7	156.9
20	2.00	0.090	40	24	1.67	0.018	0.021	25.7	1027.8	23.4	935.9
30	5.00	0.017	150	40	3.75	0.018	0.016	20.2	3027.8	17.4	2604.9
40	10.00	0.009	400	60	6.67	0.018	0.014	22.8	9135.5	18.5	7388.7

Cross section A = Width × Depth; Wetted perimeter P = Width + (2 × Depth); $R = A / P$; Resistance coefficient $n = 0.0151 D_{50}^{0.167}$; Friction factor f = $(8gn^2) / R^{0.167}$; Velocity (Darcy-Weisbach) = $\sqrt{(8gRS)} / $ f; Velocity (Manning) = $(R^{0.667} S^{0.5}) / n$; Discharge Q = Velocity × A.

Alternatively, velocity can also be calculated from the Chézy equation:

$$V = C\sqrt{RS}$$

where C is Chézy's flow resistance factor[23.1] ($C = R^{0.167}/n$).

Hydraulic radius R and velocity V are linked to a distinction between laminar and turbulent flow via the *Reynolds number Re*:

$$Re = \frac{VR}{\nu}$$

where ν is the kinematic viscosity (dynamic viscosity μ divided by the fluid density ρ_w; Table 23.4).

$Re < 500 \Rightarrow$ laminar flow; $Re > 2000 \Rightarrow$ turbulent flow; for $500 < Re < 2000$ both forms of flow are found.

Velocity V and flow depth d can be used to distinguish between *sub-critical* and *super-critical* flow regimes, defined by the *Froude number F*:

$$F = \frac{V_{\text{mean}}}{\sqrt{gd}}$$

where g is gravitational acceleration.

For tranquil or sub-critical flow regime $F \leq 1$, for rapid or super-critical flow regime $F \geq 1$. Manning's n varies with the Froude number in sand bed streams (Fig. 23.4); a range of river bed landforms develop in response to changing roughness and flow regime[23.2].

Table 23.4   Density and viscosity of water.

Temperature (°C)	Fluid density ρ_w (g cm^{-3})	Dynamic viscosity μ (g cm^{-1} s^{-1})	Kinematic velocity ν (cm^2 s^{-1})
0	1.0	0.018	0.018
20	0.998	0.00998	0.01

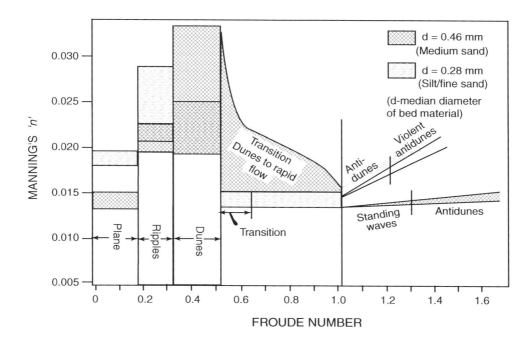

Figure 23.4   Variation of Manning's roughness coefficient n with Froude number (after Simons and Richardson 1962)[23.2].

Sediment Load

The energy available to transport sediment (i.e. *stream power*) Ω J s^{-1} is related to the discharge Q and the channel slope gradient S (i.e. the energy gradient):

$$\Omega = \rho_w \, g \, Q \, S$$

where ρ_w is the density of water (1000 kg m^{-3}).

Sediment is transported along river channels in three ways[23.3] (Fig. 23.5):

- in *solution*
- as *suspended material* within the flow
- as *bedload* rolling or bouncing along the channel floor.

The relative proportion varies with the river regime (i.e. the annual variations in river discharge, the grain size distribution and density of the sediments being transported). In Fig. 23.6 the critical fluvial erosion velocity refers to a height of 1 m above the river bed[23.4]. The two critical zones around these curves and the settling velocity for particles in water delimit the three regimes of fluvial deposition, fluvial transport and fluvial erosion[23.5].

Suspended sediment concentrations C may be estimated from:

$$C = a \, Q^{\,b}$$

Values of the coefficient a and exponent b vary for rivers in different climatic zones (Table 23.5). Note that b < 2 for ephemeral streams (i.e. arid environments), whereas b > 2 for perennial streams[23.6].

Measurement of the bedload is extremely difficult, and involves establishing a discharge/sediment load relationship at river gauging stations. Accurate sampling is virtually impossible and requires the installation of large sediment traps that can change the sediment regime they are attempting to measure.

A number of theoretical estimation methods are available for estimating bedload, although none are considered totally reliable[23.7–8]. These equations give an indication of *potential* sediment transport rather than *actual*.

Bed armouring can affect the transport rate, especially in gravel bed rivers where a coarse surface layer often protects underlying finer sediments, preventing its entrainment and removal. In such cases rivers may experience 2-phases of sediment transport:

- Phase I: a throughput load with little or no bed disturbance
- Phase II: once the armour is disrupted (e.g. by higher velocity flows) the finer material can be rapidly removed. Accelerated bed erosion can follow disruption of the bed armour e.g. during bridge or pipeline construction.

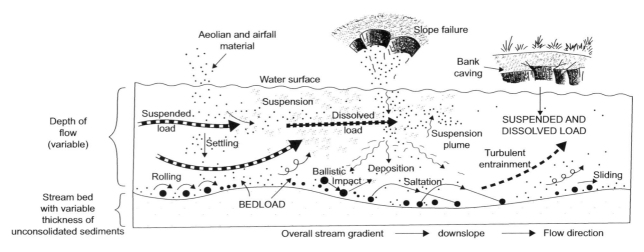

Figure 23.5   Sediment movement and local sources of material in a flowing river (based on Smithson *et al.* 2002)[23.3].

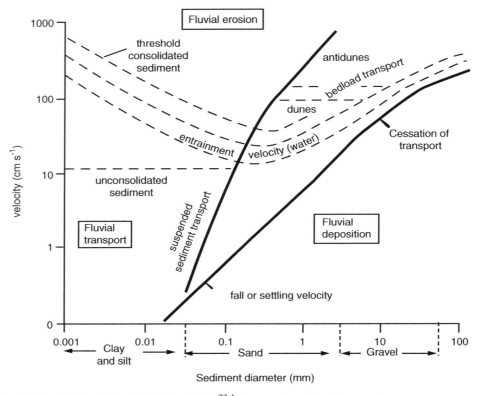

Figure 23.6   Threshold velocities (after Hjulström 1935)[23.4]; transport and bedform regimes.

Table 23.5   Examples of coefficients and exponents for estimating suspended sediment yield in different river environments[23.6].

Country	Environment	Coefficient a	Exponent b
Kenya	Arid	2570	0.512
Mexico	Arid	100	0.700
Israel	Arid	4217	0.159
Austria	Temperate, humid	0.004	2.200
USA	Temperate, sub-humid	0.01	1.600
USA	Temperate, humid	40	2.500
Germany	Temperate, humid	31	1.391

References

23.1 Chow, V. T. (1981) *Open Channel Hydraulics*. McGraw-Hill, London.

23.2 Simons, D. B. and Richardson, E. V. (1962) *The effect of bed roughness on depth-discharge relations in alluvial channels*. US Geological Survey Water-Supply Paper 1498E.

23.3 Smithson, P., Addison, K. and Atkinson, K. (2002) *Fundamentals of the Physical Environment*. 3rd Edition, Routledge, London and New York.

23.4 Hjulstrom, F. (1935) Studies of the morphological activity of rivers as illustrated by the River Fyris. *Bulletin of the Geological Institute, University of Uppsala*, **25**, 221–527.

23.5 Sundborg, A. (1956) The river Klarälven, a study of fluvial processes. *Geografiska Annaler*, **49A**, 333–43.

23.6 Reid, I. and Frostick, L. E. (1994) Fluvial sediment transport and deposition. In K. Pye (ed.) *Sediment Transport and Depositional Processes*. Blackwell, Oxford, 89–155.

23.7 Schoklitsch, A. (1933) *Uber die Verkleinerung der Geschriebe der Flusslaufen. Sitzungsberichte der Akademie der Wissenschaften in Wein*, 142, 343–66.

23.8 Bagnold, R. A. (1986) Transport of solids by natural water flow: evidence for a world-wide correlation. *Proceedings of the Royal Society*, London, A, **405**, 369–374.

24

Rivers: Channel Form

Introduction

River channel form reflects the interaction between the discharge Q and sediment load Q_s with the materials through which the river flows and the vegetation along its course. These interactions are expressed in terms of a characteristic *cross section*, *long-profile* (*thalweg*) and *planform*.

Channels that have been formed in strong bedrock reflect geological control (i.e. lithology and structure) and are described as *confined*; they are not generally prone to significant or rapid channel change. Channels formed in erodible sediments can be described as *unconfined* (alluvial or self-formed) channels and can be subject to changes in location and form over engineering time (see Chapter 25).

Channel Classification

There are four main channel patterns[24.1-2] (Figs 24.1 and 24.2):

1. *Straight*: the channel bed tends to have a sequence of regularly spaced *riffles* (shallows) and *pools* (deeps). Straight alluvial channels are rare; non-uniform flows and sediment loads result in a tendency for meandering. For example, straightened channel sections on the Lower Mississippi developed meanders during floods within 20 years of the engineering works[24.3].

2. *Meanders* (Fig. 24.3): *passive* meanders occur where a river does not have sufficient power to adjust its channel through bed scour and bank erosion; the course is diverted across the flood plain by hard-points or interlocking spurs. *Active* meanders are associated with alluvial channels that undergo almost constant re-adjustment in response to variations in discharge and sediment load, or inherent instability (see Chapters 4 and 5).

Channel patterns can be classified in terms of their *sinuosity S* (channel length/straight-line valley length).

There is a good relationship between the meander wavelength (Fig. 24.4) and other channel parameters[24.4-5]:

$$L = 12.34\, w$$
$$L = 54.3\, Q_b^{\,0.5}$$

where L = meander wavelength (m); w = mean width (m) and Q_b = bankfull discharge (m³ s⁻¹); see below.

Continued growth and elongation of meanders is a common feature of many alluvial channels, rather

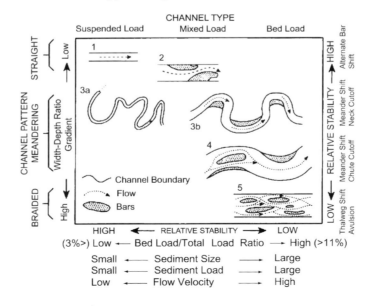

Figure 24.1 Channel classification (after Schumm 1977)[24.1].

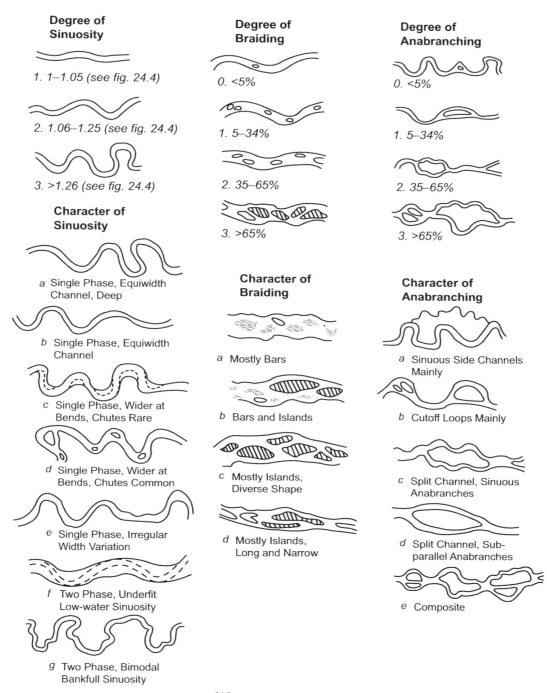

Figure 24.2 Channel patterns (after Brice 1975)[24.2].

than the maintenance of an equilibrium form (see Chapter 5).

3. *Braided* channels (Fig. 24.5): associated with high-powered rivers with abundant relatively coarse material moving as bedload, and weak erodible banks. Typically there is a sequence of sub-channels (*anabranches*) separated by *braid bars* that are flooded at bankfull discharge (i.e. at bankfull stage, the flow is in a single channel).

Braided channels resemble a string of *beads* (relatively long, wide multi-thread island reaches) separated by short, narrower single-thread reaches

(*nodes*). Braided rivers are very susceptible to rapid bank erosion and channel change (*avulsion*).

4. *Anastomosed* channels: associated with low energy rivers. The typical planform is of two or more highly sinuous channels separated by large, semi-permanent vegetated islands at a similar elevation to the surrounding floodplain (i.e. at bankfull stage the flow remains multi-thread). These tend to be relatively stable channels.

Dryland streams with permanent low flows may develop *compound* forms i.e. low flows occupy a single meander-

Figure 24.3   Meanders in a river entering a coastal plain, south Iceland.

ing channel while high flows spread across a wider braided channel.

Figure 24.6 and Table 24.1 present a system for classifying channel forms[24.6].

Regime Theory

Under stable climatic conditions, alluvial channel geometry is often assumed to be in equilibrium (i.e. in *regime*) with the prevailing flows, along with slope gradient, sediment type, bank vegetation and valley constrictions. Note, however, that climate stability is not common, even over engineering time (see Chapter 9).

Studies of channel cross-section (*hydraulic geometry*) have established key relationships:

$$w = a\,Q^b \qquad d = c\,Q^f \qquad v = k\,Q^m$$

where a, c, k, b, f, m are coefficients.

With cross-sectional area $A = w\,d$ and discharge $Q = w\,d\,v$, then

$$a\,Q^b \times c\,Q^f \times k\,Q^m = Q$$
$$\Rightarrow a \times c \times k = 1.0 \qquad \text{and} \qquad b + f + m = 1.0$$

Measured values for the numerical exponents are:

    b: 0.04–0.35 (typically 0.12)
    f: 0.33–0.56 (typically 0.45)
    m: 0.2–0.55 (typically 0.43)

Empirical equations can be used to predict channel change in alluvial channels[24.7] and relationships have been established for stable sand and gravel bed channels (Tables 24.2 and 24.3).

Dominant Discharge and Channel Form

Dominant discharge is the flow that, over time, yields the maximum bed and bank erosion, sediment transport and deposition, often assumed to be the *bankfull discharge*.

The frequency of bankfull discharge varies between rivers and can range from 1 to 32 years; its importance varies between river environments[24.8]. Bankfull is the dominant discharge where the entrainment threshold of the bed sediments is low (e.g. in sand/mud bed rivers). Relatively high frequency flows are therefore more significant than rare floods in their cumulative impact on channel form. In semi-arid and arid environments, rare flash floods are more effective.

Many channel changes initiated by large rare floods may be later removed by the action of smaller, more frequent floods. The duration of this 'healing' process is termed the *recovery time*, which varies between river environments. In deserts, the forms may be very persistent and channels may not undergo significant modification until the next major flood passes through. However, in temperate rivers the recovery time can be relatively short.

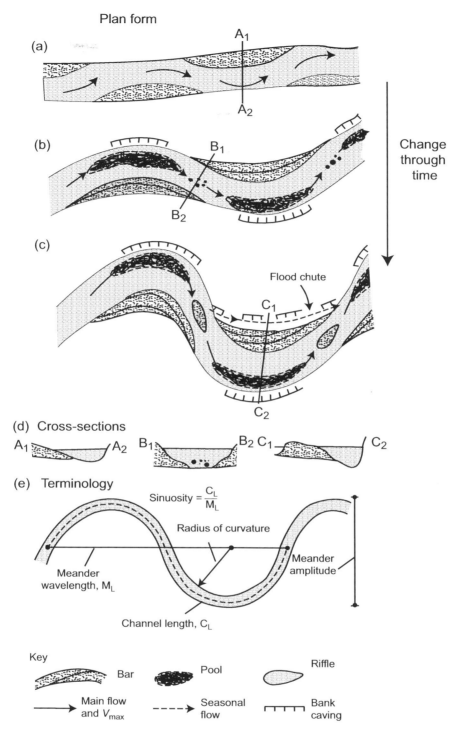

Plan form

(a)

(b)

(c)

Change through time

Flood chute

(d) Cross-sections

(e) Terminology

$$\text{Sinuosity} = \frac{C_L}{M_L}$$

Radius of curvature

Meander amplitude

Meander wavelength, M_L

Channel length, C_L

Key

Bar

Pool

Riffle

Main flow and V_{max}

Seasonal flow

Bank caving

Figure 24.4 Meander development (after Smithson *et al*. 2002)[24.5]: (a–c) plan form transition through time; (d) cross-section through the channel and (e) terminology.

Figure 24.5 Braided river system of the Brahmaputra River in China. Image ISS003-E-6632 13th October 2001 courtesy of the Image Science and Analysis Laboratory, NASA Johnson Space centre (http//eol.jsc.nasa.gov/) (width of image circa 5 km).

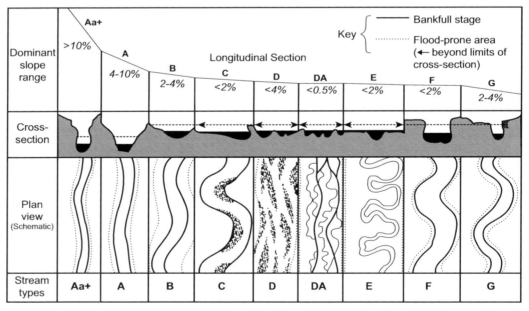

Figure 24.6 Classification of channel types (see Table 24.1) (after Rosgen 1994)[24.6].

Table 24.1 A channel classification system (see Fig. 24.6).

Stream Type	Description	Entrenchment Ratio	W/D ratio	Sinuosity (see Fig 24.4)	Slope gradient	Landform / soils / features
Aa+	Very steep, deeply entrenched, debris transport streams	<1.4	<12	1.0–1.1	>0.10	Very high relief. Erosional, bedrock or deposition features; debris flow potential. Deeply entrenched streams. Vertical steps with deep scour pools, waterfalls.
A	Steep, entrenched, cascading, step/pool streams. High energy/debris transport associated with depositional soils. Very stable if bedrock or bedrock-dominated channel.	<1.4	<12	1.0–1.2	0.04–0.10	High relief. Erosional or depositional and bedrock forms. Entrenched and confined streams with cascading reaches. Frequently spaced deep pools; associated step-pool bed morphology.
B	Moderately entrenched, moderate gradient, riffle-dominated channel with infrequently spaced pools. Very stable plan and profile. Stable banks.	1.4–2.2	>12	>1.2	0.02–0.04	Moderate relief, colluvial deposition and/or residual soils. Moderate entrenchment and width/depth ratio. Narrow, gently sloping valleys. Rapids predominate with occasional pools.
C	Low gradient, meandering, point bar, riffle-pool, alluvial channels with broad well-defined floodplains.	>2.2	>12	>1.4	<0.02	Broad valleys with terraces, in association with floodplains, alluvial soils. Slightly entrenched with well-defined meandering channel. Riffle-pool bed morphology.
D	Braided channel with longitudinal and transverse bars. Very wide channel with eroding banks.	N/A	>40	N/A	<0.04	Broad valleys with alluvial and colluvial fans. Glacial debris and depositional features. Active lateral adjustment, with abundance of sediment supply.
DA	Anastomosing narrow and deep with expansive well vegetated floodplain and associated wetlands. Very gentle relief with highly variable sinuosities. Stable streambanks.	>4.0	<40	Variable	<0.005	Broad, low gradient valleys with fine alluvium and/or lacustrine soils. Anastomosed, fine deposition with well-vegetated bars that are laterally stable with broad wetland floodplains.
E	Low gradient, meandering, riffle-pool stream with low width/depth ratio and little deposition. Very efficient and stable. High meander width ratio.	>2.2	<12	> 1.5	<0.02	Broad valley/meadows. Alluvial materials with floodplain. Highly sinuous with stable, well vegetated banks. Riffle-pool morphology with very low width/depth ratio.
F	Entrenched meandering riffle-pool channel on low gradients with high width/depth ratio.	<1.4	>12	>1.4	<0.02	Entrenched in highly weathered material. Gentle gradients, with a high width/depth ratio. Meandering, laterally unstable with high bank erosion rates. Riffle-pool morphology.
G	Entrenched gulley step-pool and low width/depth ratio on moderate gradients.	<1.4	<12	>1.2	0.02–0.04	Gulley, step-pool morphology zwith moderate slope and low width/depth ratio. Narrow valleys or deeply incised in alluvial or colluvial materials i.e. fans, deltas. Unstable with grade control problems and high bank erosion rates.

Table 24.2   Hydraulic geometry equations: sand bed channels[24.7].

Range of application	Discharge Q: 0.03–2800 m^3 s^{-1} Sediment discharge Q_s: 30–100 ppm Bed material size D: 0.1–0.6 mm Bank material type: cohesive Bedforms: ripples-dunes Planform: straight Profile: uniform
Mean width W^* (m)	$$W^* = \sqrt{\dfrac{F_{bc}}{F_s} Q}$$
Mean bed depth d^* (m)	$$d^* = \left(\dfrac{F_s\, Q}{F_{bc}^{\,2}}\right)^{0.33}$$
Slope S	$$S = \dfrac{F_{bc}^{\;0.833} F_s^{0.083} v^{0.25}}{3.63\, g\, Q^{0.166}\left(1 + \dfrac{Q_s}{2330}\right)}$$ v is the kinematic velocity m^2 s^{-1} Q_s parts per 100 000 by weight g is gravitational acceleration (9.81m s^{-2})
Bed factor F_{bc}	$F_{bc} = F_b\,(1 + 0.012\, Q_s)$ $$F_b = \dfrac{V^2}{d^*}$$ where V is the mean velocity in m s^{-1} F_b can be approximated by $0.58\ \sqrt{D_{50}}$ where D_{50} is the median particle size
Side factor F_s	$$F_s = \dfrac{V^3}{W^*}\quad \text{m}^2\text{s}^{-3}$$ 0.009 - for loam of very slight cohesiveness 0.018 - for loam of medium cohesiveness 0.027 - for loam of high cohesiveness
Meandering channels factor k	For meandering channels the slope should be multiplied by k: $k=2$ in well-developed meanders; $k=1.25$ for straight channels with alternative bars

Table 24.3   Hydraulic geometry equations: gravel bed channels[24.7].

Range of application	Discharge Q: 3.9–424 m^3 s^{-1} Bankfull sediment discharge Q_s: 0.001–14.14 kg s^{-1} Median bed material size D_{50}: 0.014–0.176 m D_{84} is the bed material size for which 84% is finer. Bank material type: composite, with cohesive fine sand, silt and clay overlying gravel Bedforms: plane Bank vegetation — Type I: 0% trees and shrubs; Type II: 1–5%; Type III: 5–50%; Type IV: > 50% Valley slope S_v: 0.00166–0.0219 Planform: straight and meandering jProfile: pools and riffles
Bankfull width (average over a channel reach) W (m)	For vegetation Type I, $\ W = 4.33\,\sqrt{Q}$
	For vegetation Type II, $\ W = 3.33\,\sqrt{Q}$
	For vegetation Type III, $\ W = 2.73\,\sqrt{Q}$
	For vegetation Type IV, $\ W = 2.34\,\sqrt{Q}$

(continued)

Table 24.3 (continued).

Bankfull mean depth (reach average) d (m)	All vegetation types $d = 0.22\,Q^{0.37} D_{50}^{-0.11}$
Bankfull slope S	All vegetation types $S = 0.087\,Q^{-0.43}\,D_{50}^{-0.09} D_{84}^{0.84}\,Q_S^{0.10}$
Bankfull maximum depth (reach average) d_m (m)	All vegetation types $d_\mathrm{m} = 0.20\,Q^{0.36}\,D_{50}^{-0.56}\,D_{84}^{0.35}$
Meander arc width z	$z = 6.31\,W$
Sinuosity p	$p = \dfrac{S_\mathrm{v}}{S}$

References

24.1 Schumm, S. A. (1977) *The Fluvial System*. Wiley, New York.

24.2 Brice, J. C. (1975) *Air photo interpretation of the form and behaviour of alluvial rivers*. Final Report to the US Army Research Office.

24.3 Winkley, B. (1982) Response of the Lower Mississippi to river training and realignment. In R. D. Hey, J. C. Bathurst and C. R. Thorne (eds.) *Gravel-Bed Rivers*. Wiley, Chichester, 659–680.

24.4 Richards, K. S. (1982) *Rivers: form and process in alluvial channels*. Methuen Press.

24.5 Smithson, P., Addison, K. and Atkinson, K. (2002) *Fundamentals of the Physical Environment*. 3rd Edition, Routledge, London and New York.

24.6 Rosgen, D. L. (1994) A classification of natural rivers. *Catena*, **22**, 169–199.

24.7 Hey, R. D. (1997) Stable channel morphology. In C. R. Thorne, R. D. Hey and M. D. Newson (eds.) *Applied Fluvial Geomorphology for River Engineering and Management*, John Wiley and Sons Ltd., 223–236.

24.8 Wolman, M. G. and Miller, J. P. (1960) Magnitude and frequency of forces in geomorphic processes. *Journal of Geology*, **68**, 54–74.

25

Rivers: Channel Change

Introduction

River channels can change over engineering time[25.1,2]. Some rivers may undergo major course changes that can have major environmental consequences:

- The Huang He (Yellow River), China, has had at least 20 major channel changes during the last 4000 years (equivalent to a course change every 200 years).
- The decline of the civilisation of Mohenjo Daro is believed to have coincided with course changes of the River Indus in modern day Pakistan.

Not all changes are as dramatic, but they can be important for engineering projects. Identifying, mapping and anticipating channel changes requires an understanding of the river system and its response to changing water and sediment inputs (Table 25.1). For engineering purposes the two main concerns will be lateral shifts in the channel location and vertical movements, predominantly downward i.e. scour (see Table 25.2).

Channel Behaviour

The causes of channel change can be complex:

1. *Long term influences* e.g. climate change. High rainfall periods such as occurred at the end of the Little Ice Age (see Chapter 9) are associated with increased erosion, lateral instability, incision and channel widening in the UK[25.3].

2. *Medium term adjustments*, frequently associated with human activities (see Chapter 14):
   - *Land use change*, through the increase in runoff and hillslope erosion e.g. the entrenchment of *arroyos* in southwest USA (see Chapter 4).
   - *Reservoirs and dams*: significant changes in both discharge regime and sediment load take place as a result of water impoundment by reservoirs[25.4]. Channel width reduces downstream; incision occurs due to the lack of sediment.
   - *Channelisation*: changes have arisen as a direct result of river works, in some cases exacerbating the problems they were meant to solve e.g. river straightening on the Lower Mississippi.
   - *Waste disposal*: mine workings in upland areas frequently dispose of spoil and waste in mounds adjacent to river channels or directly into the stream. The resulting increase in sediment load can have an impact on channel stability[25.5]. Generally, increased sediment load will lead to increased instability, tendency for braiding, aggradation and channel switching. Reduction in sediment loads following mine closure can reverse some of the channel changes.
   - *Urbanisation* can cause channel change due to increased peak flows and decreased sediment supply (see Chapter 14). Following the development of Cumbernauld New Town, UK, some

Table 25.1 Channel changes in response to changes in discharge and sediment load[25.1].

Change	River bed morphology	Change	River bed morphology
$Q_s \uparrow Q_w =$	Aggradation, channel instability, wider and shallower channel	$Q_w \uparrow Q_s =$	Incision, channel instability, wider and deeper channel
$Q_s \downarrow Q_w =$	Incision, channel instability, narrower and deeper channel	$Q_w \downarrow Q_s =$	Aggradation, channel instability, narrower and shallower channel
$Q_s \uparrow Q_w \downarrow$	Aggradation	$Q_w \downarrow Q_s \downarrow$	Processes decreased in intensity
$Q_s \uparrow Q_w \uparrow$	Processes increased in intensity	$Q_w \uparrow Q_s \downarrow$	Incision, channel instability, wider? channel

Q_s sediment discharge; Q_w water discharge; $\uparrow$ increase; $\downarrow$ decrease; = remains constant.

Table 25.2   River channel changes[25.2] (*Channel change* takes place in three-dimensions and therefore instability involves combinations of these types of changes).

Planform	*Confined or restricted migration* involves downstream translation of meandering channel with little net change in form characteristics in a reach. Movement is often constrained by terraces or valley walls. In low rates of migration, erosion occurs only on bend apexes. *Active meandering* may involve a net change in form, either an increase or decrease in channel length in a reach. Major components of change in meanders are: migration, extension, compound forms and cutoffs. Meanders may undergo an evolutionary sequence from migration of simple loops through to cutoff of compound loops. *Meander-braiding switching.* Channels may change from single to multiple and vice versa. The threshold is influenced by discharge, gradient and sediment load. Floods may cause braiding where sediment load is increased. Channels may cease to be braided when sediment supply is decreased e.g. after a period of mining. *Braiding.* Degree of braiding, indicated by number of bars and width of channel zone, may alter. Braided reaches tend to be unstable in form and are influenced by individual floods. Frequent channel switching takes place by avulsion and deposition.
Cross-sectional form	*Width* increase takes place by erosion of both banks e.g. in response to urbanisation, snowmelt, monsoon. Decrease takes place by bar deposition and formation of berms e.g. downstream of some reservoirs or at the end of the monsoon. *Depth* increase results in incision of channel and can in turn destabilise channel banks. Incision may take place in large floods and if bed armouring is removed. Decrease results from net deposition from increased sediment supply or decreased gradient.
Longitudinal form	*Steepening* results from incision and increase in slope upstream. A sharp break of slope (headcut or knickpoint) may be created and this may progress upstream over time. *Flattening* of slope results from aggradation or build-up of sediment controlled either from upstream supply or a raised local base-level downstream (e.g. sea level rise).

streams enlarged through the urban area[25.6]. Locally there were instances of extensive erosion and bank collapse; up to 10 m of vertical incision through glacial till was observed in gullies draining industrial areas.

- *Gravel extraction* increases the channel gradient through a reach, disrupts the bed armour and increases bed roughness, leading to enhanced bed scour and gravel transport e.g. the Shamkir Chay, Azerbaijan[25.7]. If the rate of removal is greater than the rate of supply from upstream, channel downcutting will occur. This incision is self-enhancing because flow becomes more confined to an entrenched channel with higher flow velocity and bed shear stresses. Entrenchment may continue long after extraction ceases, as the river channel continues to adjust to these changes.

3. *Short term responses* may follow individual extreme flood events. The 1990 and 1993 floods on the River Tay, Scotland, caused significant changes, including the creation of small sections of new channel across the floodplain[25.8].

Some changes are not related to external causes — they are an inherent feature of the natural evolution of the system e.g. meander development across floodplains i.e. inherent (*intrinsic*) instability (see Chapters 4 and 5).

Hazards

- Loss of land and property due to the lateral shifting of a channel. For example, Majuli island on the Brahmaputra River (Assam, India) reduced in size from c. 1250 km^2 in 1950 to c. 880 km^2 by 1993 (see http://majuli-assam.tripod.com/).
- Bed scour around bridge piers e.g. in 1987 four people died when channel scour and erosion around the piers of a railway bridge at Glanrhyd, Wales, led to the bridge collapsing under the weight of a train[25.2].
- Bed scour and exposure of buried pipelines e.g. the exposure of a gas pipeline crossing the Shamkir Chay gravel-bed river, Azerbaijan; bed lowering rates of around 1 m yr^{-1} are due to downstream gravel extraction[25.7].
- Reduction in channel cross section due to deposition of sediment, leading to increased flood hazard. For example, 5000–8000 tonnes of gravel have to be removed from the River Usk, Wales, every year as part of the maintenance of a flood defence scheme[25.9].

Hazard assessment: Channel Migration and Bank Erosion

Historical topographical maps, charts and aerial photographs provide a record of the former positions of channels (see Chapter 40). For example, survey maps and aerial photographs from the 1950s, together with 1989 SPOT satellite imagery were used to assess bank migration as part of the Brahmaputra River Training Study, Bangladesh[25.10]. The river was sub-divided into a series of reaches, and point measurements of right and left bank erosion were made at 0.5 km intervals. The survey revealed that bank erosion was more severe on the right bank, where the average annual rate was 90 m (in some reaches the rate was double this figure).

Hazard Assessment: Scour

Hydraulic geometry equations (Chapter 24) can be used to estimate the channel geometry under design flood conditions. The results can be converted to the potential average *scour depths* if evidence of maximum flood flow heights can be deduced from the channel morphology.

In ephemeral streams, the channel geometry reflects the effects of high magnitude low frequency flood flows. The maximum height of river terraces and braid channel bars lie close to the maximum flood height and can be used to establish the channel width during the design flood.

Multiplying factors can be applied to the mean depth to obtain the maximum scour depths (Table 25.3):

$$\text{Scour depth} = (\text{Mean depth} \times \text{Multiplying Factor}) - \text{Mean depth}$$

This approach has been used to estimate scour depth on an ephemeral stream in southeast Spain[25.12] (Fig. 25.1). The results help identify suitable crossing sites i.e. avoiding areas of channel instability and provide an estimated discharge for a 1:500 year flood that had passed through the channel three years before the mapping. Using empirical hydraulic geometry equations (Table 24.3), the discharge of this 1:500 year flood and the flow depths were estimated based on mapping the identifiable maximum stage height. The analysis for one of the cross sections indicated in Fig. 25.1 is presented in Table 25.4. In this example, the predicted scour depth of 1.7 m would exceed the depth of alluvial infill at the site, indicating that a major flood would be most likely to scour out all sedimentary channel infill in the area.

These approaches can be used as a *guide* to the potential severity of the problems, rather than a source of detailed predictions. The use of these techniques must be considered carefully and only after a detailed geomorphological investigation of the channels (Table 22.2).

Channel Change and Nature Conservation

Channel changes can be vital to the maintenance of both channel and floodplain habitats. River engineering on mobile river reaches can involve destruction and removal of vegetation and habitats, particularly the loss of pools and riffles, vertical eroding banks and sinuous courses.

Changes in fish populations can result from the destruction of natural riffle-pool sequences. Fish require sheltered water in fast flowing rivers, these conditions may be absent where a meandering stream has been artificially straightened.

Soft engineering designs, which work *with* natural processes, are likely to be the most cost-effective solutions, requiring less maintenance and minimising the environmental impact (see http://www.environment-agency.gov.uk/). Soft engineering solutions can also help restore or rehabilitate reaches that have been damaged by previous channelisation works[25.13].

Table 25.3 Multiplying factors for maximum channel scour depth[25.11].

Location	Multiplying factor	Maximum scour depth (for a 10m deep flow)
Straight channel reach	1.25	2.5
As above with mobile bed dunes	1.5	5
Moderate channel bend	1.5	5
Severe channel bend	1.75	7.5
Right-angle abrupt channel bend	2.0	10

Table 25.4 Example flood discharge and scour calculation for gravel bed channel, Ugijar district, southeast Spain.

Channel capacity $= 21.4$ m^2

Measured channel width $B = 18.4$ m

Grading of channel deposits: $D_{84} = 0.09$ m; $D_{50} = 0.0038$ mm

Channel planform $=$ relatively straight

From the empirical hydraulic geometry equations, calculation of discharge Q based on the measured width for Type II bank vegetation $= 30.5$ m^3 s^{-1}

Mean depth $= 1.4$ m; maximum mean depth over a reach $= 6.7$ m

Scour factor $= 1.25$

Estimated maximum local scour within reach $= 8.4$ m below maximum flow height $= 1.7$ m

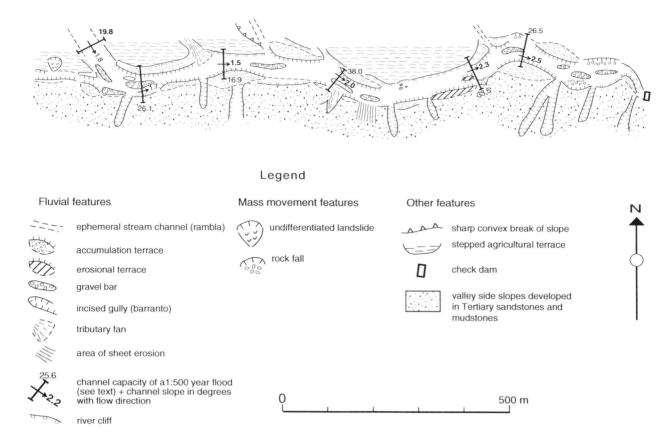

Legend

Fluvial features

- - - - ephemeral stream channel (rambla)

accumulation terrace

erosional terrace

gravel bar

incised gully (barranto)

tributary fan

area of sheet erosion

25.6 / 2.2   channel capacity of a1:500 year flood (see text) + channel slope in degrees with flow direction

river cliff

Mass movement features

undifferentiated landslide

rock fall

Other features

sharp convex break of slope

stepped agricultural terrace

check dam

valley side slopes developed in Tertiary sandstones and mudstones

N

0   500 m

Figure 25.1 Simplified geomorphological map of Rambla Seca, southeast Spain (from Griffiths *et al.* 2004)[25.12].

References

25.1 Schumm, S. A. (1977) *The Fluvial System.* Wiley, New York.

25.2 Lee, E. M. (1994) *The Occurrence and Significance of Erosion, Deposition and Flooding in Great Britain.* HMSO.

25.3 Higgett, D. and Lee, E. M. (eds.). (2001) *Geomorphological Processes and Landscape Change: Britain in the last 1000 years.* Blackwell, Oxford.

25.4 Petts, G. E. (1984) *Impounded Rivers.* Wiley & Sons.

25.5 Macklin, M. G. and Lewin, J. (1989) Sediment transfer and transformation of an alluvial valley floor: the River South Tyne, Northumbria, UK. *Earth Science Processes and Landforms,* **14**, 233–246.

25.6 Roberts, C. R. (1989) Flood frequency and urban-induced channel changes: some British examples. In K. Beven and P. A. Carling (eds.) *Floods: Hydrological, Sedimentological and Geomorphological Implications.* Wiley and Sons, Chichester, 57–82.

25.7 Walker, C., Skipper, C. and Bettes R. (2004) AGT Pipelines Project Overview: Engineering design and construction of significant river crossings in Azerbaijan and Georgia. In M. Sweeney (ed.) *Terrain and Geohazard Challenges Facing Onshore Oil and Gas Pipelines,* Thomas Telford, 480–491.

25.8 Gilvear, D. J., Davies, J. R. and Winterbottom, S. J. (1994) Mechanisms of floodbank failure during large flood events on the rivers Tay and Earn, Scotland. *Quarterly Journal of Engineering Geology,* **27**, 319–332.

25.9 Sear, D. A. and Newson, M. D. (1992) *Sediment and gravel transportation in rivers including the use of gravel traps.* NRA project report 232/1/T.

25.10 Thorne, C. R., Russell, A. P. G. and Alam, M. K. (1993) Planform pattern and channel evolution of the Brahmaputra River, Bangladesh. In J. L. Best and C. S. Bristow (eds.) *Braided Rivers,* Geological Society Special Publication **75**, 257–276.

25.11 Neill, C. R. (1973) *Guide to Bridge Hydraulics.* Roads and Transportation Association of Canada, University of Toronto Press.

25.12 Griffiths, J. S., Mather, A. E. and Stokes, M. (2004) Construction design data provided by the investigation of geomorphological processes and landforms. R. J. Jardine, D. M. Potts and K. G. Higgins (eds.) *Advances in Geotechnical Engineering: the Skempton Conference,* Thomas Telford, London, 1292–1303.

25.13 Brookes, A. B. (1988) *Channelized rivers: perspective for environmental management.* Wiley and Sons, Chichester.

26

Rivers: Flooding

Introduction

Flooding is the temporary inundation of land by surface water (see http://www.fema.gov/hazard/flood/ or http://www.environment-agency.gov.uk/):

- Worldwide, flooding is the most important cause of economic and loss of life from a natural event.
- Flooding is an increasing problem as the population expands to occupy areas that were formerly left uninhabited or only temporarily occupied because of flood risk.
- Flooding is a natural phenomena and evidence of former flood events are often recorded in the landscape which can be identified by geomorphological investigations.
- Floods tend to be episodic in nature; geomorphological evidence can provide some indication of the frequency of occurrence of floods of various magnitudes i.e. the *recurrence interval* or *return period* (see Chapter 11).

The concept of return period is widely used in civil engineering. The design requirements for drainage works such as bridges or dam spillways being the capacity to withstand a flood discharge with specified return period (e.g. 50 or 100 years; see Chapter 11).

The most common causes of flooding are rainfall and snowmelt, although floods are influenced by a variety of factors:

- *Catchment area* affects the total volume of streamflow i.e. the larger the catchment the greater the potential rainfall input.
- *Slope characteristics* influence the amount of runoff produced by an event i.e. slope angle, bedrock geology, soil type, land use and vegetation cover (see Chapter 15).
- *Network characteristics* influence the speed at which water is transmitted through the channel system. *Dendritic networks* (Fig. 22.2) tend to produce a marked concentration of flow in the lower catchment as floodwaters are delivered down the major tributaries at similar speeds; *trellised networks* tend to produce a more muted response. Lakes, reservoirs or other storage areas can reduce the size of a

flood. Artificial drainage (e.g. field drains) speeds up the movement of water towards the channel network.

- *Channel characteristics* influence the ability of the channel to carry a flood flow. Channel capacity is not constant; deposition of eroded sediments can significantly reduce the channel depth and cross section. Debris (e.g. uprooted trees) trapped behind structures may cause 'backing-up', as occurred during the August 16th 2004 floods in Boscastle, UK[26.1] (discharge = 100 m³ s⁻¹).
- *Antecedent conditions* determine the amount of the catchment that is saturated prior to a rainstorm or snowmelt event and, hence, the amount of runoff that is generated. The river or stream level prior to an event is also critical as this will influence whether the channel system can carry the additional runoff.

World Flood Zones

Figure 26.1 presents a global classification of flood zones[26.2]:

- *High latitudes*: depressions are a key factor in the occurrence of floods; the tracks taken by depressions are controlled by the position and strength of the jet streams. Changes in atmospheric circulation can lead to changes in flood frequency; more frequent floods in the Upper Mississippi valley during the late 1800s and since 1950 were due to a weak westerly circulation in mid latitudes[26.3].
- *Low latitudes*: floods and/or drought events are associated with tropical storms, the monsoon and the El Niño Southern Oscillation (ENSO; see Chapter 9). The Intertropical Convergence Zone (ITCZ) produces marked seasonality of rainfall in the tropics (the *monsoon*). The zone is not constant; it can extend further northwards, triggering flooding in normally arid areas (e.g. Bangladesh in 1987 and 1988[26.4]) or contract southwards causing droughts (e.g. the 1981–85 Ethiopian drought[26.5]).

An important consequence of short-term climatic variation is that the probability of flood events may not be constant over time. Many of the world's major rivers

Figure 26.1   Flood climate regions (after Hayden 1988)[26.2]. Baroclinicity describes the air stratification in which surfaces of constant pressure intersect surfaces of constant density e.g. winter temperature gradients on the east coast of the US. Barotropy refers to surfaces of constant density or temperature coinciding with surfaces of constant pressure i.e. zero baroclinicity.

experienced both 'humid' and 'dry' periods throughout the last century[26.6]. Caution is needed when using return period statistics; they rely on the assumption that the probability of a flood event of a given magnitude has remained constant over the historical record.

Flood Types

1. *River Floods*

These are generally the result of relatively long periods (days or weeks) of intense and/or persistent rain over large areas, sometimes combined with snowmelt. The flood builds up gradually as the ground within the catchment becomes saturated and the capacity of the soil to store water is exceeded (a similar effect occurs when the ground is frozen and water cannot infiltrate the soil).

In floodplains and broad open valleys, large areas can be affected (Fig. 26.2). In narrow valleys only a narrow strip maybe inundated, but water depths will increase and flow velocities tend to become higher.

- In 2000 there was extensive flooding in Mozambique, particularly along the Limpopo, Save and Zambezi valleys. Half a million people were made homeless and 700 lost their lives. The floods destroyed crops and overwhelmed water and sanitation infrastructure in many areas[26.7] (see http://www.oxfam.org.uk/).
- Over 110 million people live within the relatively unprotected Jumna floodplain in Bangladesh; monsoon-generated flooding occurs regularly over 20–30% of the country's total land area. In 1988, floods inundated 46% of the land area and killed an estimated 1500 people[26.4].

2. *Flash Floods*

These are localised and episodic river floods, associated with small, steep catchments and semi-arid or arid regions. They are the result of intense rainfall over a limited area, typically when rainfall intensity exceeds the infiltration rate. Flash floods have an extremely rapid onset and are often marked by the passage of a storm wave passing down the river channel.

Flash floods are very difficult to predict because of the limited spatial extent and extremely short lead times. They have enormous destructive potential:

- 25 adults and 9 children were killed, 90 houses destroyed and 130 cars swept away during the Lynmouth floods of August 1952 (see

Figure 26.2   Flood area of the Amazon rain forest in Brazil, June 1996. Image STS078-751-94 courtesy of the Image Science and Analysis Laboratory, NASA Johnson Space Centre (http://eol.jsc.nasa.gov/) (width of image circa 2 km).

http://www.exmoor-nationalpark.gov.uk/). An intense summer storm of around 225 mm fell over Exmoor and the steep Lyn catchment, one of the heaviest 24-hour totals ever recorded in the UK (discharge = 511 m^3 s^{-1}).

- Catastrophic flash flooding of the Albuñol stream, Spain, during October 1973, caused the loss of 46 lives, the destruction of 91 houses and serious damages to 141 in La Rabita[26.8]. A heavy rainstorm (600 mm) generated a discharge of 2580 m^3 s^{-1}.
- The El Arish flood of 1975 in the Sinai desert (discharge of 1650 m^3 sec^{-1}). The floodwaters destroyed a railway bridge and deposited a delta on the Mediterranean shoreline[26.9].

3. *Dam failures*

Dam failures have also resulted in disastrous flash flood events:

- Over 2200 people died as a result of the Johnstown Flood in 1889, Pennsylvania. The old South Fork Dam collapsed during heavy rains, sending 20 Mtons of water down through the town at 40 miles per hour (see http://www.jaha.org/).
- In 1963 a landslide moving into a partially filled reservoir created a massive flood wave (a lake *tsunami*) which overtopped the Vaiont dam, northern Italy. The dam survived the event, but the flash flood created as the waters moved downstream killed over 3000 people[26.10].

4. *Glacial lake outburst floods (Jökulhlaup)*

Proglacial lakes impounded behind moraines or an ice dam can present a major flood risk to downstream communities. Failure of the moraine or ice dam can lead to a catastrophic outburst flood (*aluvión, débâcles* or *glacier lake outburst floods*) with the discharge reaching up to several thousand cubic metres per second[26.11], often accompanied by large volumes of mobile sediment. An outburst flood in 1725 killed 1500–2000 people in the village of Ancash, in the Cordillera Blanca of Peru[26.12] (see http://pubs.usgs.gov/prof/p1386i/peru/hazards. html).

In Iceland some jökulhlaups may be triggered by volcanic activity (see http://www.raunvis.hi. is/~alexandr/glaciorisk/). Failure of *landslide dams* can produce similar devastation (see Chapter 20).

References

26.1 Barham, N. (2004) A flash flood assessment. An insurance perspective on the Boscastle floods of 16 August 2004. *Journal of Meteorology*, **29**, 334–339.

26.2 Hayden, B. P. (1988) Flood climates. In V. R. Baker, R. Kochel and P. C. Patton (eds.) *Flood Geomorphology*, John Wiley, New York, 13–26.

26.3 Knox, J. C. (1988) Climatic influence on Upper Mississippi Valley. *Annals of the Association of American Geographers*, **77**, 224–244.

26.4 Brammer, H. (1990) Floods in Bangladesh. 1 Geographical background to the 1987 and 1988 floods. *Geographical Journal*, **156**, 12–22.

26.5 Seleshi, Y. and Zanke, U. (2004) Recent changes in rainfall and rainy days in Ethiopia. *International Journal of Climatology*, **24**, 8, 973–983.

26.6 Probst, J. L. and Tardy, Y. (1987) Long range streamflow and world continental runoff fluctuations since the beginning of this century. *Journal of Hydrology*, **94**, 289–311.

26.7 Vitart, F., Anderson, D. and Stockdale, T. (2003) Seasonal Forecasting of Tropical Cyclone Landfall over Mozambique. *Journal of Climate*, **16**, 23, 3932–3945.

26.8 Romero Cordón, E., Alvarez Vigil, E., Garrido Perez, L., De Las Heras Martín, F., Duran Valsero, J. J. and Madrid García, R. (2003) The catastrophic flood event of the 18th of October, 1973, in the rambla of Albuñol (La Rabita, Granada, Spain). In V. R. Thorndycraft, G. Benito, M. Barriendos and M. C. Llasat (eds.) *Palaeofloods, Historical Floods and Climatic Variability: Applications in Flood Risk Assessment* (Proceedings of the PHEFRA Workshop, Barcelona, 16–19th October, 2002).

26.9 Gilead, D. (1975) *A preliminary hydrological appraisal of the Wadi El-Arish flood 1975*. Mimeograph Report of the Israel Hydrological Service, Jerusalem.

26.10 Hendron, A. J. Jr. and Patton, F. D. (1985) *The Vaiont Slide: a geotechnical analysis based on new geologic observations of the failure surface*. Waterways Experiment Station Technical Report, US Army Corps of Engineers, Vicksburg, Mississippi.

26.11 Reynolds, J. M. (1992) The identification and mitigation of glacier-related hazards: examples from the Cordillera Blanca, Peru. In G. J. H. McCall, D. J. C. Laming and S. C. Scott (eds.) *Geohazards*. Chapman and Hall, London, 143–157.

26.12 Lliboutry, L., Arnao, B., Morales, A., Pautre, A. and Scneider, B. (1977) Glaciological problems set in the control of dangerous lakes in the Cordillera Blanca, Peru. Part 1 Historical failures of moraine dams, their causes and prevention. *Journal of Glaciology*, **18**(79), 239–254.

27

Rivers: Flood Hazard and Risk

Introduction

The level of flood damage varies according to the event (depth, duration, velocity, sediment load), the land use affected and the action taken by the occupants of the flooded area.

Loss of life: Table 27.1 shows the most disastrous flood events in Europe in terms of fatalities during the 1990s. The number of deaths associated with flooding is closely related to the nature of the event (rapidly rising waters, deep flood waters, objects carried by the rapidly flowing water) and the behaviour of victims[27.1]:

- In the 1998 flood in Sarno, Italy, 147 people were killed by a river of mud that rapidly destroyed an urban area.
- The 1927 Mississippi River floods resulted in more than 700 000 displaced people, 246 deaths, nearly 70 000 km^2 of flooded land, and an estimated $400 million worth of damage (see http://www.pbs.org/wgbh/amex/flood/).

Economic losses: flooding is the most common natural disaster, and the most costly in economic terms (Table 27.2). Between January and July 2002, Europe suffered eight major floods that killed 93 people and affected 336 000, and caused damage of around US $480M. Over the period 1991 to 2001, losses amounting to US$ 250 billion had to borne by societies all over the world[27.1].

Intangible damages ranging from anxiety and stress to ill health related to the general inconvenience caused by the event.

Hazard Assessment: Flood Flow Estimation

The standard methods of flood estimation are listed in Table 27.3. These methods can be used where comprehensive data exist on stream flow, flow stage heights, precipitation, and catchment characteristics[27.2–4].

A fluvial geomorphological field survey (Table 22.2) can identify evidence of flood stage heights: trash lines (debris trapped in trees, bushes and structures; see Fig. 27.1), high water marks, and depositional or erosional terraces (*strath lines*) related to contemporary, recent, or historical flood behaviour. These observations can then be

Table 27.1 Disastrous European flood events in the 1990s.

Location	Year	Fatalities
Italy	1998	147
Romania	1991	108
Spain	1996	86
Turkey	1995	78
Turkey	1995	70
Italy	1994	64

Table 27.2 Flood damages in the USA during the 1990s.

Year	Flood Damages (Billion)
1990	$2.12
1991	$2.15
1992	$0.93
1993	$19.45
1994	$1.29
1995	$5.73
1996	$6.71
1997	$9.15
1998	$2.57
1999	$5.45

used to provide estimates of flood flow discharges, especially in ephemeral streams where accurate field measurements of channel dimensions and boulder size are possible.

1. *Slope/area method*: based on field data that establish the stage height reached after a flood wave has passed. The basis is the identification of a length of stream (a stream reach) with a uniform cross section, free from obstructions or any backwater effects (e.g. near a tributary), length $\geq$ 75 times the mean depth of flow, length $\geq$ 5 times the mean width, $\geq$ 300 m in length, and with a surface fall of $\geq$ 0.15 m over the reach.

   The cross section is divided into sub-sections (1, 2, ...n) with a constant Chézy resistance coefficient C_j (Fig. 27.2). The cross-sectional area A_j and

Table 27.3 Standard methods of flood estimation.

Technique	Field Data Required	Analytical Work
Stream flow analysis – long record of river discharge (e.g. 25 years +, minimum of 10 years)	Regular discharge and water level measurements on river channel, preferably at the site of the structure but can be interpolated between river flow gauging stations	Development of stage height to discharge relationship; frequency analysis of discharge data; interpolation ($\times$ 50 years of data) or extrapolation ($\times$ 50 years of data) of design flood estimate – use of extreme value distributions (e.g. Gumbel)
Rainfall analysis and synthetic unit hydrographs*	Good spatial and temporal coverage of rainfall data within catchment; geomorphological and geological characteristics of the catchment; river cross-sections and gradient	Frequency analyses of rainfall and development of rainfall-intensity and area relationships; development of rainfall intensity–runoff relationships; if runoff data is limited synthesis of unit hydrographs; develop stage–discharge relationship; design flood estimation based on the rainfall recurrence interval
Rainfall–runoff analysis and development of actual unit hydrographs; possible development of the Probable Maximum Precipitation (PMP)	Full hydro-meteorolgical data set for river catchment; full geomorphological assessment of catchment characteristics; river cross-sections and channel gradient	Frequency analyses of rainfall and development of rainfall-intensity and area relationships; development of rainfall intensity–runoff relationships; creation of unit hydrographs; develop stage–discharge relationship; design flood estimation based on the rainfall recurrence interval. Use of the rainfall data and catchment characteristics to calculate the Probable Maximum Flood (PMF) that could be expected to occur
Flood estimation formulae (not recommended unless there is no other option). Various formulae have been developed for many different situations and countries	Input to formulae highly variable and can involve a range of catchment, rainfall and channel characteristics	Manipulation of the flood formula; relating the result of the flood formula to the design requirements (e.g. flood frequency, stage height etc)
Computer modelling of flow regime. Many different models have now been developed.	An extensive data set is required, covering all hydro-meteorological conditions, catchment, and channel characteristics. The longer the data record the better	Use of field data to calibrate the model followed by modelling the design flood based on the engineering requirements

*A unit hydrograph is the flow hydrograph of direct runoff (i.e. excluding base flow) resulting from 1 cm of effective rainfall (i.e. after allowing for evapo-transpiration) generated uniformly in space and time over the catchment in unit time T (an arbitrary period such as 1 hour).

hydraulic radius R_j are determined for each sub-section. The discharge Q_j through each is then estimated from the *Chézy equation* (see Chapter 23):

$$Q_j = VA_j = C_j A_j \sqrt{R_j S}$$

where Q_j is the discharge of each sub-section; V is the mean velocity of each sub-section; S is the mean slope through the overall reach; $C_j A_j R_j^{0.5}$ is the *conveyance* K_j for each sub-section, thus the discharge for the entire cross-section is:

$$Q = \sqrt{S} \sum K_j$$

sub-section j to n.

If the cross-sections differ from the beginning to the end of the reach, the geometric mean conveyance is used:

$$K_j = \sqrt{K_{j1} K_{j2}}$$

The *Darcy-Weisbach equation* can also be used to estimate velocity and discharge through each of the sub-sections (see Table 23.2). The Darcy-Weisbach friction factor f can be estimated from field measurements of bed particle size[27.6]:

$$\frac{1}{\sqrt{f}} = 2 \, \log\left(\frac{d}{D_{84}}\right) + 1.16$$

where d = water depth (m).

Figure 27.1   Flood damage and debris (trash lines) indicating the extent of a flood.

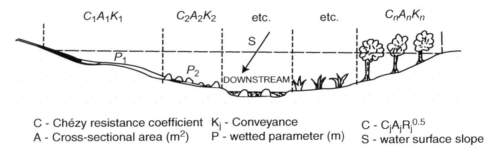

Figure 27.2   Channel cross-section for flood calculations (after Gardiner and Dackombe 1983)[27.5].

2. *Maximum boulder size*[27.7]: the maximum boulder sizes moved by flood flows can be used to give an approximate estimation of the highest flood discharges, or probable maximum flood (PMF). A number of methods have been developed; the method presented here (Table 27.4) is appropriate for both active and palaeochannels.

A crude approximation of the critical velocity V_c (m s^{-1}) needed to move the maximum sediment size in a channel can be obtained from:

$$V_c = 0.18 \ d^{0.49}$$

where d = particle diameter (mm)[27.8].

Hazard Assessment: Flood Area Zonation

Geomorphological surveys can define the nature and extent of flood-prone areas. Hazard zonation studies should incorporate all available flood data and historical records in order to establish the spatial extent of floods of given frequency. This technique has been employed by the German insurance industry for example, who have established a GIS-based rating system (Table 27.5) for the whole country that defines flood exposure according to three classes[27.9]. This approach is also used in flood engineering[27.10] (see Fig. 27.3).

Flood Risk Assessment

One approach is to develop a *flood damage curve* that relates flood losses associated with a flood event to the event probability (Fig. 27.4). Potential flood damage can be determined from past events or through the use of standard flood depth/damage data[27.11]. As any of the flood events could occur in a given year it is necessary to calculate the *average annual damage*, taking account of every possible combination of event probability and loss (Table 27.6). On Fig. 27.4 the average annual risk equates to the area below the damage curve, calculated as a series of slices between scenario probabilities.

Flood Risk Management

- *Do nothing*: accept the consequences associated with flooding.
- *Avoiding flood risk areas*: e.g. ensuring that the potential losses in risk areas are not increased by the siting of new development. In the UK, for example, government planning guidance[27.12] advises local authorities of the need to take flood risk into account during the land use planning process. This approach is supported by the preparation of flood risk maps by the Environment Agency (www.environment-agency. gov.uk/subjects/flood).

Table 27.4 Flood flow estimation based on maximum boulder size.

Property	Formula	Example calculation
A – boulder major axis (m) B – boulder intermediate axis (m) C – boulder minor axis (m) σ – boulder's density (kg m^{-3}) w – channel width (m) β – bed slope angle (radians/degree)	Basic concepts: ρ – fluid density = 1150 kg m^{-3} μ – coefficient of sliding for a cubic boulder = 0.675 μ – coefficient of sliding for a round boulder = 0.225 g – gravitational acceleration (9.81 m s^{-2})	A = 0.5 m B = 1.0 m C = 0.75 m σ = 2700 kg m^{-3} β = 0.052 radians (3 degrees) w = 10 m
Nominal diameter of boulder D	$D = (A\,B\,C)^{0.33}$	$D = (1 \times 2 \times 3)^{0.33} = 0.721$ m
Mass of a cubic boulder M_B	$M_B = \sigma D^3$	M_B = 1013 kg
Mass of a spherical boulder M_B	$M_B = \sigma \dfrac{\pi}{6} D^3$	M_B = 530 kg
Resisting force F_R (Newtons)	$F_R = M_B \dfrac{\sigma - \rho}{\sigma}\, g\, (\mu\cos\beta - \sin\beta)$	F_R = 3545 for a cubic boulder F_R = 515 for a round boulder
The critical force is the minimum force that can be applied to the boulder in the direction of stream flow that will initiate movement F_C	Set $F_C = F_R$ (i.e. the critical force is assumed to be equal to the resisting force)	
The drag force is dependent on the flow conditions and the shape of the boulder F_D. It is a function of the lift C_L and drag C_D coefficients of the boulder	$F_D = \dfrac{C_D F_C}{C_L + C_D}$ C_L = 0.178 for the cubic boulder and 0.20 for the round boulder C_D = 1.18 for the cubic boulder and 0.20 for the round boulder	F_D = 3081 N for the cubic boulder and 257 N for the round boulder
The critical velocity V_C is equivalent to the competent bottom velocity at a height of about 1/3 of a particle diameter above the bed at the condition of incipient movement (m s^{-1})	$V_C = \sqrt{\dfrac{2F_D}{C_D \rho A_B}}$ A_B = cross-sectional area of the boulder	A_B = 0.52 m^2 for a cubic boulder and 0.408 m^2 for a round boulder V_C = 2.95 m s^{-1} for a cubic boulder and 2.34 m s^{-1} for a round boulder The average V_C = 2.65 m s^{-1} for both forms
The average velocity of stream flow, V_{avg} is $1.2 \times V_C^{27.8}$	$V_{avg} = 1.2\, V_C$	V_{avg} = 3.18 m s^{-1}
Mannings roughness coefficient n as a function of channel slope (in degrees)	$n = 0.295\,(\tan\beta)^{0.377}$	n = 0.0971
Mean flow depth: for channel flows with high width to depth ratio, hydraulic radius is approximately equal to mean flow depth, i.e. $d = R$ (m), and Manning's equation can be used to calculate mean flow depth	$d = \left(\dfrac{V_{avg} n}{\sqrt{\tan\beta}}\right)^{3/2}$	d = 1.56 m
Discharge Q (m^3 s^{-1})	$Q = w\, d\, V_{avg}$	$Q = (20 \times 2.99 \times 4.98) = 50$ m^3 s^{-1}

- *Spread the losses*: e.g. emergency aid or disaster relief, such as the Hazard Mitigation Grant Program (HMGP) in the USA. The HMGP is coordinated by the Federal Emergency Management Agency (FEMA) and assists States and local communities in implementing long-term hazard mitigation measures following a declaration of major disaster (www.fema.gov).
- *Plan for losses* through insurance and the establishment of reserve funds. For example, in the USA the Flood Mitigation Assistance (FMA) Program (coordinated by FEMA) provides funds to assist States and communities in implementing measures to reduce or eliminate flood damage to buildings and homes. The US National Flood Insurance Program (NFIP) is the primary federal programme to reduce flood costs to the nation (www.fema.gov).
- *Mitigate the losses through early warning systems*: in the UK, the Environment Agency provide flood warnings online 24 hours a day; the information is updated every 15 minutes (www.environment-agency.gov.uk/subjects/flood/floodwarning/).

The flood warning system consists of the following codes:

Flood Watch	Flooding possible. Be aware! Be prepared! Watch out!
Flood Warning	Flooding expected affecting homes, businesses and main roads. Act now!
Severe Flood Warning	Severe flooding expected. Imminent danger to life and property. Act now!
All Clear	An all clear will be issued when flood watches or warnings are no longer in force.

Table 27.5 A flood rating system for risk assessment.

Rating	Class	Definition
I	Small exposure	Areas that are affected by floods less than once per 50 years on average; objects are insurable without restriction
II	Moderate exposure	Areas that are affected by floods in the recurrence interval 10–50 years; objects are basically insurable
III	High exposure	Areas on flood plains that are affected by floods with recurrence interval up to 10 years; objects in these areas are generally not insurable

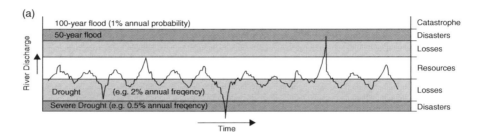

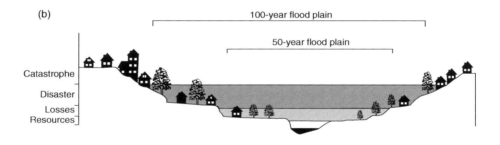

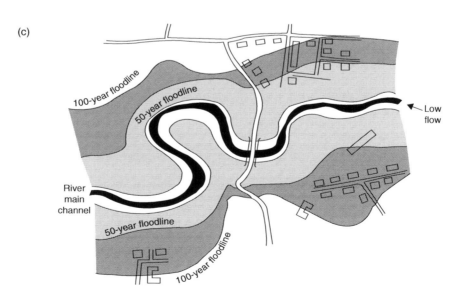

a) River discharge over time (50 year and 100 year low and high flows indicated).
b) Cross-section through floodplain.
c) Plan showing extent of the 50 year and 100 year floods.

Figure 27.3 Example of a flood zonation scheme (after Hewitt 1997)[27.10].

Table 27.6 Typical flood damage data (damage in £M).

Return period (RP) of flood event (years)	1	10	50	100	500
Probability (flood event)	1.000	0.100	0.020	0.010	0.002
Residential property	0.25	1	5	10	25
Industrial property	0	0	1	5	5
Indirect losses (e.g. tourism)	0	0	0	20	25
Traffic disruption	0.1	0.2	0.5	1	5
Emergency services	0.1	0.5	1	2	5
Other	0	0	0	0	0
Total damage £M	0.45	1.7	7.5	38	65
Area (damage × frequency)		0.97	0.37	0.23	0.41
Average annual risk (area beneath curve) £M					1.98

Area (damage × frequency) for between the 1 in 1 (Scenario 1) and 1 in 10 (Scenario 2) year events is calculated as:
Area (damage × frequency) = (Prob. 1 – Prob. 2) × (Total Damage 2 + Total Damage 1)/2.
Average annual risk =Σ Area (damage × frequency).

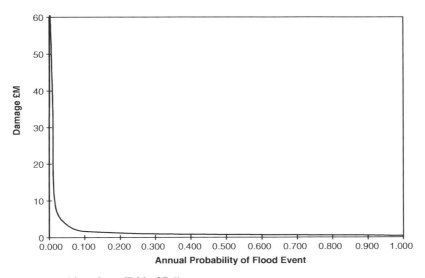

Figure 27.4 Flood damage curve (data from Table 27.6).

• *Floodplain management*: risks to property owners can be reduced by incorporating specific flood-proofing measures into the building design. For example, elevate property above a prescribed design flood level (e.g. on stilts or an earth bund). In the UK Royal Borough of Windsor and Maidenhead, for example, new residential properties on the Thames floodplain must have an internal ground floor level 0.15 m above the 1947 flood level[27.13]. The risks to residents of single storey properties can be reduced by requiring a means of escape such as a dormer roof window to be incorporated into the building design. In some countries, boats are placed on roofs.

• *River channel management*: obstructions to flow or inappropriate river channel/bankside works can increase flooding problems. In England, consent is needed from the Environment Agency for any structure in, over or under a river, which might obstruct flow in a watercourse[27.13].

• *Flood defence engineering*: engineering measures represent the traditional response to flood risks (Table 27.7). Defences can involve *passive measures* (e.g. floodbanks and flood alleviation channels) and *active measures* (e.g. flood gates and barriers, storage and diversion channels which can be operated in times of anticipated flooding).

Flood defences only *reduce* the risk; they cannot *eliminate* it. Much depends on the design life of the structure and the degree of risk that is acceptable (see Chapter 12).

Table 27.7   Summary of flood defence options.

Flood regulation	Reservoirs for temporary storage of floodwater; barriers to control tidal surges
Flood relief	Construction of by-pass and diversion channels to carry some of the excess floodwater away from the protected area
Channel improvements	Enlarging channel capacity by straightening, widening or deepening
Flood protection	Embankments and walls to confine the floodwaters
River beach and point bar management structures	Groynes, breakwaters and strongpoints to promote a natural defence; beach nourishment

References

27.1   World Health Organisation (2002) *Flooding: Health Effects and Preventive Measures*. Fact sheet 05/02 Copenhagen and Rome, 13 September 2002.

27.2   Kron, W. (2002) Flood risk = hazard × exposure × vulnerability. In Wu et al (eds.) *Flood Defence 2002*, Science Press, New York, 1–16.

27.3   Farraday, R. V. and Charleton, F. G. (1983) *Hydraulic Factors in Bridge Design*. Hydraulics Research, Wallingford.

27.4   Transport Research Laboratory (TRL) (1997) *Principles of low cost road engineering in mountainous regions*. Overseas Road Note 16, Transport Research Laboratory, Crowthorne.

27.5   Gardiner, V. and Dackombe, R. (1983) *Geomorphological Field Manual*. Allen and Unwin, London.

27.6   Limerinos, J. T. (1969) *Relation of the Manning coefficient to measured bed roughness in stable natural channels*. US Geological Survey Professional Paper 650-D 215–221.

27.7   Clark, A. O. (1996) Estimating probable maximum floods in the upper Santa Ana Basin, Southern California, from stream boulder size. *Environmental and Engineering Geoscience*, **II**(2), 15–182.

27.8   Costa, J. E. (1983) Palaeohydraulic reconstruction of flash-flood peaks from boulder deposits in the Colorado Front Range. *Geological Society of America Bulletin*, **94**, 986–1004.

27.9   Kron, W. and Willems, W. (2002) Flood risk zoning and loss accumulation analysis for Germany. In: Proceedings of the First International Conference on *Flood Estimation*, March 6–8, Berne.

27.10 Hewitt, K. (1997) *Regions of Risk: A Geographical Introduction to Disaster*, Longman, London.

27.11 Penning-Rowsell, E., Johnson, C., Tunstall, S., Tapsell, S., Morris, J., Chatterton, J., Coker, A. and Green, C. (2003) *The benefits of flood and coastal defence: techniques and data for 2003*, Flood Hazard Research Centre, Middlesex University.

27.12 Department for Transport, Local Government and the Regions (DTLR) (2001) *Planning Policy Guidance note 25: Development and Flood Risk*. HMSO.

27.13 Lee, E. M. (1994) *The Investigation and Management of Erosion, Deposition and Flooding in Great Britain*. HMSO.

Engineering Geomorphology

Key to Colour Section

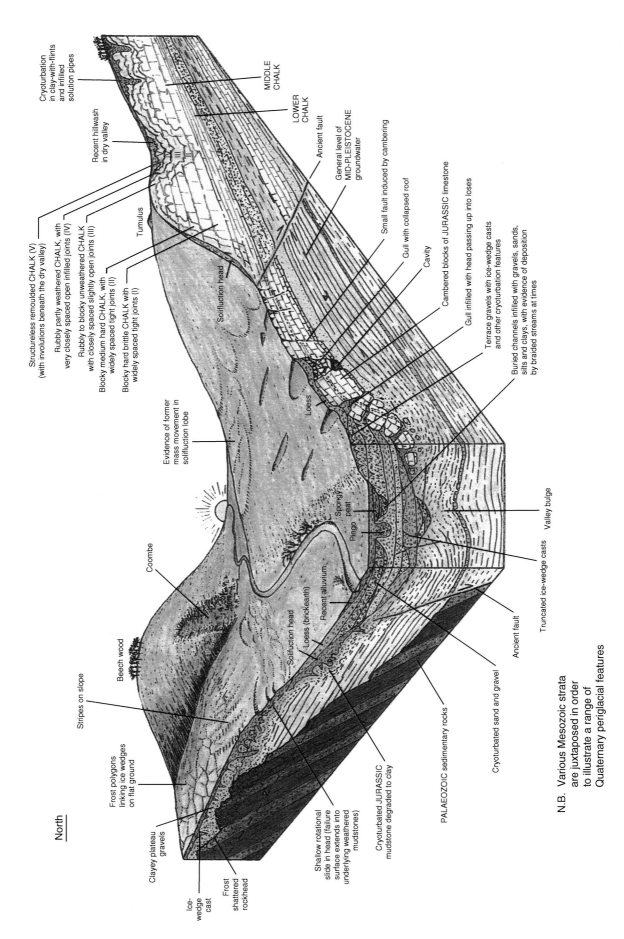

Figure 2.6   Relict periglacial terrain model in southern England (after Fookes 1997)[2.13].

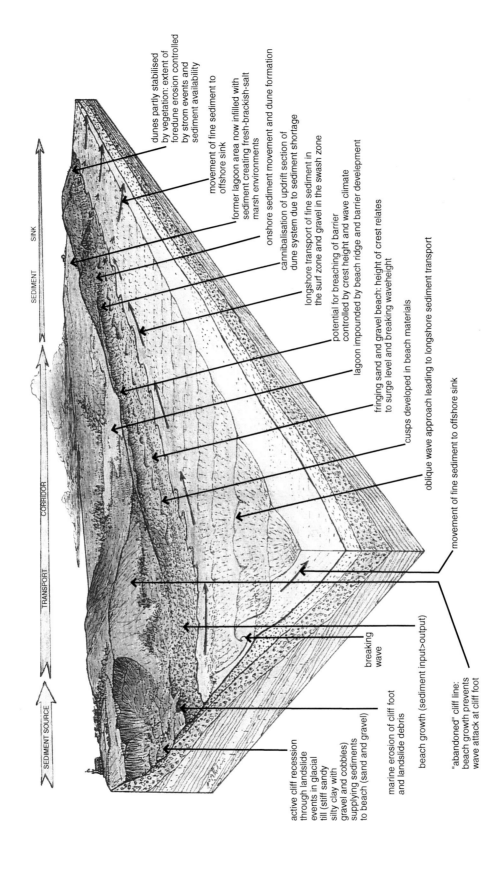

dunes partly stabilised by vegetation: extent of foredune erosion controlled by strom events and sediment availability

movement of fine sediment to offshore sink

former lagoon area now infilled with sediment creating fresh-brackish-salt marsh environments

onshore sediment movement and dune formation

cannibalisation of updrift section of dune system due to sediment shortage

longshore transport of fine sediment in the surf zone and gravel in the swash zone

potential for breaching of barrier controlled by crest height and wave climate

lagoon impounded by beach ridge and barrier develepment

fringing sand and gravel beach: height of crest relates to surge level and breaking waveheight

cusps developed in beach materials

oblique wave approach leading to longshore sediment transport

movement of fine sediment to offshore sink

breaking wave

beach growth (sediment input>output)

active cliff recession through landslide events in glacial till (stiff sandy silty clay with gravel and cobbles) supplying sediments to beach (sand and gravel)

marine erosion of cliff foot and landslide debris

"abandoned" cliff line: beach growth prevents wave attack at cliff foot

SEDIMENT SOURCE

TRANSPORT

CORRIDOR

SEDIMENT

SINK

Figure 29.2 A sediment transport cell model.

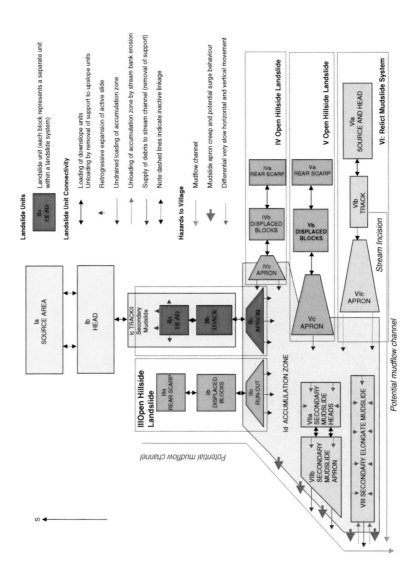

Figure 37.1 Schematic landslide model of a mudslide system, Georgia, highlighting interconnectivity between landslide systems.

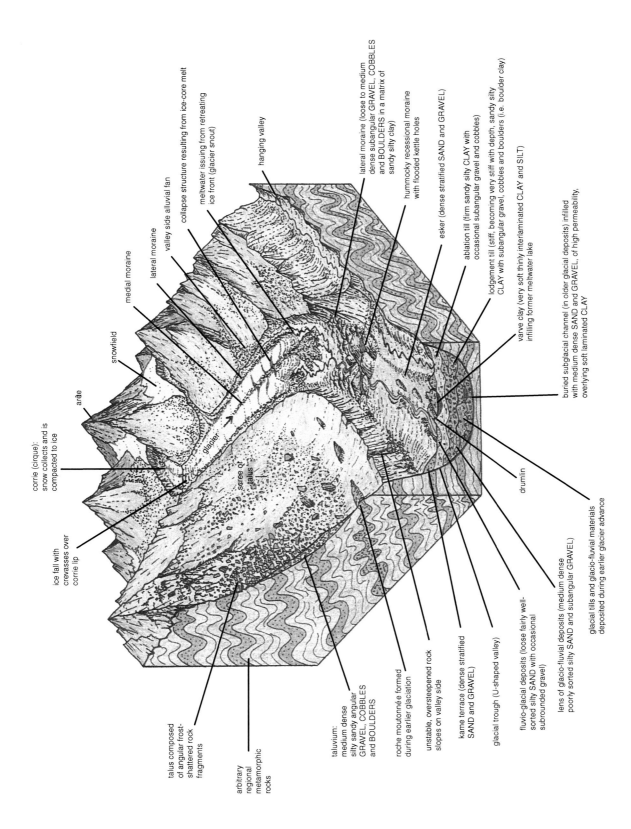

Figure 38.2   Glacial valley terrain model.

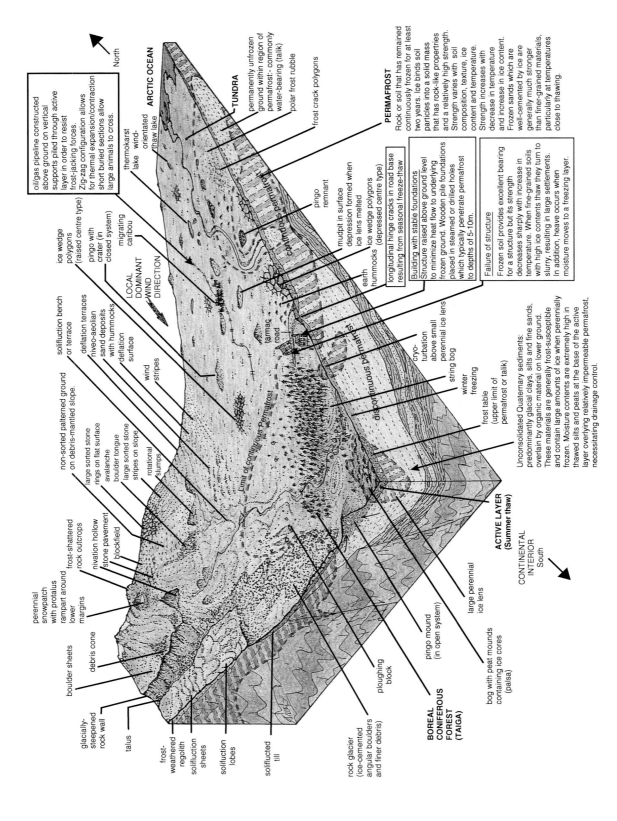

Figure 38.3 Periglacial terrain model.

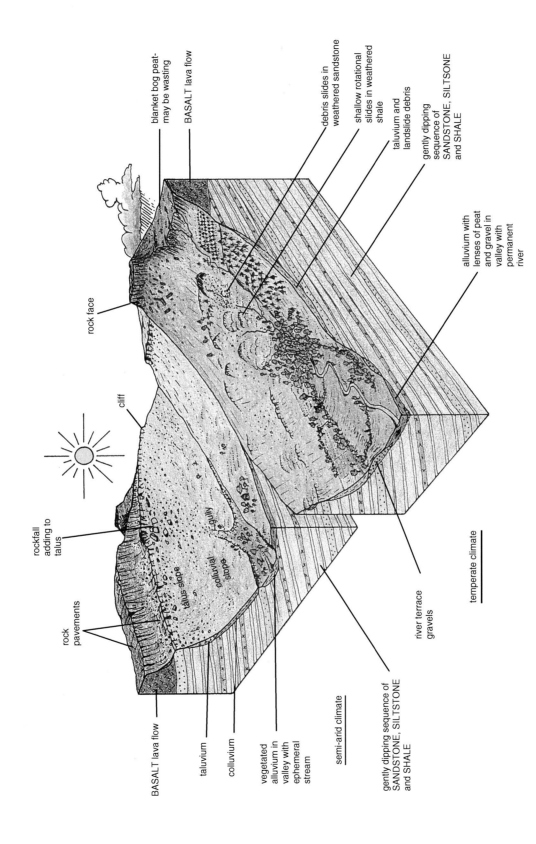

Figure 38.4 Semi-arid and temperate terrain models (after Fookes 1997)[38.1].

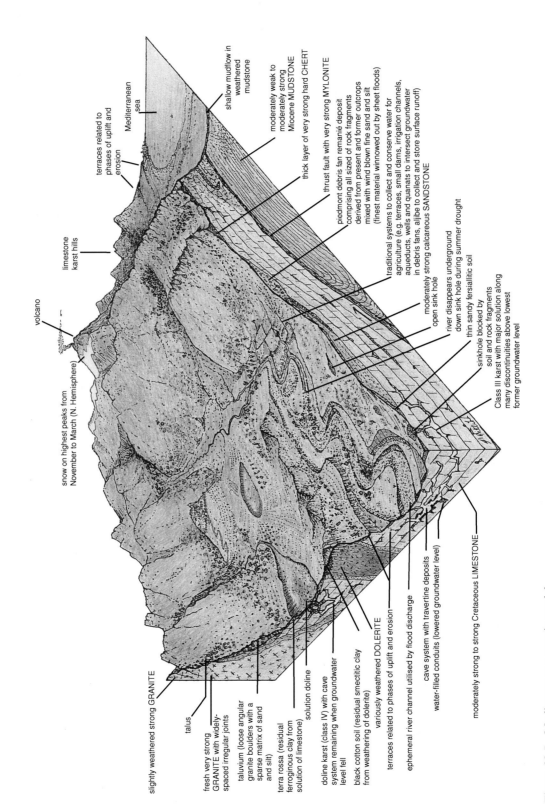

Figure 38.5   Mediterranean terrain model.

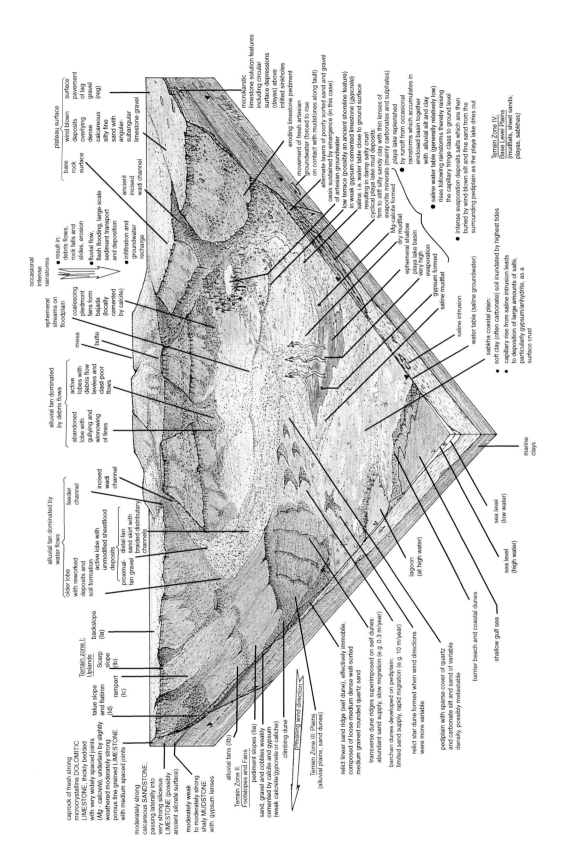

Figure 38.6 Desert terrain model (after Fookes 1997).[38.1]

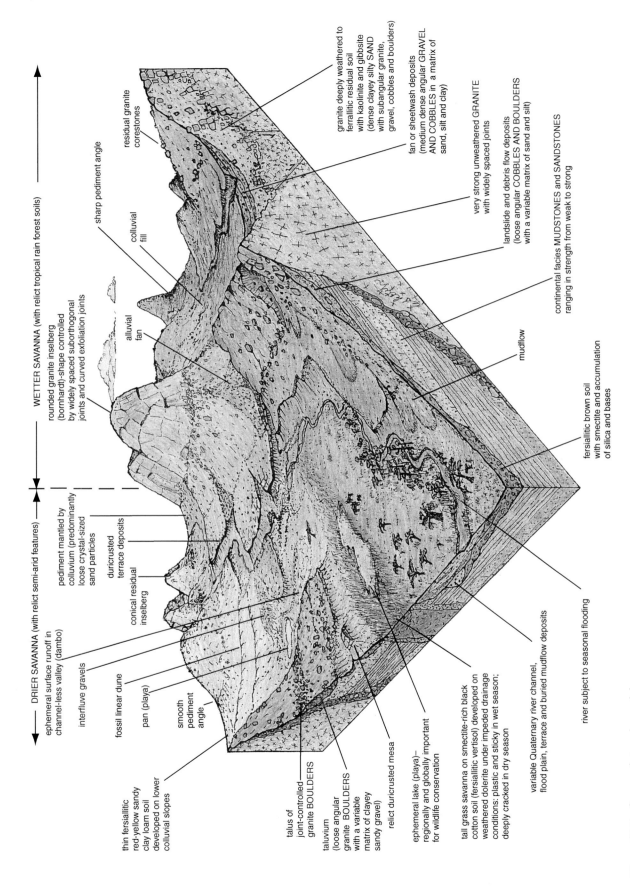

residual granite corestones

sharp pediment angle

colluvial fill

alluvial fan

WETTER SAVANNA (with relict tropical rain forest soils)

rounded granite inselberg (bornhardt)-shape controlled by widely spaced suborthogonal joints and curved exfoliation joints

pediment mantled by colluvium (predominantly loose crystal-sized sand particles)

duricrusted terrace deposits

conical residual inselberg

DRIER SAVANNA (with relict semi-arid features)

ephemeral surface runoff in channel-less valley (dambo)

interfluve gravels

fossil linear dune

pan (playa)

smooth pediment angle

granite deeply weathered to ferrallitic residual soil with kaolinite and gibbsite (dense clayey silty SAND with subangular granite, gravel, cobbles and boulders)

fan or sheetwash deposits (medium dense angular GRAVEL AND COBBLES in a matrix of sand, silt and clay)

very strong unweathered GRANITE with widely spaced joints

landslide and debris flow deposits (loose angular COBBLES AND BOULDERS with a variable matrix of sand and silt)

continental facies MUDSTONES and SANDSTONES ranging in strength from weak to strong

mudflow

fersiallitic brown soil with smectite and accumulation of silica and bases

thin fersiallitic red-yellow sandy clay loam soil developed on lower colluvial slopes

talus of joint-controlled granite BOULDERS

taluvium (loose angular granite BOULDERS with a variable matrix of clayey sandy gravel)

relict duricrusted mesa

ephemeral lake (playa)—regionally and globally important for wildlife conservation

tall grass savanna on smectite-rich black cotton soil (fersiallitic vertisol) developed on weathered dolerite under impeded drainage conditions: plastic and sticky in wet season; deeply cracked in dry season

variable Quaternary river channel, flood plain, terrace and buried mudflow deposits

river subject to seasonal flooding

Figure 38.7 Savannah terrain model.

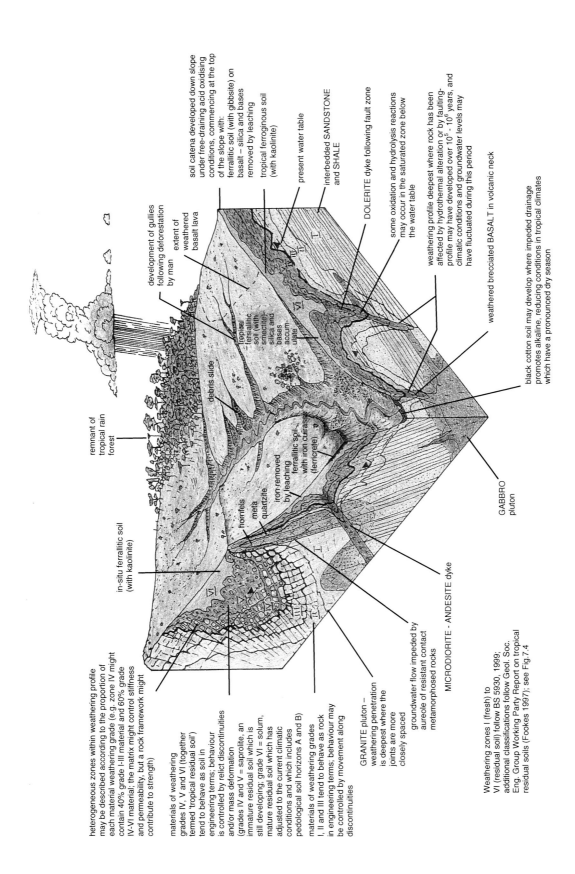

heterogeneous zones within weathering profile may be described according to the proportion of each material weathering grade (e.g. zone IV might contain 40% grade I-III material and 60% grade IV-VI material; the matrix might control stiffness and permeability, but a rock framework might contribute to strength)

materials of weathering grades IV, V and VI (together termed 'tropical residual soil') tend to behave as soil in engineering terms; behaviour is controlled by relict discontinuities and/or mass deformation (grades IV and V = saprolite, an immature residual soil which is still developing; grade VI = solum, mature residual soil which has adjusted to the current climatic conditions and which includes pedological soil horizons A and B)

materials of weathering grades I, II and III tend to behave as rock in engineering terms; behaviour may be controlled by movement along discontinuities

GRANITE pluton – weathering penetration is deepest where the joints are more closely spaced

groundwater flow impeded by aureole of resistant contact metamorphosed rocks

MICRODIORITE - ANDESITE dyke

Weathering zones I (fresh) to VI (residual soil) follow BS 5930, 1999; additional classifications follow Geol. Soc. Eng. Group Working Party Report on tropical residual soils (Fookes 1997); see Fig.7.4

remnant of tropical rain forest

in-situ ferrallitic soil (with kaolinite)

development of gullies following deforestation by man

extent of weathered basalt lava

debris slide

tropical ferrallitic soil (with smectite) - silica and bases accumulate

hornfels

meta quartzite

iron removed by leaching

ferrallitic soil with iron cuirass (ferricrete)

GABBRO pluton

soil catena developed down slope under free-draining acid oxidising conditions, commencing at the top of the slope with:
ferrallitic soil (with gibbsite) on basalt – silica and bases removed by leaching

tropical ferroginous soil (with kaolinite)

present water table

interbedded SANDSTONE and SHALE

DOLERITE dyke following fault zone

some oxidation and hydrolysis reactions may occur in the saturated zone below the water table

weathering profile deepest where rock has been affected by hydrothermal alteration or by faulting- profile may have developed over 10^5 - 10^6 years, and climatic conditions and groundwater levels may have fluctuated during this period

weathered brecciated BASALT in volcanic neck

black cotton soil may develop where impeded drainage promotes alkaline, reducing conditions in tropical climates which have a pronounced dry season

Figure 38.8 Wet tropical terrain model (after Fookes 1997)[38.1].

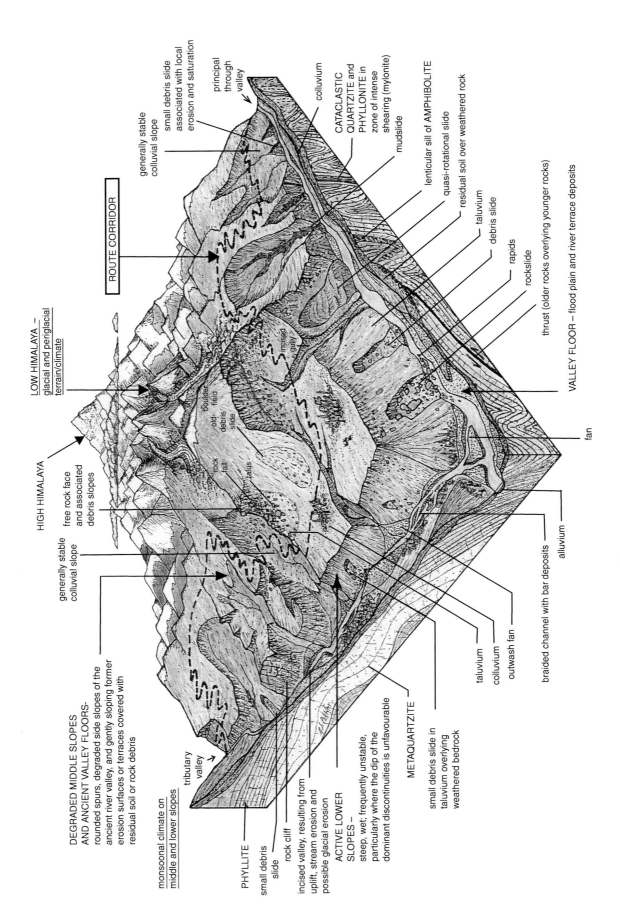

Figure 38.9 Mountain terrain model: Himalayas (after Fookes 1997)[38.1].

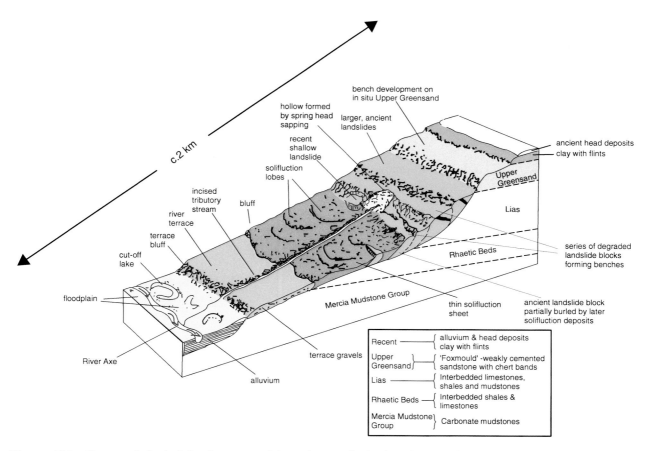

Figure 42.1 Geomorphological landscape model used as a basis for the morphological and derivative maps in Fig. 42.2 (after Croot and Griffiths 2001)[42.1].

A) Morphological Map

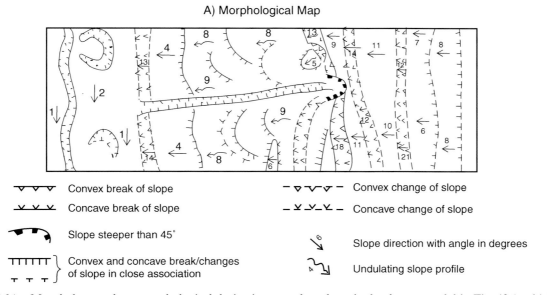

Figure 42.2A Morphology and geomorphological derivative maps based on the landscape model in Fig. 42.1, which gives the approximate scale.

B) Morphographic Map

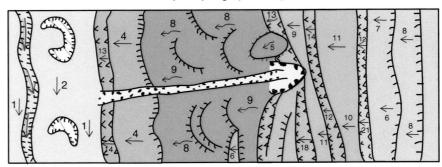

▢ Planation surface/Upper Greensand dip slope with superficial cover	◿ Upper Greensand cuesta scrap face
▢ Degraded blocks and benches in ancient rotational landslides developed on Lias & Rhaetic beds	⬭ Recent landslide TR translational
Hollow developed by spring head sapping	▢ Solifluction apron with intermediate and frontal lobes
▢ River terrace with terrace bluff	Incised stream
▢ Floodplain ⬿ Meander cut-offs	Contemporary River (with flow direction)

C) Morphochronological Map

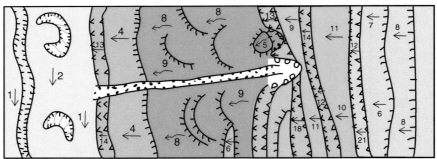

▢ Tertiary planation surface developed on Cretaceous Upper Greensand with Late Tertiary clay-with-flints and early to mid Pleistocene 'head' deposits	▢ Late to early post-glacial landslide deposits developed in Jurassic Lias Rhaetic beds
▢ Devensian, solifluction apron	▢ Ipswichian river terrace deposits overlying Jurassic Mercia Mudstones
▢ Post-glacial river area floodplain	▢ Recent landslide
▨ Recent meander cut-offs	

Figure 42.2 (continued).

D) Morphogenetic Map

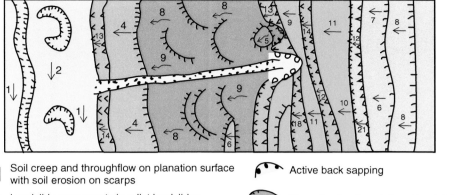

☐ Soil creep and throughflow on planation surface with soil erosion on scarps

☐ Landslide movements in relict landslides and soil erosion on scarps

☐ Stable solifluction apron with relict shear surfaces and soil water throughflow

☐ River terrace with soilwater throughflow and minor instability off terrace bluff

Active back sapping

Recent active translational landslide

Seasonal flooding with deposition on floodplain

Actively eroding gully

E) Resource Map

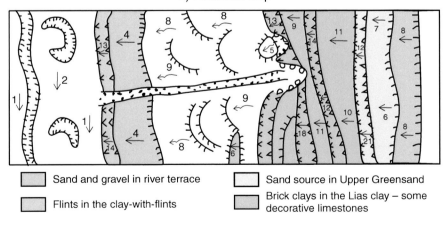

☐ Sand and gravel in river terrace

☐ Flints in the clay-with-flints

☐ Sand source in Upper Greensand

☐ Brick clays in the Lias clay – some decorative limestones

Figure 42.2 (continued).

F) Hazard Map

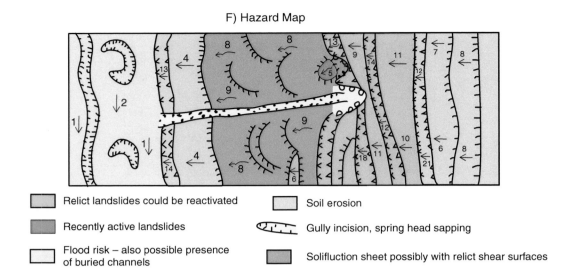

Relict landslides could be reactivated

Recently active landslides

Flood risk – also possible presence of buried channels

Soil erosion

Gully incision, spring head sapping

Solifluction sheet possibly with relict shear surfaces

Figure 42.2   (continued).

28

The Coast: Energy Inputs

Energy Gradients and Erosional/Depositional Settings

The coastline is subject to energy inputs from three main *forcing* processes: *waves* (wind waves, tides, tsunami), *wind* and *river mouth flows*. This energy is reflected back, dissipated by the frictional drag exerted by the shoreline or used to move sediment (from high energy to low energy settings; Table 28.1). The forcing processes interact with the inherited morphology (i.e. topography and bathymetry) to generate energy gradients along and across the nearshore and shoreline, including high energy zones of *convergence* (e.g. at headlands) and low energy zones of *divergence* (e.g. within bays).

Depending on the availability of mobile sediment, characteristic landforms develop in specific settings (*niches*) within these energy gradients. Fine-grained sediments (i.e. *muds*) tend to accumulate in sheltered, low energy environments (e.g. estuaries with their intertidal mudflats and saltmarshes; see Chapter 31) whereas coarse sediments can be found on the open coast where the energy inputs are higher (e.g. sand and shingle beaches; see Chapters 33 and 34). The highest energy environments are often characterised by rock cliffs (i.e. reflective rock barriers; see Chapter 36).

The combination of variable energy inputs and mobile sediment lead to on-going *morphological adjustments* of the shoreline, ranging from beach profile changes over the course of a single storm to long-term changes (e.g. hundreds to thousands of years) in response to factors such as relative sea-level rise or changes in sediment availability. If the energy niche changes then the landform will either be left as a relict form, replaced by a form that is more suited to the new setting (e.g. a change in sediment size or profile shape) or lost.

Wind Waves

Waves generated by wind blowing across the water surface (*wind* or *gravity* waves; see Table 28.2) supply enormous amounts of energy to the shoreline:

- *Storm waves*: waves developed in storm areas (e.g. depressions or tropical storms). They directly affect a particular shoreline typically less than 10 times a year and only last for a few days.
- *Swell waves*: waves emanating from storms some distances away are a regular occurrence. They travel beyond the area of generation; wave energy generated by the storm is spread out and not significantly reduced.

Coastlines can be classified in terms of the wave energy inputs[28.2]:

- *Storm wave environments*: a significant proportion of the waves are generated in local waters by storms (short, high energy waves of varying directions), with swell waves forming a persistent background (e.g. the north Atlantic coasts).
- *Swell wave environments*: dominated by consistent patterns of long, low waves (e.g. the west coast of Africa, southern Australia, the Gulf of Mexico).
- *Protected environments*: seas where there is little ocean swell penetration and outside the main temperate storm belts (e.g. the Mediterranean).

It is the *wave form* (Fig. 28.1) not the water mass that moves across the surface. The individual water particles

Table 28.1 Classification of coasts [28.1].

Type of coast	High energy		Low energy
	Open coast	Bay	Estuary
Sediment type	Solid	Non cohesive	Cohesive
Foreshore	Rock platform	Shingle/sand beach	Mudflat
Upper shore	Rocky/cliff	Shingle ridge-foredunes	Marsh
Supra tidal	Cliff	Cliff-dune	Reclaimed marsh

Table 28.2 Beaufort scale of wind and significant wave characteristics.

Beaufort Wind Force	Wind speed (m s⁻¹)	Description	Sea description	Approx significant wave height H_s (m)	Approx significant wave period T_s (s)
0	0–0.2	Calm	Like a mirror	0	1
1	0.3–1.5	Light air	Rippled	0.025	2
2	1.6–3.3	Light breeze	Small wavelets	0.1	3
3	3.4–5.4	Gentle breeze	Large wavelets	0.4	4
4	5.5–7.9	Moderate breeze	Small waves	1	5
5	8.0–10.7	Fresh breeze	Moderate waves	2	6
6	10.8–13.8	Strong breeze	Large waves	4	8
7	13.9–17.1	Moderate gale	Sea heaped up with white foam breakers	7	10
8	17.2–20.7	Fresh gale	Moderately high waves with spindrift	11	13
9	20.8–24.4	Strong gale	High waves with streaks of foam	18	16
10	24.5–28.4	Whole gale	Very high waves with overhanging crests	25	18
11	28.5–32.7	Storm	Exceptionally high waves with extensive foam patches	35	20
12	> 32.7	Hurricane	Sea completely covered with foam, driving spray	40	22

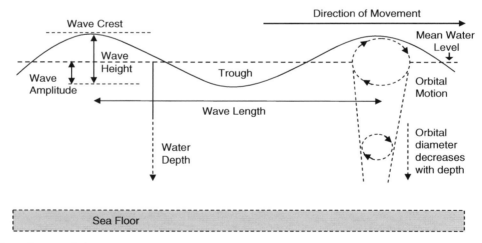

Figure 28.1 Wave characteristics.

rotate around closed orbits that decrease in size with water depth. The wave climate is often described in terms of the *significant wave height* H_s, the mean of the highest third of waves in a series.

Key equations from the *Airy wave theory* are presented in Table 28.3:
- Wave energy E is related to the square of the wave height H.

- The rate at which energy is transported across the sea E_f (energy flux) is related to the velocity C of the wave (*celerity*).
- Wave speed in deep water is a function of the wavelength L and period T. Long period waves travel faster than short period waves. If a storm produces a variety of waves, the longer waves will reach a distant shoreline first and lose less

Table 28.3 Airy wave theory equations.

	Deep water: $h/L > 0.25$	Shallow water: $h/L < 0.05$
Wave phase velocity C (m s^{-1}) (individual wave)	$C = \dfrac{gT}{2\pi}$	$C = \sqrt{gh}$
Wavelength L (m)	$L = \dfrac{gT^2}{2\pi}$	$L = T\sqrt{gh}$
Wave group velocity C_g (m s^{-1})	$C_g = \dfrac{C}{2}$	$C_g = C$
Energy density, per wavelength, for each unit wave crest E (joules)	$E = \dfrac{\rho g L}{2}\left(\dfrac{H}{2}\right)^2$	$E = \dfrac{\rho g L}{2}\left(\dfrac{H}{2}\right)^2$
Energy flux (wave power) for each unit of wave crest E_f (joules)	$E_f = \dfrac{\rho g}{4}\left(\dfrac{H}{2}\right)^2 C$	$E_f = \dfrac{\rho g}{4}\left(\dfrac{H}{2}\right)^2 C$

T is the wave period (seconds); g is the gravitational acceleration; H = wave height (m); ρ = density of water (kg m^{-3}); h = water depth (m); L = wavelength (m).

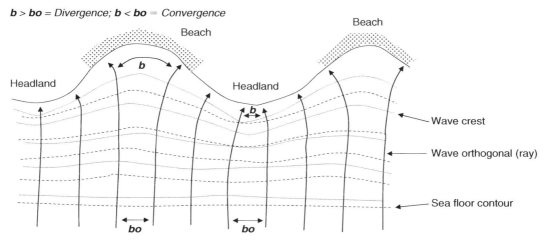

Figure 28.2 Wave convergence and divergence.

energy than the slower moving short period waves.

• Deep water waves are *transformed* as they enter shallow water. As water depth h decreases so both the wavelength L and wave phase velocity C decrease. However, the rate at which energy arrives at the coast (wave power) E_f does not change, as wave height H increases to offset the reduction in velocity and conserve the potential energy of the wave.

Waves are *refracted* in shallow water and each wave crest will become parallel to the seabed contours (*bathymetry*) and the shoreline. This occurs because wave velocity is related to water depth i.e. a wave crest arriving obliquely to the shore will be travelling slower in the shallow water than in the deep water. As a result the wave crest becomes curved to maintain an equal velocity. This process generates zones of convergence where wave height increases (i.e. at headlands) and zones of divergence (in bays) where wave height decreases (Fig. 28.2). The combination of shoreline

currents and convergence/divergence zones results in the establishment of *coastal cells* (see Chapter 29).

Wave transformation in shallow water results in changes to the orbit of the water particles in the wave. The orbits change from circular in deep water to elliptical and finally horizontal.

Much of the wave energy is dissipated during wave breaking. The point at which waves break is defined by the ratio between water depth h and wave height H i.e. wave height within the surf zone is depth-limited. The ratio ranges from 0.6 to 1.2, with a mean value of 0.78. This would appear to limit the potential variability of wave energy arriving at the shoreline, as all surf zone waves (storm or swell) will be the same height at a particular point in the nearshore. However, beach processes are actually driven by oscillations generated at the shoreline by a combination of incident and reflected waves (e.g. *infragravity* and *edge waves*) which lead to the generation of currents:

• *Cross shore currents*, acting normal to the shoreline (rip currents).

• *Longshore currents,* acting obliquely to the shoreline at a velocity proportional to the wave approach angle. These currents generate *longshore drift.*

Swell waves generate *swash* containing suspended sand which surges up the beach. As the velocity of the swash is greater than that of the returning (*backwash*) water there is net transport of material onshore and longshore. A large range of sediment sizes are moved shorewards, but only the finer grains are moved offshore. This results in an overall coarsening towards the shoreline. *Storm waves* generate steeper spilling waves and high returning velocities. As a result, material is transported offshore, creating a nearshore bar (see Chapter 33).

Tides

Tides are related to the balance of the *gravitational force* generated by the moon and the sun and the *centrifugal force* generated by the Earth's rotation. Water is pulled towards these bodies (mainly the moon), forming a bulge on both sides. On the side closest to the moon, the gravitational force is greater than the centrifugal force, while the reverse is true on the opposite side. These bulges move through the oceans as tidal waves ($C = \sqrt{gh}$). High and low tide are the crest and trough of the tidal wave, which arrives once every 12 hrs 25 mins.

Variations in the positions of the moon and the sun relative to the earth result in a 14 day cycle of:

• higher, *spring tides* (i.e. upwelling tides), when the moon and sun act together, and
• lower, *neap tides* when the moon and the sun act at 90° to each other.

The ocean tidal wave breaks against the continents and is reflected and dissipated. The combination of the reflected wave and the later, oncoming, wave result in the formation of a *standing wave*. Rather than travelling around the ocean surface as a progressive wave, the standing wave simply heaves up and down around a line where there is no change in water level (the *nodal line*). As the earth is spinning on its axis the Coriolis force causes the standing wave to rotate around a nodal point (*amphidromic point*). This rotation is anti-clockwise in the Northern Hemisphere, clockwise in the Southern Hemisphere. The tidal wave sweeps around the amphidromic point every 12 hrs 25 mins; *co-tidal lines* around this point define locations which have high tide at the same time (Fig. 28.3).

The tidal range experienced at a shoreline depends on (Table 28.4):

• The distance from the amphidromic point: tidal wave height increases away from this point.

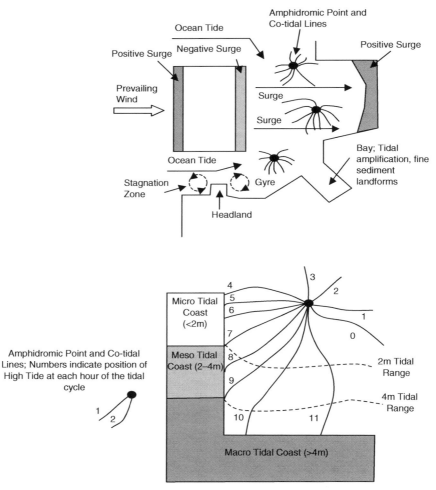

Figure 28.3 Schematic plan view of tidal systems.

Table 28.4 Tidal heights.

Abbreviation	Tidal height	Description
HAT	Highest astronomical tide	Highest which can be predicted to occur under any combination of astronomical circumstances
MHWS	Mean high water springs	Average (19-year) of high water occurring on spring tides
MHW	Mean high water	Average (19-year) of high water
MTL	Mean tidal level	Average of MHW and MLW
MLW	Mean low water	Average (19-year) of low water
MLWS	Mean low water springs	Average (19-year) of low water occurring on spring tides
LAT	Lowest astronomical tide	Lowest which can be predicted to occur under any combination of astronomical circumstances; often used as Chart Datum.

- The water depth: wave height increases in shallow water, as the wave velocity decreases.

Ocean tidal ranges rarely exceed 2.5 m even in shallow coastal waters. However, in enclosed seas, bays and inlets the range can increase dramatically (12 m in the Bristol Channel, 16.8 m in the Bay of Fundy, Canada), because of *resonance*; this occurs when the standing wavelength is 2 times the length of an enclosed basin, or 4 times that of an open basin.

As the tidal wave moves along the coastline it generates currents. Sediment tends to move in the direction of the net residual current. Headlands create *tidal gyres* (zones of circulating current) and areas of *stagnation,* leading to the development of shoals.

Tides generate enormous amounts of energy. Much of this energy is dissipated in internal friction, but some of the energy is dissipated as the tidal wave is affected by the frictional drag exerted by gently sloping shallow banks (i.e. mudflats and saltmarshes). The wave may ultimately break and form a *tidal bore* (e.g. on the River Severn, UK).

The tidal wave height varies with barometric pressure – sea surface level changes by around 0.1 m for every millibar difference. Extreme sea surface level changes can occur in the form of:

- *Negative tidal surges* i.e. lower water levels which can present a threat for shipping in estuaries and harbours.
- *Positive tidal surges* i.e. increased flood risk, as occurred during the 1953 Southern North Sea coastal floods when over 2000 people died. Possibly over 100 000 may have been killed in the 'Grote Mandrenke' (drowning) of 1362, including more than half the populations of the North Sea coasts of Jutland and Schleswig[28.3].

Tidal range is an important control on the types of landforms present on a shoreline; it controls the rate that wave energy is applied to the foreshore:

- *Micro-tidal* ($<$ 2 m tidal range) i.e. open coasts with little tidal amplification. These coasts are dominated

by wind waves and coarse sediments (e.g. fringing sand and shingle beaches, spits and barrier islands).

- *Macro-tidal* ($>$ 4 m tidal range) i.e. sheltered bays and inlets where the tidal range is amplified. These areas are dominated by tides and characterised by forms that dissipate tidal energy (e.g. mudflats and saltmarshes).

The intermediate *meso-tidal* (2–4 m tidal range) coasts are composite shorelines with both wind wave and tidal wave related landforms.

Combinations of tide level and wave height (i.e. joint probabilities) are important in the design of engineering structures. Severe storms superimposed on an extreme high tide level can have more significance than frequent lower-energy events at neap-tide positions.

Tsunami

Seismic sea waves or *tsunami* (Japanese for 'great wave in harbour') are large ocean waves with very long wavelengths (often over 100 km from crest to crest) and high velocities (reaching in excess of 500 km hr^{-1} in deep water). In 1896 an earthquake-triggered tsunami (the Great Japanese Sanriku Tsunami) hit the coast of Japan. Over 10 000 houses were washed away, around 27 000 people killed and 9000 injured.

Although they are often referred to as tidal waves, tsunamis are not caused by tides. They are most common in the Pacific Ocean, because the region is surrounded by active plate boundaries. However, they are also known to have occurred on many other shorelines around the world. They are caused by[28.4]:

- *Fault displacement on the seabed*, causing water to rush in from the surrounding sea to restore the water level. A wave travels out from the sea area above the zone of seabed displacement. For example, the December 26th 2004 (Boxing Day) tsunami was caused by a major earthquake (magnitude 9.3) 320 km west of Medan, Sumatra. It is possible that over 200 000 people were killed, with widespread devastation in Sumatra, Sri Lanka, India, Thailand and the east coast of Africa (see http://walrus.wr.usgs.gov/ tsunami/).

- *Landslides*, either submarine failures (e.g. of the continental shelf margin) or a landslide entering the sea. The Storegga landslide occurred around 7200 BP on the continental slope west of Norway. The resulting tsunami hit the east coast of Scotland and extended as far south as Lindisfarne; at Inverness the waves may have been 8 m high[28.5].
- *Volcanic activity*, with the uplift or settlement of the volcano flank triggering a shock wave. Tsunami may also be caused by the actual explosion of submarine or shoreline volcanoes – the 1883 Krakatoa eruption triggered a tsunami which was 40 m high when it hit the coasts of Java and Sumatra (see http://vulcan.wr.usgs.gov/Glossary/Tsunami/).

Most tsunamis do not hit the shore as a huge breaking wave – they arrive like rapidly rising tides. Occasionally the tsunami wave can develop an abrupt, steep front (i.e. a bore). The 1960 tsunami that hit Hilo, Hawaii had a 10 m high bore. The extent of tsunami *run-up* onshore is influenced by the reflection of the wave seawards and amplification by later incoming waves (see the Pacific Tsunami Warning Centre at: http://www.prh.noaa.gov/ptwc/).

Most tsunamis deposit a spread of sediment inland, often as a *tapering sediment wedge*. There is a progressive landward decrease in material size, from boulder spreads to sheets of finer grained sediment (sands and coarse silts) that occur as sets of fining upwards sequences. The 1992 tsunami that hit the island of Flores, Indonesia deposited an extensive sheet up to 1 m thick. The upper limit of run-up is often marked by a zone of stripped vegetation and soil[28.6].

Tsunamis can also be associated with major shoreline change, especially to barrier beaches and spits. For example, the 1755 Lisbon tsunami (caused by a magnitude 8.5 earthquake) resulted in extensive modification of the Rio Formosa barrier island system on the Algarve coast, Portugal[28.7].

Wind

Onshore winds move sediment (around 0.006–2 mm grain size) from exposed intertidal sand sheets to form sand dunes, by saltation (bouncing), surface creep and in suspension (see Chapter 18). The wind applies a shear force to the surface sand grains, generally expressed as the *wind shear velocity* u^*. The amount of sand transport q is related to the cube of the wind shear velocity (i.e. a rise in wind speed will result in a significant increase in sand transport).

Winds crossing a sea or basin can pile water up on the windward coast (positive surge) and cause a marked negative surge on the leeward coast (Fig. 28.3).

River Mouth Flows

The interaction between the strong currents entering into and emerging from river mouths (estuaries) and the processes operating on the open coast create complex local energy gradients. *Deltas* form where the supply of sediment delivered by the river at its mouth exceeds the amount that is mobilized and removed by the waves and tidal currents (see Chapter 32).

In estuaries, the balance between the river flows and the tide and wave forces controls the position and type of the various bedforms that develop at the mouth[28.8] (Fig. 28.4):

- river dominated environments: prograding deltas with distributaries flanked by levées
- tidal dominated environments: macro-tidal estuaries and tide-dominated deltas
- wave dominated environments: barriers (shore parallel sand/shingle bodies), strand plains and lagoons.

A general model of an estuary mouth is presented in Fig. 28.5; this highlights the development of a strong, steady and unidirectional jet at the mouth. Turbulent vortices form at the margins and front of the jet, drawing

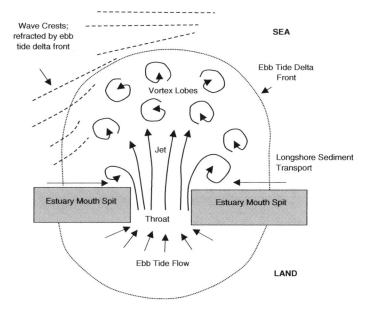

Figure 28.4 Schematic plan view of estuary mouth currents and sediment movement (after Carter 1988)[28.8].

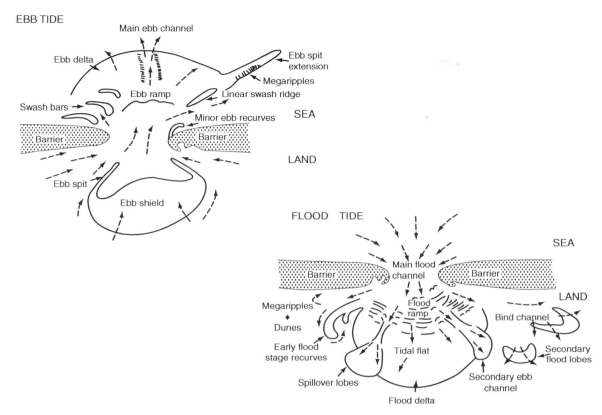

Figure 28.5 Typical ebb and flood tide delta features (after Carter 1988)[28.8].

sediment alongshore towards the inlet and leading to the development of lobate shoals and delta ramparts. A *flood-tide delta* forms inside the inlet and an *ebb-tide delta* forms on the seaward side, often comprising wave-worked structures. Excess ebb-tide delta material often accumulates as marginal swash bars that are driven shoreward to feed the adjacent beaches (see Chapter 31).

Management Issues

Shoreline management must be based on an understanding of how the coastline works, especially the way it adjusts to variations in energy inputs (i.e. energy gradients):

- The use of hard structures to prevent erosion or reduce flood hazard disrupts the energy gradients established on the natural shoreline. The adjacent shoreline will adjust its morphology to reflect these changes. As a result, seawalls can result in beach loss, both in front of the structure (scour) and downdrift (increased energy resulting in increased sediment transport). Offshore breakwaters reduce the energy arriving at the shoreline and can lead to sediment accumulation in the wave shadow. However, this interruption of the energy gradient along the shoreline can result in a reduction in longshore sediment transport past the structure and sediment starvation downdrift.
- Landforms can only be successfully recreated in the appropriate energy niches. Beach recharge schemes need to be designed to fit the particular setting (i.e. profile, planshape and range of sediment grain sizes). Saltmarsh recreation needs to be accompanied by the development of mudflats that dissipate tidal and wave energy before it reaches the marsh.
- Sea-level rise (see Chapter 10) will result in increased water depths and, hence, gradual changes in energy gradients. Coastal landforms will respond by changing their location – either onshore or alongshore – so as to maintain their relative energy level position.

References

28.1 Pethick, J. S. and Burd, F. (1994) *Coastal Defence and the Environment*. MAFF Publications, London.

28.2 Davies, J. (1972) *Geographical Variation in Coastline Development*. Oliver and Boyd.

28.3 Lamb, H. H. (1991) *Historic storms of the North Sea, British Isles and Northwest Europe*. Cambridge University Press.

28.4 Dawson, A. G., Long, D. and Smith, D. E. (1991) Tsunamis. *Disaster Management*, **3**, 155–160.

28.5 Dawson, A. G., Long, D. and Smith, D. E. (1988) The Storegga slides: evidence from eastern Scotland for a possible tsunami. *Marine Geology*, **82**, 271–276.

28.6 Dawson, A. G. (1994) Geomorphological effects of tsunami runup and backwash. *Geomorphology*, **10**, 83–94.

28.7 Dawson, A. G., Hindson, R., Andrade, C., Freitas, C., Parish, R. and Bateman, M. (1995) Tsunami sedimentation associated with the Lisbon earthquake of 1 November AD 1755: Boca do Rio, Algarve, Portugal. *The Holocene*, **5.2**, 209–215.

28.8 Carter, R. W. G. (1988) *Coastal Environments*. Academic Press.

29

The Coast: Sediment Cells and Budgets

Introduction

Beaches are stores of gravel and sand supplied from source areas on the adjacent coastline or offshore. They can be viewed as parts of a larger system, a *coastal cell* or *sediment transport cell*, within which a range of sediment transfers takes place (see Fig. 29.1). Beach building material might be supplied from the seabed, moved onshore by wave energy, or from rivers and eroding cliffs. This material is then redistributed along the shoreline by waves (*longshore drift*), unless prevented by barriers such as headlands, groynes or harbour arms. Some of the material can be lost to the cell around the seaward end of the barriers or offshore, particularly during large storms.

Longshore Drift

Longshore drift and offshore/onshore exchanges of coarse sediment are driven by wave energy (see Chapter 28). Storms arrive at infrequent and irregular intervals, removing the upper beach berm and placing the material offshore in the form of a bar. Subsequent smaller waves sweep the material back onto the beach. At this time, sediment moving onshore from the bar can be driven alongshore if the waves are approaching obliquely to the coastline (if waves arrive normal to the coast no longshore drift occurs). Longshore drift generally occurs in pulses, during and immediately after storms. While the sediment is within the beach berm, calm weather waves have little influence on it.

Over the course of a year, waves will arrive at a beach from a range of directions. As the wave approach angle changes so the direction of the longshore drift will change:

- The *gross drift* is the total volume of longshore movement irrespective of direction.
- The *net* drift is the difference between the volumes moving in the two opposite drift directions.

Numerous equations are available for calculating the longshore drift rate; one of the more widely used methods is the CERC equation [29.1]:

$$Q = \frac{K P_{ls}}{(s-1)\rho_{w} g a}$$

where:

Q = volumetric longshore transport rate ($m^3\ s^{-1}$)

P_{ls} = longshore energy flux = $0.05\rho\ g^{3/2} H_{50}{}^{5/2}$ $(\cos \alpha_{o})^{1/4} \sin \alpha_{o}$

ρ_{w} = water density ($1000\ kg\ m^{-3}$)

g = gravitational acceleration

H_{50} = significant wave height offshore (m)

α_{o} = angle between wave crest (offshore) and the shoreline

K = constant (0.32 for a sand beach, 0.025 for a shingle beach)

s = specific gravity of the beach material (e.g. 2.65)

a = ratio of particle volume to total volume of the sediment (usually taken as 0.6)

Coastal Cells

The shoreline comprises a series of inter-linked systems within which sediment is moved along energy gradients (Fig. 29.2) (see colour section), from high energy sediment *source* areas (e.g. eroding cliffs, offshore sand banks), along sediment *transport pathways* (e.g. beaches) to low energy sediment *stores* (e.g. dunes, offshore banks). Suspended sediment can be carried thousands of miles around the coast and seas.

Boundaries of coarse sediment cells occur where there is no breaking wave or the angle of wave approach is zero. They can be:

- *Fixed* (e.g. headlands, river mouths): geology and topography are major controls on cell size and form. Headlands can change over time (e.g. the gradual emergence and loss of eroding drumlin headlands on the Nova Scotia coast, Canada[29.2]).
- *Free* (i.e. energy convergence and divergence zones): the cell boundary position along a shoreline can vary as waves of different height, period and approach angle reach the shoreline (Fig. 29.3).

Longshore Sediment Transport Power

The size and shape of a shoreline embayment and the available materials influence the patterns of sediment circulation. The longshore component of wave power is

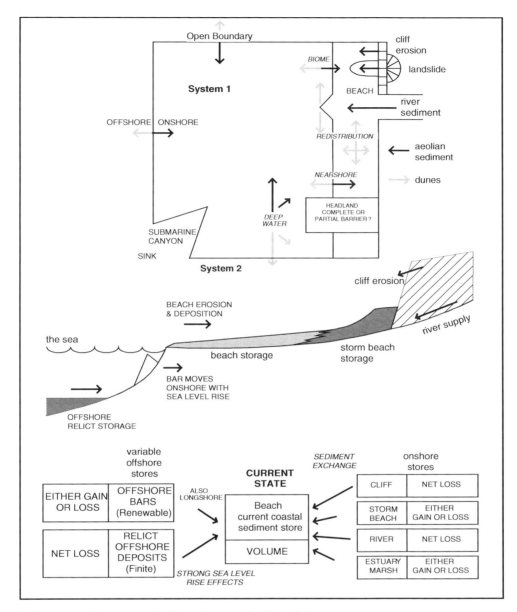

Figure 29.1   Sediment transfers on a typical beach: coastal cell model.

related to the wave approach angle relative to the shoreline; longshore wave power P_L is defined by[29.3]:

$$P_L = \frac{EC_b}{2}\sin 2\beta$$

where β = angle between the wave crest and the shoreline; E = wave energy density and C_b = breaking wave crest velocity = $\sqrt{2gH_b}$.

Considering a headland-bay unit that is not in equilibrium (Fig. 29.4), a number of points define the nature of the sediment transport system (Table 29.1). A positive longshore power gradient (i.e. increasing P_L) indicates shoreline erosion, whereas a negative gradient (i.e. decreasing P_L) indicates deposition. These trends will act to re-shape the shoreline with erosion between points A and C, and deposition between C and E. Considering any cross-shore profile around the bay, if the sediment

transport rate Q is greater than at the adjacent updrift profile (i.e. increasing dQ/dX) then more sediment will be moved away from the profile than is supplied to it, i.e. shoreline erosion occurs. Deposition occurs if the sediment transport rate is lower than at the adjacent updrift profile (i.e. decreasing dQ/dX). Maximum erosion occurs where dQ/dX reaches its peak (point B), and maximum deposition occurs where dQ/dX is at its minimum value (point D).

Shoreline Alignments

On many coasts, the dominant waves arrive from a narrow band of directions resulting in a relatively fixed distribution of wave energy along the shore. Coasts tend to become orientated in relation to the dominant wave direction:

• *Drift-aligned coasts*, where the shoreline is adjusted to maximise sediment transport (P_L = maximum).

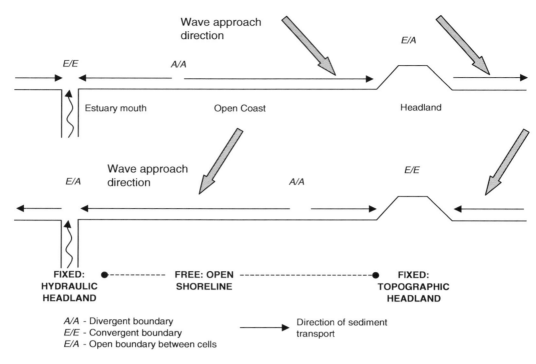

Figure 29.3   Coastal cell boundaries.

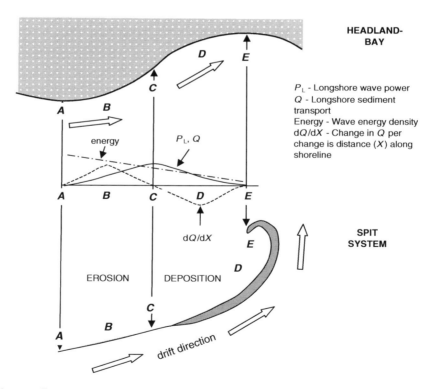

Figure 29.4   Longshore sediment transport power in a headland-bay (top) and spit system (bottom) (after May and Tanner 1973)[29.3].

Table 29.1 Longshore power gradients and sediment transport [29.3].

Point (Figure 29.4)	Wave Energy E	Longshore wave power P_L	dQ/dX	Beach trend
A	Maximum	Minimum	Zero	No change
B			Maximum	Erosion
C		Maximum	Zero	No change
D			Minimum	Deposition
E	Minimum	Minimum	Zero	No change

dQ/dX is the change in sediment transport Q, dX is the change in distance X along the shoreline. Where dQ/dX is zero, there is no change in the amount of sediment in transit.

These cells are open systems (i.e. elongated beaches) and associated with a continuous sediment supply from updrift. Changes in the wave approach angle cause a reduction in the transport rate and lead to deposition, so that the alignment is restored (i.e. *dynamic equilibrium*).

- *Swash-aligned coasts*, where scarcity of sediment results in a tendency for plan-form adjustment to match the refraction pattern of the dominant wave system; sediment transport potential is minimised (P_L = minimum). Eventually the readjustment leads to a stable bay (i.e. pocket beaches) within which longshore sediment transport potential is zero at all points (i.e. *static equilibrium*).

Stable bay planform (Fig. 29.5) is controlled by the wave approach angle (i.e. wave crest angle β) and the shoreline orientation[29.4]. Key design parameters are:

- The maximum bay indentation width a is defined by:

$$a = (0.014\,\beta - 0.000094\,\beta^2)\,R_o$$

where R_o is the length of the *control line* between the two headlands that define the boundaries of the bay.

- The longshore position of the maximum indentation a is defined by the angle θ_c:

$$\theta_c = 63 + 1.04\,\beta$$

Values of the indentation ratio a/R_o and θ_c for a range of wave crest angles β are presented in Table 29.2.

Sediment Budgets

Over time, the balance between sediment inputs and outputs (i.e. the *sediment budget*) within the system will determine whether the beach experiences growth, decline or has remained constant in overall size.

A contemporary shingle budget for part of the West Dorset coast, UK, is presented in Table 29.3. The inputs into the various cells are limited to cliff erosion inputs, especially from the cliffs between Lyme Regis and Golden Cap[29.5]. Although the net drift direction is eastwards (from cell 1 towards cell 4), the presence of headlands at Golden Cap and Doghouse Hill restricts the longshore movement of sediment. Landslide debris on the foreshore

has prevented sediment exchange between the cells since 1850 (cells 2 and 3) and 1962 (cells 1 and 2).

Sediment Scarcity and Cannibalisation

Sea-levels during the last glaciation (Devensian in the UK; see Chapter 10) were in excess of 100 m lower than present day, exposing a gently sloping coastal plain. This plain would have been partly covered by extensive spreads of river sands and gravels (i.e. floodplain, delta and estuary deposits, together with glacial sediments in high-latitudes). During the early Holocene, rapidly rising relative sea level resulted in onshore movement of large volumes of this sand and gravel onto the coast (e.g. the formation of Chesil Beach, UK[29.6]).

Beach development in the Holocene was probably controlled by the interaction of sea-level rise, sediment supply and longshore transport. As the rate of sea-level rise decreased, sediment supply and longshore transport declined substantially (see Chapter 8). Sediment supply to many mid- to high-latitude beaches has probably declined substantially since around 5–6000 BP, providing a long-term restriction on beach development i.e. *sediment scarcity*[29.7] (Fig. 29.6).

Beaches respond to scarcity of sediment by *cannibalisation* i.e. the re-working of sediment within a beach or barrier structure. The updrift sections of the beach experience erosion, feeding the continued supply of the downdrift sections. In time, this internal redistribution of sediment may transform a beach from a drift-aligned to swash-aligned structure or can involve the subdivision of the beach into a series of sediment sub-cells (Fig. 29.7).

Management Issues

Management decisions should be based on a sound understanding of coastal processes and their interactions with the existing coastal defences and assets. Coastal cells provide the framework for ensuring that wider-scale issues are taken into account when planning engineering works at a specific site.

Often it is not practical to use coastal cells as a basic shoreline planning and management unit e.g. they do not correspond with administrative boundaries. A

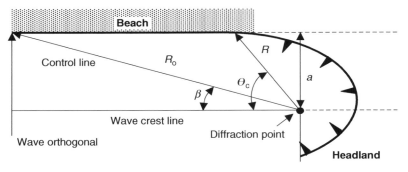

Figure 29.5   Stable equilibrium bay definition diagram (after Hsu *et al.* 1989)[29.4].

Table 29.2   Static equilibrium bay design parameters[29.4].

Wave crest angle β	Indentation ratio a/R_o	Intercept angle θ_c
10	0.13	73.4
20	0.19	78.6
30	0.34	94.2
40	0.41	104.6
50	0.47	115
60	0.5	125.4
70	0.52	135.8
80	0.52	146.2
90	0.5	156.6

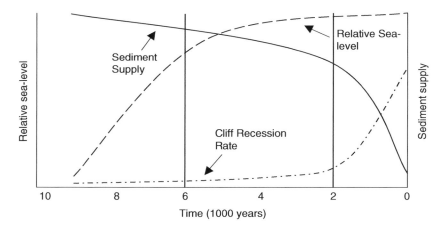

PERIOD I	PERIOD II	PERIOD III
Onshore sediment transfer Rapid beach growth Very low cliff recession rates	Declining onshore sediment Inputs from cliff recession Beach decline Low cliff recession rates	Cessation of onshore transfer Reduction in cliff inputs Marked beach decline Accelerated cliff recession

Figure 29.6   Holocene cliff/beach evolution periods, defined by sea-level rise and sediment supply (after Jennings *et al.* 1998; Brunsden and Lee 2004)[29.7–8].

Table 29.3 Current Shingle Budgets for the West Dorset coast, UK [29.5].

Cells	Gravel inputs					Gravel outputs							
	Cliff recession	Rivers	Littoral drift	Onshore transport	TOTAL	Attrition	Littoral drift	Entrapment	Offshore transport	TOTAL	Balance (input–output)	Beach mining	Total balance
1. Lyme Regis To Golden Cap	6.2	0.2	0	NS	6.4	1.3	0	0.2	NS	1.5	4.9	0.6 (up to 1956)	4.3
2. Golden Cap To Doghouse Hill	0.1	0	0	NS	0.1	0.1	0	0.05	NS	0.15	−0.15	2.1 (up to 1987)	−2.9
3. Doghouse Hill To West Bay Piers	NS	0	0	NS	NS	0.1	0	0.05	2.7	2.85	−2.85	0.4 (up to 1986)	−3.2
4. West Bay Piers To The Isle Of Portland	NS	NS	0	NS	NS	0.2	1.2*	0	NS	1.4	−1.4	16 (up to 1986)	−17.4
TOTAL	6.3	0.2	0	NS	6.5	1.6	1.2	0.3	2.7	5.8	0.7	19.2	−18.5

* Dredged from West Bay harbour channel; all figures in 1000's of m^3 yr^{-1}; NS = not significant.

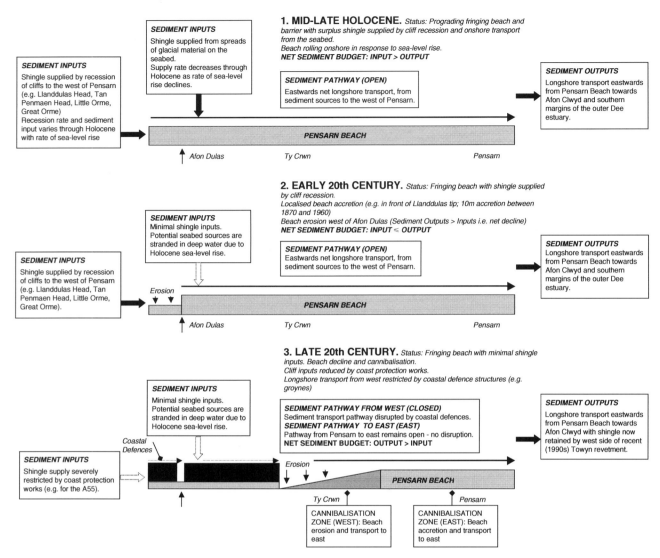

Figure 29.7   Conceptual evolutionary model of Pensarn Beach, North Wales, since the mid-Holocene.

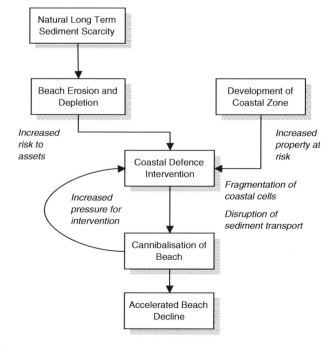

Figure 29.8   A summary of the 'knock-on' effects of coast protection works on shingle beaches (after Lee 2001)[29.9].

more pragmatic approach is to identify *management units* i.e. lengths of shoreline with similar characteristics of coastal processes and assets at risk that can be managed efficiently, within a broader coastal cell framework.

The plan for individual management units needs to be compatible across the coastal cell and they need to take account of:

- the operation of coastal processes within a coastal cell
- the potential impact of the management policies on the operation of coastal processes elsewhere within the coastal cell.

Coastal engineering works can disrupt the operation of coastal cells and threaten the integrity of many of the landforms within the cell (Fig. 29.8) by causing:

- *Reduction in sediment supply* from eroding cliffs. On the North Yorkshire coast, UK, between Whitby and Sandsend, current sediment inputs from the eroding glacial till cliffs are around 35% of the inputs prior to the onset of coast protection works in the late 19th and early 20th centuries[29.8] (see Table 14.2).
- *Disruption of longshore sediment transport pathways* e.g. by the construction of breakwaters and groynes (e.g. at the port of Map Ta Phut, Thailand).
- *Disruption of cross shore sediment exchanges* e.g. the construction of seawalls to prevent dune erosion (see Chapter 35).

The result tends to be beach depletion downdrift and may lead to the need for further coastal defence works, again with impacts on the sediment transport system operating in the cell (see Fig. 33.4). In time, transport pathways become broken and sediment stores may begin to erode.

Static equilibrium, swash-aligned bays may be reproduced by the construction of artificial headlands (*strong points*). These structures can be used to protect the shorelines in their 'wave shadow', while allowing erosion to continue over the intervening frontage. A series of such structures can be used to break up the coastline into a number of cells and thereby create a series of 'stable' bays[29.10].

References

29.1 CERC (1984) *Shore Protection Manual*. Coastal Engineering Research Centre, US Corps of Engineers, 4th Edition.

29.2 Orford, J. D., Carter, R. W. G. and Jennings, S. C. (1996) Control domains and morphological phases in gravel-dominated coastal barriers of Nova Scotia. *Journal of Coastal Research*, **12**, 589–604.

29.3 May, J. P. and Tanner, W. F. (1973) The littoral power gradient and shoreline changes. In D. R. Coates (ed.) *Coastal Geomorphology*, Binghampton University, 43–60.

29.4 Hsu, J. R. C., Silvester, R. and Xia, Y. M. (1989) Generalities on static equilibrium bays. *Coastal Engineering*, **12**(4), 353–369.

29.5 Lee, E. M. and Brunsden, D. (2001) Sediment budget analysis for coastal management, West Dorset. In J. S. Griffiths (ed.) *Land Surface Evaluation for Engineering Practice*, Geological Society Special Publication 18, 181–187. Geological Society Publishing House, Bath.

29.6 Carr, A. P. and Blackley, M. W. L. (1974) Ideas on the origin and development of Chesil Beach, Dorset. Proceedings *Dorset Natural History and Archaeological Society*, **95**, 9–17.

29.7 Jennings, S., Orford, J. D., Canti, M., Devoy, R. J. N. and Straker, V. (1998) The role of relative sea-level rise and changing sediment supply on Holocene gravel barrier development: the example of Porlock, Somerset, UK. *The Holocene*, **8**(2), 165–181.

29.8 Brunsden, D. and Lee, E. M. (2004) Behaviour of coastal landslide systems: an inter-disciplinary view. *Zeitschrift fur Geomorphologie*, **134**, 1–112.

29.9 Lee, E. M. (2001) Estuaries and Coasts: Morphological adjustment and process domains. In D. Higgett and Lee E. M. (eds.) *Geomorphological processes and landscape change: Britain in the last 1000 years.* Blackwell, 147–189.

29.10 Silvester, R. and Hsu, J. R. C. (1993) *Coastal stabilisation: innovative concepts.* Prentice Hall.

30

The Coast: Hazard and Risk Assessment

Introduction

The coast is a dynamic environment. Coastal processes can present significant hazards to engineering projects and development, especially flooding and erosion (Fig. 30.1). Over the past decade there has been an increasing emphasis on *risk-based* methods of evaluating and managing the problems (i.e. considering the probability of the hazard and adverse consequences).

Problems can arise as a result of progressive loss of land (e.g. cliff recession), extreme conditions (e.g. a storm surge or tsunami) or defence failure (e.g. *overtopping* by water levels higher than the design standard or *breaching*). Risk assessment needs to take account of all possible scenarios (i.e. a combination of circumstances) that might generate adverse consequences (see Chapter 12).

Failure Scenarios

Coastal hazards often arise as a result of failure of defence systems e.g. flood embankments, seawall or a barrier beach. Developing an understanding of the potential failure modes and pathways (i.e. the way in which an initial failure develops into a hazard) is an essential aspect of risk assessment[30.1]. Failure of the Lake Pontchartrain levees in New Orleans during Hurricane Katrina (August 2005) led to flooding of about 80% of the city, causing around $75 billion in damages (see http://en.wikipedia.org/wiki/ Hurricane_Katrina).

Figure 30.2 presents a possible *failure mode* of a stabilised coastal cliff protected by a seawall at the cliff foot[30.2]. A seawall breach (e.g. as a result of high wave loadings during a storm) results in renewed landsliding on the coastal slopes and subsequent renewal of cliff top recession.

For a shingle barrier there may be several failure modes leading to breach and flooding of the backshore area[30.3] (Fig. 30.3; and Table 30.1). The failure modes and pathways can be modelled through event tree analysis (see Chapter 12).

Figure 30.1   Example of a typical coastal urban development at Gibraltar that has to incorporate coastal geomorphological processes in the infrastructure design.

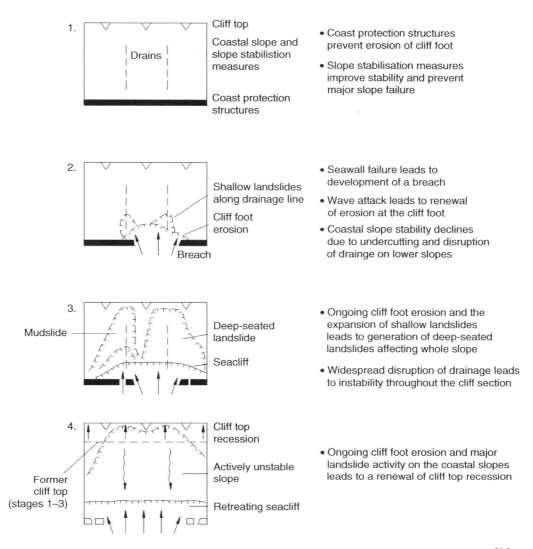

Figure 30.2   A four-stage cliff recession scenario (plan view): Whitby, UK (after Lee and Sellwood 2002)[30.2].

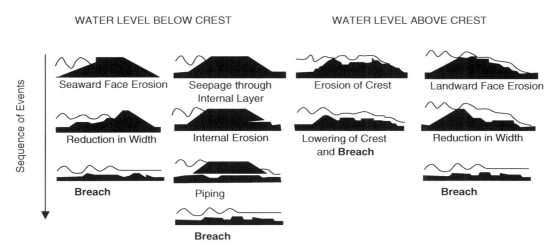

Figure 30.3   Shingle barrier failure modes (see Table 30.1; after Meadowcroft *et al.* 1996)[30.3].

System Loadings (Initiating Events)

Hazard scenarios (failure mode-pathway-outcome sequences) are often triggered by extreme conditions (e.g. wave energy or high groundwater levels – *initiating events*, see Chapter 12). For example, many coastal landslides are triggered by periods of heavy rainfall or wet year sequences. In Lyme Regis, UK, the annual probability of wet years/high groundwater levels was estimated from the historical trends of annual rainfall[30.4] (1868–1998). There have been 8 wet year sequences in 130 years (i.e. annual probability of 1 in 16 or 0.06).

Combinations of extreme loadings can also be modelled, such as the *joint probability* of wave heights and sea water level[30.5]. This approach can be used to identify a range of conditions and their likelihood under which defences can be overtopped by floodwaters. Values can be expressed as lines of equal probability (Fig. 30.4).

Estimates of the probability of defence failure can be made from field inspections of the defence condition. An alternative approach is to consider the *fragility* of the structure[30.6]. The fragility is the probability of its failure conditional on its loadings (e.g. wave and water levels). Fragility curves are plots of the conditional probability of failure over the complete range of loadings (Fig. 30.5); the overall probability of failure can be determined by integrating the fragility curve over the loading distribution (i.e. the area beneath the curve in Fig. 30.5).

Table 30.1 Possible shingle barrier failure modes, pathways and outcome (see Fig. 30.3).

Failure Mode	Failure Pathway	Outcome
Barrier erosion	Crest narrowing	Breach
Overtopping	Erosion of crest; Crest narrowing	Breach
Overtopping	Erosion of rear face; Crest narrowing	Breach
Seepage	Internal erosion; Washout of barrier material	Breach

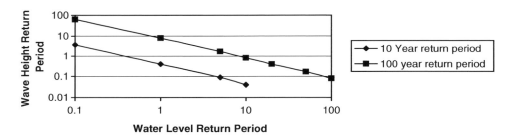

Figure 30.4 Estimates of joint probability of waves and water levels: Scarborough, UK.

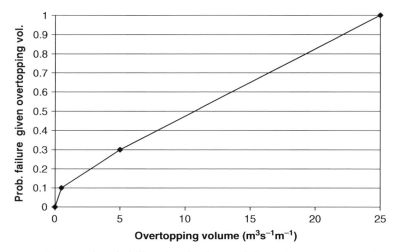

Figure 30.5 Fragility curve: estimates of probability of embankment failure given an overtopping volume.

Hazard Models

Two approaches are available for modelling the hazard and outcomes:

- *Deterministic analysis* of a system: a unique set of input parameters is entered into the hazard model, in order to obtain a unique value for the system response/outcome.
- *Probabilistic analysis*; a full range of input values is tested in the model, each one weighted by the probability of encountering it. A range of probabilistic methods can be used, including *event trees* and *simulation models*.

Simulation methods are a good way of accounting for uncertainty in system behaviour. The approach involves establishing a quantitative model of the system and proceeding through numerous analyses, each one using randomly sampled input parameters (e.g. wave height, shear strength) drawn from a probability distribution (i.e. *Monte Carlo* simulation). The higher the probability, the more likely the value will be sampled. Each analysis is essentially deterministic, but repeated simulation allows a probability distribution to be produced for the results (i.e. an output distribution). The accuracy of the output distribution generated by Monte Carlo simulation depends on the number of runs (i.e. trials) conducted.

Table 30.2 presents a frequency distribution for shingle barrier annual retreat rates. A very simple Monte Carlo simulation model of the barrier retreat over 25 years is presented in Table 30.3. After 10 trials (i.e. repeat runs of the model) the mean 25-year retreat distance is 8.9 m, with a standard deviation of 3.5 m. The more trials that are conducted (e.g. 1000) the more reliable the mean retreat distance will be.

Hazard Zones

Hazard zones can be defined in a number of ways:

- Projection of erosion rates (either as single value or a probability distribution) to establish erosion contours or zones[30.7] (Fig. 30.6). This approach can also be used on defended clifflines, recognising that future recession is conditional on defence failure and renewal of cliff foot erosion (i.e. it is a conditional probability).
- Establishing the flood extents and depths associated with particular scenarios (e.g. the 1 in 100 year). This involves the horizontal projection of water levels across low-lying areas for a given flood event. Flood area is very sensitive to water level in low-lying areas. Accurate topographic data is needed for reliable zonation at a detailed level; GIS and digital terrain models (DTM) are useful tools in flood modelling. The implications of relative sea-level change can be modelled by progressively raising the flood level across the DTM surface.

Flood discharge is usually limited by the high-tide duration i.e. flows are *time-dependent*. On defended coasts flood levels will be conditional on defence failure; this needs to be taken into account in the modelling of flows away from a breach[30.8].

Adverse Consequences

For coastal erosion, losses tend to be *irreversible* i.e. once the land has been eroded, or the cliff top is considered too close for the property to be used safely, then the land or property cannot be regained. In areas where property is affected by slow coastal landslide movement (e.g. the Portuguese Bend landslide in the Palos Verdes Hills, California[30.9]), damage may be repairable; differing magnitudes of ground movement can be associated with particular damage levels (e.g. 'unaffected' to 'write-off'), each of which can be assumed to be equivalent to specific percentages of the average property value.

Flood damage can be estimated from past events or through the use of standard flood depth/damage data. In the UK, flood depth-damage relationships are available for a range of residential and commercial property types[30.10]. Inundation by sea water can result in a 10–20% increase in damage compared with freshwater flooding (e.g. salt impregnation of floor timbers, leading to an increased tendency for dry rot; rusting and damage to central heating systems).

Risk Estimation

Risk = Probability (Event) × Adverse Consequences

- Table 27.6 presents a model that can be used to estimate the average annual damage associated with a range of flood scenarios in an area.
- Table 30.4 presents a risk model for a major 'one-off' landslide (estimated annual probability of 0.0025 i.e. 1 in 400) on a protected cliffline.

Table 30.2 Example frequency distribution for shingle barrier retreat classes.

Retreat class	Annual retreat (m)	Average retreat (m)	Relative frequency (Number of times retreat class occurs out of 25)
1	0	0	10/25
2	0–0.5	0.25	8/25
3	0.5–1.0	0.75	4/25
4	1.0–1.5	1.25	2/25
5	1.5–2.0	1.75	1/25

Table 30.3 Simple Monte Carlo simulation model: Shingle barrier retreat over 25 years.

Yr	Annual retreat (m)	Trial number and simulated annual recession distance (m)									
		1	2	3	4	5	6	7	8	9	10
1	0	0	0.75	0	0	0.25	0.25	0	0	0	0.25
2	0	0.25	0	0.75	0.75	0.25	0	0.75	0.25	0.25	1.25
3	0	1.75	0.75	0.25	0	0	0	0.25	0	0	0.25
4	0	0.75	0	0	0	0	1.25	0	0.75	0	1.75
5	0	0.75	0.25	0	0.25	0.25	0.25	0	0.25	0.75	1.25
6	0	0.75	0	0.25	0	0.75	0	0.75	0.75	0	1.25
7	0	0	0.75	0.25	0	0	0	0.25	0.75	0.25	0.25
8	0	0.25	0	0	0.25	0	0.25	0	0	0	1.25
9	0	0	0.25	0.25	0	0.75	0.25	0.25	0	0.25	0.75
10	0	0.75	0	0.25	0.25	0.25	0.75	0	0	0	1.25
11	0.25	0.25	0	0.25	1.25	0	0.25	0.75	0.25	0	0
12	0.25	0.75	1.25	0	0	0	0	0	0.25	0	0.25
13	0.25	0	1.25	1.25	0	0.75	0.75	0	0.25	1.75	0.75
14	0.25	0	0	0.25	0	0.25	0.75	0	0	0.25	0
15	0.25	1.25	0	0	0.25	0.25	0.25	1.75	0.25	0.75	1.25
16	0.25	1.25	0	0.75	0	0.25	0	0.25	0.75	1.25	0.25
17	0.25	0.75	0	0	0.25	0.75	0	1.75	0.25	0.75	0.25
18	0.25	0	0	0.75	0	0	0.25	0	0.25	0	0.75
19	0.75	0	0.25	0	0.25	0.25	0	0	0.25	0.25	0.75
20	0.75	1.75	0	0.25	0.25	0.25	0.75	0.25	0	0.25	0.25
21	0.75	0	1.75	0	1.25	0.25	0	0.25	0	0	0
22	0.75	0.25	1.75	0.25	0	1.25	0	1.25	0.25	0.25	0.25
23	1.25	0.25	0.25	0	0.25	0.25	0.75	0.25	0.25	0	0.25
24	1.25	0	0.75	0	0	0.25	0	0.25	0	0.25	1.25
25	1.75	0.75	1.75	0	0	0.25	0	0.25	0.75	0.25	0.25
Total Retreat		12.5	11.75	5.75	5.25	7.5	6.75	9.25	6.5	7.5	16

1. Column 2 represents the frequency distribution i.e. 8 times out of 25 the retreat distance is 0.25m (see Table 30.2).

2. Column 3 represents the retreat distance each year from Year 1 to 25. Each retreat value is randomly sampled from the frequency distribution in Column 2. In Excel this can be achieved by: = VLOOKUP(INT(RAND()*25+1),$A$3:$B$27,2) where $A$3 is Year 1 etc. INT(RAND()*25+1) finds an integer random number between 1 and 25 i.e. go to a randomly selected row in Column 1. VLOOKUP then selects the retreat distance from the adjacent cell in Column 2.

3. Columns 4–12 show the retreat distance each year from Year 1 to 25 for different trials/simulations, using the same formula as Column 3.

4. Total retreat = the sum of the simulated retreat distances from Year 1 to 25 in each separate trial.

Table 30.5 presents an approach to estimating the risk associated with breaching of a shingle barrier (assets at risk = £25M). The probability of the breach increases exponentially over time from 0.1 in Year 0 until Year 10 by which time failure is expected to have occurred. The change in probability P with the remaining "life" T years is represented by:

$$P = \exp(a + b \ln T)$$

As it was assumed that the barrier had a 0.99 probability of failure in Year 10 i.e. the end of its effective life, then:

$$a = \ln 0.99 = -0.01005$$

$$b = \frac{\ln P_0 + a}{\ln L} = 0.99564$$

where P_0 is the initial probability of failure (0.1) and L is the remaining barrier 'life' (a further 10 years).

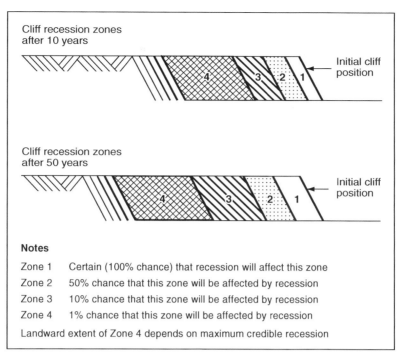

Figure 30.6   Cliff recession zones (after Lee and Clark 2002)[30.7].

Table 30.4   Annual risk calculation: Major landslide event[30.11].

Probability major landslide	Direct losses £M	Indirect losses £M	Annual risk £M
0.0025	3.75	5	0.022

Annual risk = Probability (Landslide event) × (Direct losses + Indirect losses).

Table 30.5   Shingle barrier breach risk model.

Year T	Probability (breach)	Probability that breach			
		Occurs in year T	Has not occurred by year T	Has occurred by year T	Loss due to breach (£M)
0	0.100	0.100	0.900	0.100	2.50
1	0.111	0.100	0.800	0.200	2.50
2	0.125	0.100	0.700	0.300	2.50
3	0.143	0.100	0.600	0.400	2.50
4	0.166	0.100	0.500	0.500	2.50
5	0.199	0.100	0.401	0.599	2.49
6	0.249	0.100	0.301	0.699	2.49
7	0.332	0.100	0.201	0.799	2.49
8	0.496	0.100	0.101	0.899	2.50
9	0.990	0.100	0.001	0.999	2.51
10	1.000	0.001	0.000	1.000	0.03

1. In Excel, Column 2 = EXP(a+b*(LN(L-Year T))).
2. Column 3 = Probability (Breach, Year T) × Probability (Has occurred by Year T–1).
3. Column 4 = Probability (Has not occurred, Year T–1) – Probability (Occurs, Year T).
4. Column 5 = 1 – Probability (Has not occurred, Year T).
5. Column 6 = Probability (Occurs in Year T) × 25.

References

30.1 Hall, J. W., Lee, E. M. and Meadowcroft, I. C. (2000) Risk-based assessment of coastal cliff recession. Proceedings of the ICE: *Water and Maritime Engineering*, **142**, 127–139.

30.2 Lee, E. M. and Sellwood, M. (2002) An approach to assessing risk on a protected coastal cliff: Whitby, UK. In R. G. McInnes and J. Jakeways (eds.) *Instability, Planning and Management*. Thomas Telford, 617–624.

30.3 Meadowcroft, I. C., Reeve, D. E., Allsop, N. W. H., Diment, R. P. and Cross, J. (1996) Development of new risk assessment procedures for coastal structures. In *Advances in Coastal Structures and Breakwaters '95*. Thomas Telford.

30.4 Brunsden, D. and Lee, E. M. (2004) Behaviour of coastal landslide systems: an inter-disciplinary view. *Zeitschrift fur Geomorphologie*, **134**, 1–112.

30.5 HR Wallingford and Lancaster University (2000) *The Joint Probability of Waves and Water Levels*: JOIN-SEA. Report SR 537. Wallingford.

30.6 Dawson, R. J. and Hall, J. W. (2002) Improved condition characterisation of coastal defences. Proceedings of the ICE Conference on *Coastlines, Structures and Breakwaters*, 123–134. Thomas Telford.

30.7 Lee, E. M. and Clark, A. R. (2002) *The Investigation and Management of Soft Rock Cliffs*. Thomas Telford.

30.8 Bates, P. D. and De Roo, A. P. J. (2000) A simple raster-based model for flood inundation simulation. *Journal of Hydrology*, **236**, 54–77.

30.9 Merraim, R. (1960) The Portuguese Bend landslide, Palos Verdes, California. *Journal of Geology*, **68**, 140–153.

30.10 Penning-Rowsell, E. C., Johnson, C., Tunstall, S. M., Tapsell, S. M., Morris, J., Chatterton, J. B., Coker, A. and Green, C. (2003) *The Benefits of Flood and Coastal Defence: Techniques and Data for 2003*. Middlesex University Flood Hazard Research Centre.

30.11 Lee, E. M. and Jones, D. K. C. (2004) *Landslide Risk Assessment*. Thomas Telford.

31

The Coast: Estuaries, Mudflats and Saltmarshes

Settings

Estuaries are the tidal mouths of rivers, but they do not behave as rivers[31.1]:

- The size and shape of a river is a function of the discharge (i.e. controlled by catchment size and characteristics). In an estuary the tidal discharge is controlled by the size of the channel into which the tidal water moves; changes to the channel form are reflected in changes in tidal discharge.
- Rivers tend to develop a shape and pattern which *minimises* the frictional drag of water on the bed thus optimising the transport of water and sediment to the sea. The opposite is true of estuaries which tend to *maximise* frictional effects in order that the tidal energy is fully dissipated within the channel.
- The velocity of flow in rivers is greatest during high flows (i.e. flood conditions). In estuaries maximum velocities are achieved at mid-tide when the channel is part-full (Fig. 31.1).

Most estuaries have developed over the Holocene (see Chapter 9), as the post-glacial rising seas flooded pre-existing river valleys or glacial troughs. These drowned valleys quickly began to infill with sediment carried by the rivers and from marine sources.

The interaction between tides and river flows create strong seaward currents. Wave processes can also influence estuary form, creating *barriers* at the mouth; total closure of the embayment results in the formation of *lagoons*. *Deltas* are large accumulations of sediment (both fluvial and marine) which has infilled a river mouth, forming a part-submerged structure that extends onto the continental shelf (see Chapter 32).

The location and character of *mudflats* and *saltmarshes* are controlled by:

- the availability and nature of silt and clay material (*mud*), supplied as suspended sediment from river catchments, eroding cliffs and offshore mud deposits
- the tidal regime which controls the location and extent of intertidal wetlands within an estuary (see Chapter 28)
- wind waves which determine the location of the mudflat/saltmarsh boundary (see below).

In the sub-tropics the upper sections of the intertidal mudflats are colonised by mangrove trees (e.g. the *Big Swamp* of Northern Australia[31.2]).

Forms

Typical groupings include *bar-built estuaries* (e.g. the Gulf of Mexico), *drowned river valleys* (e.g. the Thames, UK), *rias* (drowned incised valleys e.g.

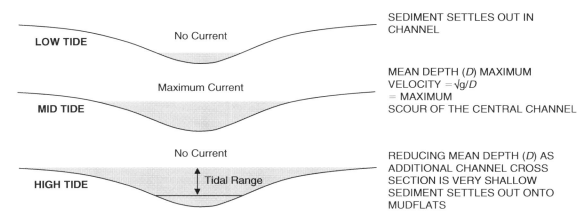

Figure 31.1 The relationship between flow velocity and tidal stage.

the Tamar in Devon, UK) and *fjords* and *fiards* (drowned glacier-scoured valleys and depressions e.g. Milford Sound, New Zealand). These types simply reflect different combinations of evolution history and environmental controls, rather than being the product of contrasting estuarine processes. The key controls on estuary form are the interaction between Holocene sea-level change, the drowned river valley (i.e. the prior topography) and the relative balance between river discharge and tidal energy.

Estuary morphology tends to maximise the frictional drag exerted by the bed and mudflats/saltmarshes (i.e. dissipation of tidal energy). The tidal energy is reduced to zero at the estuary head; width/depth ratios are generally an order of magnitude greater than for rivers with similar discharge. Over time the form is adjusted to the tidal energy and the strength of the bed/bank materials. The process of channel adjustment is determined by the critical erosion and deposition bed shear stress, itself a property of velocity and water depth. If the channel section is oversized, velocity is reduced, depth increased and bed shear falls to give deposition, so that the channel size (i.e. cross section area) decreases. If the channel is undersized, bed shears increase and erosion leads to increased cross section area. Both changes (erosion and deposition) act to restore the steady state form.

Tidal resonance models can be used to predict equilibrium (steady state) forms. The planform is a reflection of the available tidal energy:

$$C = \sqrt{gD}$$

$$L = T\sqrt{gD}$$

where C = tidal wave speed; L = wavelength; T = tidal period (44 640 seconds); D = *mean* depth and g = gravitational acceleration.

Estuary infill results in a decrease in mean depth D (average cross-section depth) and reduces the tidal wavelength L. Continued deposition occurs until the tidal wavelength L becomes around 4 times the estuary length L_e:

$$L_e \approx 0.25L$$

At this point *resonance* is established. This results in a resonant tide with a standing wave characterised by a *nodal point* at the estuary mouth (maximum velocity) and an *anti-nodal point* at the estuary head (zero velocity). The position of these points establishes the energy gradient within the estuary (from high at the mouth to low at the head, where deposition will be greatest).

The tidal resonance model can be used to predict the *rate of width decrease* inland; this is related to the tidal wave speed C:

$$W_x = W_0 \exp(-x\sqrt{gD})$$

where W_0 = width of estuary mouth, W_x = width at point x from the mouth and x = distance from the mouth.

This highlights the control exerted by mean depth D (controlled by the tidal range) on intertidal sedimentation and, hence, the distribution of mudflats and saltmarshes. There can be considerable variations from this 'ideal' form, due to the effects of controls such as tidal range, sediment supply and accommodation space.

Regime models (see Chapter 24) can also be used to predict steady state estuary morphology. The landward decrease in width results in an exponential decrease in the *tidal prism* – the tidal discharge through a given section on a single tide (i.e. the volume of water entering a section between LW and HW). The cross section A_m at any point at the mouth or within the tidal basin is related to the tidal prism P by a *regime* equation[31.3]:

$$A_m = cP^n$$

where c and n are constants. For example, for Hamford Water, UK:

Tidal prism (m^3)	c	n	Predicted cross sectional area (m^2)
17.7×10^6	6.6×10^{-5}	1	1168

This empirical relationship reflects the balance between flow velocity and bed shear stress; the cross section is adjusted to provide a velocity and bed shear stress at or immediately below a critical threshold (around 1.0 m s^{-1} for muds).

There is a continuum of estuarine settings, influenced by river, tidal and wave energy. However, three major tidal types can be distinguished[31.4]:

- *Micro-tidal*: processes are dominated by freshwater discharge, with wind-driven waves tending to produce spits and barrier islands which enclose a very shallow lagoon (e.g. Galveston Bay, Texas; Christchurch and Poole harbours, UK). Types of lagoons systems include:
  - *deficit lagoon*, where relative sea-level rise > sediment supply i.e. increasing lagoon volume (e.g. Laguna Madre, Texas)
  - *surplus lagoon*, where relative sea-level rise < sediment supply i.e. lagoon infill (e.g. Coila Lake, Australia).
- *Meso-tidal* or *wave-influenced estuaries* (Fig. 31.2): dominated by a combination of waves and tidal currents, although the limited tidal range means that tidal flow does not extend far upstream (e.g. the St Johns River estuary in northeast Florida). Tidal deltas and barriers form around the estuary mouth (Fig. 28.5). There may be progressive decrease in water area and mean depth due to sediment infill (Fig. 31.3).
- *Macro-tidal* or *tide-dominated estuaries* (Fig. 31.4): strong tidal currents may extend far inland, producing trumpet-like forms with extensive creek networks (e.g. the Hooghly estuary, India and the Severn, UK). Tidal deltas are generally absent, replaced by long linear sand bars parallel with the tidal flow. In the tropics, macro-tidal estuaries may

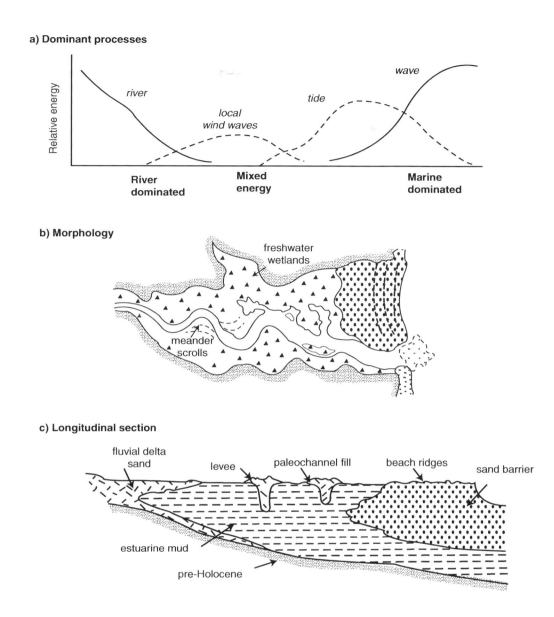

a) Dominant processes

b) Morphology

c) Longitudinal section

Figure 31.2   Features of wave dominated estuaries (after Woodroffe 2002)[31.4].

be dominated by river flows during the monsoon season e.g. northern Australia[31.6].

Three distinct zones are often present in macrotidal estuaries:

- *upstream* river dominated zone: relatively straight channel, with net seaward movement of sediment
- *central* mixed energy zone: sinuous channel, zone of sediment convergence influenced by marine and river processes
- *seaward* marine dominated zone: relatively straight channel, funnel shaped; net landward sediment movement.

Mudflats typically comprise a gently sloping (around 1 in 1000) upper profile with a marked break of slope at about mid tide level, below which the surface steepens towards the low water mark (Fig. 31.5). The lower

sections of the intertidal flats are often developed in sandy muds and sands i.e. sediments become coarser away from the shoreline.

Saltmarshes occur high in the intertidal zone and are largely covered by halophytic vegetation. They generally develop below the level of the highest astronomical tide, where wind wave attenuation by the mudflat provides low energy conditions suitable for vegetation growth. They often have a convexo-concave profile, comprising a convex seaward margin, a flat central section and a concave landward edge. A small cliff may be present at the mudflat/saltmarsh boundary. Saltmarshes have a distinctive creek system, with branching and meandering channels. Drainage densities are typically very high. Vertical accretion of saltmarshes is often able to keep pace with rates of sea-level rise of up to 10 mm yr^{-1}, provided that an adequate sediment supply is maintained.

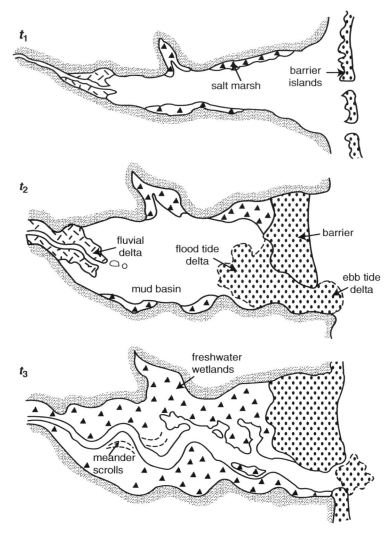

Figure 31.3 Long-term evolution of a barrier estuary, through transgression (t_1) sea-level stabilisation (t_2) and stillstand conditions (t_3) (after Roy 1984)[31.5].

Behaviour

Throughout the Holocene, river-mouth embayments have continued to be infilled by river and marine sediments. Some embayments, such as the Shoalhaven River, Australia, have almost completely infilled[31.5]; others appear to have developed equilibrium forms that reflect a balance between the sediment availability and the wave/tide/river energy.

Stages in the Holocene infill of the South Alligator River, Northern Territory, Australia were[31.6]:

1. 8000–6800 BP: inundation of a river valley and deposition of marine muds
2. 6800–5300 BP: extensive mangrove growth on the river plain (the *Big Swamp* phase)
3. 5300 BP to present: alluvial floodplain development, with abandonment and infill of palaeochannels, meander development and cut-off.

The transition between stages 2 and 3 was due to a depletion of marine sediment sources.

Over shorter timescales (e.g. the 50–100 year 'engineering time'), many estuaries oscillate between two forms[31.1]:

1. *Flood dominance*: initially, estuaries would have been relatively deep and wide allowing the crest of the tidal wave to travel faster than the trough which progresses slower in the shallower water:

$$C_{crest} = \sqrt{g(D+0.5H)}$$

$$C_{trough} = \sqrt{g(D-0.5H)}$$

where C = wave speed; D = mean depth and H = wave height.

This leads to an asymmetry in the tidal wave, with a steep flood limb of short duration (i.e. higher velocity), while the ebb duration increases (i.e. lower velocity). The estuary becomes a *sediment sink* with rapid deposition rates on the intertidal areas, which gradually increase in elevation. For example, the estuarine plain formed at the mouth of the South Alligator River, Northern Australia is dominated by marine sediments[31.6].

2. *Ebb dominance*: as intertidal deposition proceeds over time, the channel cross section defines an

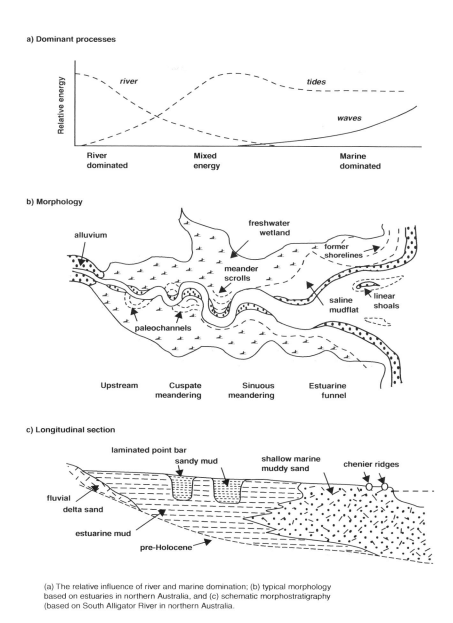

a) Dominant processes

b) Morphology

c) Longitudinal section

(a) The relative influence of river and marine domination; (b) typical morphology based on estuaries in northern Australia, and (c) schematic morphostratigraphy (based on South Alligator River in northern Australia.

Figure 31.4 Features of tide dominated estuaries (after Woodroffe *et al.* 1989)[31.6] (a) The relative influence of river and marine domination; (b) typical morphology based on estuaries in northern Australia, and (c) schematic morphostratigraphy (based on South Alligator River in northern Australia).

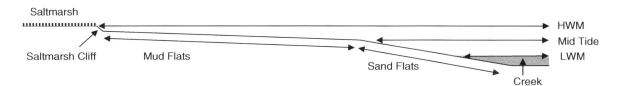

Figure 31.5 Typical saltmarsh and mudflat morphology.

estuary in which the mean water depth D is greater at low tide than at high tide (the high tide cross section is dominated by shallow water over intertidal flats, whilst the low tide cross section is confined to the deeper sub-tidal channel). The asymmetry of the tidal wave is reversed, with the passage of the tidal wave crest slower than the trough. An ebb dominant estuary acts as a *sediment source* and is characterised by erosion of the intertidal areas.

Erosion of the *ebb dominant* estuary leads to an increase in mean depth, gradually returning the channel to a *flood dominant* form. These oscillations involve phases of saltmarsh/mudflat erosion during periods of ebb dominance, with accretion occurring during flood dominance (i.e. cyclic change; see Chapter 5). Erosion does not necessarily imply long-term degradation. Changes in *mean* channel depth and intertidal area can result in morphological adjustments within the estuary:

- If mean depth D increases (e.g. as a result of sea-level rise, dredging of a navigable channel or land reclamation – which removes intertidal areas and, hence, effectively increases mean depth) the tidal wavelength L will increase causing the nodal point to migrate seaward $L = T\sqrt{gD}$. This reduces the slope of the energy gradient between the mouth and the head and, hence, velocities at any given point in the channel, allowing increased intertidal deposition and restoration of the regime form.

- Changes in tidal prism, often associated with land reclamation (i.e. decrease in intertidal area) lead to shrinkage of ebb tide deltas and partial closure of the inlet by spits, bars and dune systems (Fig. 31.6).

Mudflats and saltmarshes form an integrated system, similar to the beach-dune system, in that they respond to wind waves by adjusting their form (Fig. 31.7). However, erosion thresholds are higher for consolidated fine sediments than for sands and so mudflats and saltmarshes are only sensitive to the larger, higher energy wind waves.

High energy storms cause wind wave erosion of the upper mudflat surface and recession of the saltmarsh cliff edge, allowing the mudflat profile to extend landward. The erosion releases large volumes of sediment which is deposited on the lower mudflat surface. The result is a longer, flatter profile which increases wave attenuation. Recovery occurs during calmer periods with rapid deposition on the upper mudflat surface

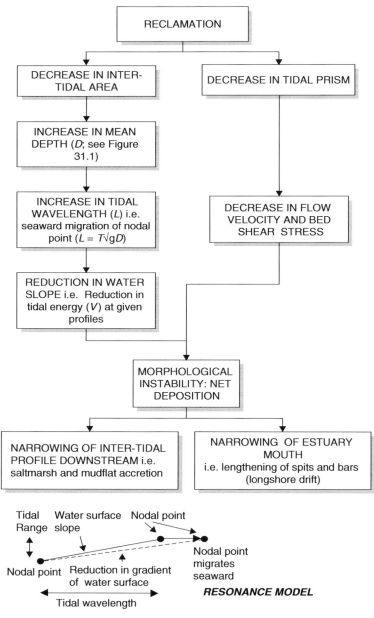

Figure 31.6   The effects of land reclamation on inter-tidal wetlands.

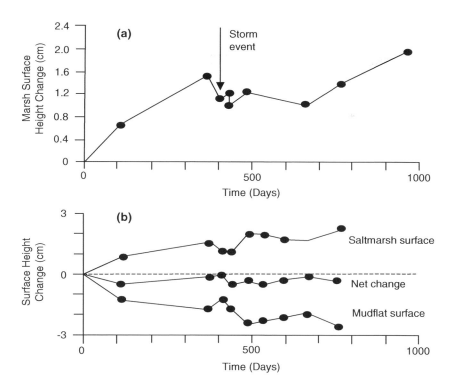

(a) Response of an Essex saltmarsh surface to a storm event i.e. erosion and recovery
(b) the close association between vertical changes to a mudflat and saltmarsh surface, Essex

Figure 31.7   Mudflat and saltmarsh behaviour (after Pethick 1994)[31.1]: (a) response of an accreting Essex saltmarsh surface to a storm event i.e. erosion and recovery and (b) the close association between vertical changes to a mudflat and saltmarsh surface, Essex, UK.

and saltmarsh accretion (note that the marsh cliff is an erosional/deposition feature that can advance forward as well as retreat).

Materials

The Holocene sediment infill created extensive alluvial plains which are traversed and re-worked by the contemporary channels. There can be extremely complex sequences of materials and a variety of problem ground conditions, including:

- buried channels at different depths and different dimensions (can be critical for tunnels, piles and bridge piers etc.)
- rapid vertical and horizontal changes in sediment types, the general stratification and whether the deposits are *under, normally* or even *over-consolidated* (Chapter 7)
- varying piezometric conditions between adjacent layers associated with variable permeabilities of the different materials, especially problematic for caissons
- metastable sediments subject to liquefaction (see Chapter 19)
- irregularities in the rockhead profile (again critical for tunnels, but will also have an important effect on end-bearing pile and pier design).

Hazards

Estuaries provide natural harbours and transport routes and are the focus of considerable port activity. Sedimentation can cause major problems for many ports, with dredging required to keep the port facilities open:

- *Maintenance dredging*: routine dredging to keep the navigable channel open
- *Capital dredging*: dredging of new channels or berths, deepening or changing existing channels to ensure that the port can be competitive by handling larger vessels. In 2004, the Port of Brisbane, Australia, removed about 100 000 m³ of material from the main harbour (A\$400 000) restoring navigation depths for large cargo vessels.

The effect of dredging is dependent on the relative size of the channel to its intertidal area. Where a deep navigation channel is dredged in the bed of a formerly wide shallow natural estuary, the effect can be to produce *ebb asymmetry* (i.e. mean depth is greatest at low water), causing a loss of sediment from the intertidal zone. In a narrow confined channel the effect of dredging can be to produce *flood asymmetry* setting up a movement of sediment into the deepened channel from marine sources. In the Thames estuary, UK:

- Dredging in the *outer* estuary has resulted in a deep narrow sub-tidal channel, with no effect on the

Table 31.1 Effects of various freshwater inputs upon tidal levels in the Humber Estuary, UK (m).

Location	Tidal state	Freshwater input			
		No input	285 m³ s⁻¹	570 m³ s⁻¹	855 m³ s⁻¹
Keadby	HW	+ 3.81m	+ 4.12	+ 4.21	+ 4.25
	LW	– 0.64	+ 0.18	+ 0.70	+ 1.25
Owston Ferry	HW	+ 3.69	+ 4.45	+ 4.63	+ 4.75
	LW	+ 0.18	+ 1.83	+ 2.71	+ 3.48
Gainsborough	HW	+ 3.38	+ 4.82	+ 5.09	+ 5.73
	LW	+ 0.5	+ 2.98	+ 4.55	+ 5.67

intertidal features. This has resulted in ebb dominance and a self maintaining channel.

- Dredging in the *inner* estuary has affected both the sub-tidal and intertidal zone, increasing the mean depth. Rapid deposition occurs in the flood dominant channel, requiring constant maintenance dredging to keep the navigable channel open.

Estuarine floods occur when the seaward flow of freshwater is impeded by the rising tide. Particular problems can arise with the higher spring tides. On the Humber in the UK, for example, the freshwater river input can have a significant effect on high water (Table 31.1), although the importance declines towards the estuary mouth. Flood embankments can confine the channel and cause a reduction in the width/depth ratio which in turn leads to an increase in the tidal amplitude and thus flood risk. High tides on the River Thames, UK, have risen by around 4 m over the last 150 years, following the construction of embankments and land settlement[31.8].

Management

Any changes to estuary morphology, either natural or man-induced, are transmitted throughout the system and particularly affect the intertidal zone. Management requires an understanding of the whole estuary system, not a site-by-site approach.

- Upstream engineering projects such as river diversion, dredging and dam construction may result in changes in the behaviour of estuaries, because they cause changes in river flows and sediment loads.
- Foundation conditions may be extremely variable over short distances as a result of the complex dynamics of the system and the depositional history.
- Relatively small changes (e.g. construction of piers or retaining walls, dredging of new channels) may result in dramatic large-scale changes in estuarine dynamics, with consequences that may not be entirely predictable.

Estuaries are often of major national and international importance for migrant and wintering waterfowl and breeding birds. Many have been designated or otherwise identified under a variety of national and international measures, including Ramsar sites (designated wetlands of international significance) and Special Areas of Conservation (SAC). Nature conservation will form a key component of most estuary management plans.

Mudflats and saltmarshes need to be managed as an integrated system. An understanding of the broader estuary dynamics is central to effective management of intertidal wetlands:

- Mudflats can only respond to high energy storms by eroding landwards into the saltmarsh. Where the marsh is absent (e.g. because of reclamation) mudflats can suffer severe erosion during storms and will take longer to recover; some mudflats will suffer long-term degradation.
- Flood defences confine the estuary channel and prevent the landward migration of saltmarshes as relative sea-level rises. This results in the loss of saltmarsh habitats (i.e. '*coastal squeeze*') and makes the mudflats more susceptible to long-term erosion.
- Intertidal wetlands are effective in dissipating wave energy and form important parts of flood defence schemes. For example, where defences in the Gwent Levels, south Wales, are fronted by healthy saltmarsh they comprise an earth embankment with crest heights around 8.8–9.5 m above Ordnance datum (AOD); where there is no marsh, the defence levels are at 10.5 m AOD, involving a near vertical wall capped by a wave return wall and fronted by an 8 m high rock slope[30.9].
- Mudflat erosion does not necessarily imply long-term degradation. It may be in response to high-energy waves and will be followed by a period of profile recovery. It may reflect longer-term morphological changes within the estuary e.g. the oscillation between flood and ebb dominance.
- Saltmarshes need to be fronted by efficient wave attenuating surfaces that provide the low energy conditions needed for vegetation growth. Saltmarshes cannot be recreated if adequate mudflats are not present.
- Mudflats and saltmarshes influence the dynamics of the whole estuary (i.e. the tidal wavelength and tidal prism). Changes in their distribution through managed retreat or land reclamation

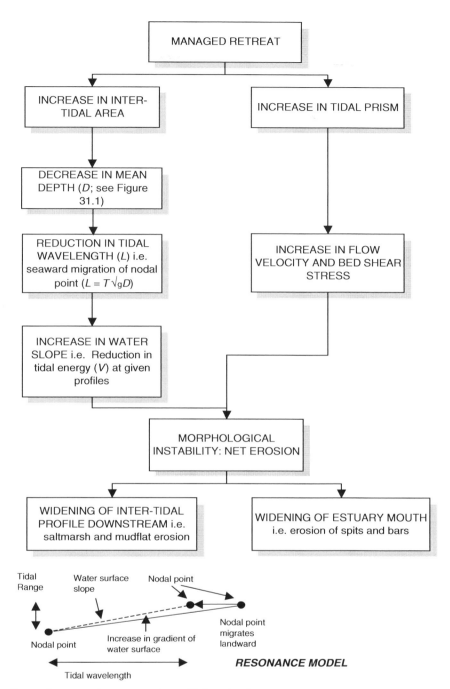

Figure 31.8 The effects of managed retreat on intertidal wetlands.

can have significant short-term impacts on the intertidal wetlands elsewhere within an estuary (Fig. 31.8).

• It has proved difficult to develop reliable models to predict the short to medium term response to changing environmental controls. *Top-down models* (e.g. the tidal resonance and regime models[31.10]) ignore processes and concentrate on defining a steady-state form. Bottom-up *hydrodynamic models* that integrate estuarine processes over a limited number of tidal cycles have been largely unsuccessful in predicting estuary wide morphological change. Combined, *hybrid models* are in development and may prove successful in overcoming the problems[31.11].

References

31.1 Pethick, J. S. (1994) Estuaries and wetlands: form and function. In R. A. Falconer and P. Goodwin (eds.) *Wetland Management*, Thomas Telford, 75–87.

31.2 Crowley, G. M. (1996) Late Quaternary mangrove distribution in Northern Australia. *Australian Systematic Review*, **9**, 219–225.

31.3 O'Brien, M. P. (1972) Equilibrium Flow Areas of Inlets on Sandy Coasts. Proceedings of the *Thirteenth Coastal Engineering Conference*, July 10–14, Vancouver, BC, Canada, American Society of Civil Engineers, New York, NY, Vol II, pp 761–780.

31.4 Woodroffe, C. D. (2002) *Coasts: Form, Process and Evolution*. Cambridge University Press.

31.5 Roy, P. S. (1984) New South Wales estuaries: their origin and evolution. In R. G. Thom (ed.) *Coastal Geomorphology in Australia*. Academic Press, Sydney.

31.6 Woodroffe, C. D., Chappell, J., Thom, R. G. and Wallensky, E. (1989) Depositional model of a macrotidal estuary and floodplain, South Alligator River, Northern Australia. *Sedimentology*, **36**, 737–756.

31.7 Chappel, J. and Woodroffe, C. D. (1994) Macrotidal estuaries. In R. W. G. Carter and C. D. Woodroffe (eds.) *Coastal Evolution: Late Quaternary Shoreline Morphodynamics*. Cambridge University Press, 187–218.

31.8 Horner, R. W. (1979) The Thames Barrier project. *Geographical Journal*, **154**, 242–253.

31.9 Brampton, A. (1992) Engineering significance of British saltmarshes. In J. R. L. Allen and K. Pye (eds.) *Saltmarshes: Morphodynamics, Conservation and Engineering Significance*, Cambridge University Press, 115–122.

31.10 Pethick, J. and Lowe, J. (2000) Regime models in estuarine research. In EMPHASYS Consortium, *A Guide to Prediction of Morphological Change within Estuarine Systems*, HR Wallingford, TR114 (Available from: http://www.estuary-guide.net/pdfs/emphasys% 20guide.pdf).

31.11 Whitehouse, R. J. S. (2001) Predicting estuary morphology and process: An assessment of tools used to support estuary management. In M. L. Spaulding (ed.) *Estuarine and Coastal Modeling*, 2001, 344–363. ASCE.

32

The Coast: Deltas

Settings

Deltas are large accumulations of sediment (both fluvial and marine) which have infilled a river mouth, forming a part-submerged structure that extends onto the continental shelf. Deltas build seaward where:

- The supply of sediment delivered by the river at its mouth exceeds the amount that is reworked and removed by the waves and tidal currents.
- The supply of sediment to the sub-aerial portion of the delta more than compensates for any subsidence-related crustal down-warping and/or compaction of previously-deposited deltaic sediments.

Deltas are generally associated with large, high sediment yield catchments, low nearshore gradients and relatively low wave and tidal energy settings (Table 32.1). Wave action restricts the offshore movement of river sediment, creating a *littoral energy fence* i.e. sediment is retained within the nearshore zone. Significant offshore transport requires:

- *River mouth bypassing*: rivers in flood jet sediment out to sea due to the extremely high discharge (hyperpycnal flows).
- *Estuary mouth bypassing*: strong ebb tide flows carry sediment offshore.
- *Shoreface bypassing*: storm wave conditions create strong offshore currents. On very exposed coasts the fluvial sediment supply is dispersed by wave action i.e. sand and gravels accumulate on adjacent beaches, and muds are carried away in suspension.

Deltas are the product of the post-glacial sea-level rise; worldwide delta expansion occurred in the mid Holocene, coinciding with the deceleration in the rate of sea-level rise (*stillstand*; see Chapter 10).

Forms

Deltas comprise two main components: a *sub-aerial deltaic plain* and a *subaqueous delta* (Fig. 32.1).

The *sub-aerial deltaic plain* (fluvial sediments) contains a suite of low-lying plains and terraces:

- The *upper deltaic plain*, above the tidal influence (i.e. no active mangroves or saltmarshes). It can include both active and abandoned channels, mean-

der belts, lakes, backswamps, and marshes (e.g. the Danube delta). The features result from migrations or abandonment of both main and distributary channels, overbank flooding, and periodic breaks (*crevasses*) in the levees that flank the channels. Sediments are deposited in very low-lying areas between the levees of active or abandoned channels (*interdistributary basins*).

- The *lower deltaic plain*, which is exposed to both river and tidal/wave processes. There is a dense network of bifurcating and/or anastomosing channels with levees. Marshes, swamps, and lakes occur in the interdistributary basins (e.g. the lower Mississippi Delta; Fig. 32.2).
- The *marginal deltaic plain*, on the flanks of the lower deltaic plain. The plain is dominated by complex patterns of saltmarshes and active or abandoned wave-created beach ridges. Beach ridges can form seaward of the marsh edge, creating semi-landlocked salt bays and lagoons. Occasionally the rapid growth of a *delta lobe* will almost completely cut off a large area adjacent to the lower deltaic plain, thus creating extensive brackish-water lakes (e.g. Lake Pontchartrain in the Mississippi delta).

The *subaqueous delta* is that portion of the delta that lies below sea level and comprises:

- The *delta front* (fluvial and marine sediments), a gentle slope (<0.5° for the Mississippi and Rhone deltas) extending from the seaward margin of the sub-aerial delta into deep water. In general, the sediments tend to be coarse-grained near the mouths of the *distributaries*, becoming finer with greater distance from the shore. *Channel mouth bars* form as localized arcuate seabed structures just offshore of the mouths of both the main channel as well as the distributaries. In general they consist of ridges of the most coarse-grained sediments transported by the streams.
- The *prodelta area* (marine sediments), an almost flat surface extending beyond the delta front, which receives only the very finest grain sizes (e.g. fine silt and flocculated clays).

Table 32.1 Major river deltas.

Drainage basin	Area (10³ km²)	Mean annual discharge (m³ s⁻¹)	Sediment load (10⁶ t yr⁻¹)	Wave Energy	Tidal range
Amazon	6300	200 000	900	Moderate	Macro
Ganges-Brahmaputra-Megha	1650	30 000	1670	Low (surges)	Macro/meso
Mississippi	3300	15 600	349	Very low	Micro
Nile (Pre-Aswan Dam)	2700	1480	111	Moderate	Micro
Yangtse	1800	28 500	478	Moderate	Macro/meso
Yellow River	860	1550	1080	Moderate	Micro
Zaire	3800	1250	43	Moderate	Micro

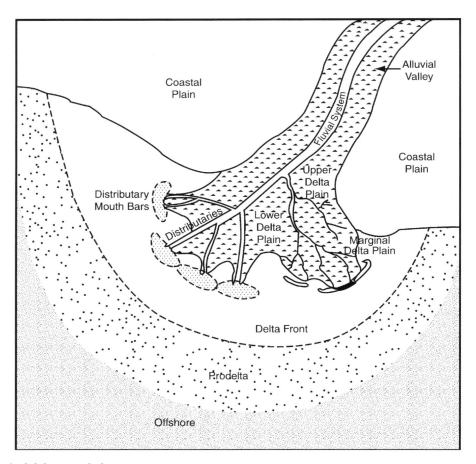

Figure 32.1 Typical delta morphology.

Sediment supply and wave energy for reworking the deposited material are important controls on delta planform (Fig. 32.3):

1. *River-dominated*: *digitate* (birdsfoot) deltas with a branching outline (e.g. the Mississippi delta; see http://gulfsci.usgs.gov/missriv/), associated with low wave energy, tidal dominance and abundant fluvial sediment supply.
2. *Wave-dominated*:
   - *Cuspate* with smoother outlines (e.g. the Danube delta), associated with higher wave energy. Wave action shapes the sediment into a series of cuspate salients.
   - *Lobate* (e.g. the Niger delta) where coasts are exposed to high energy ocean waves.
   - *Blunted* (e.g. the Sao Francisco delta, Brazil) where wave energy is very high; sediment is rapidly transported onto adjacent beaches. The delta grows as a series of prograding beach ridges.
3. *Tide-dominated*: distributary channels with linear river-mouth bars (e.g. the Ganges-Brahmaputra-Megha delta, India and Bangladesh).

Delta planform can change over time; the Mississippi delta changed from lobate to digitate as it extended to the edge of the continental shelf. Since construction of the Aswan dams in the 20th Century, the fluvial

Figure 32.2   The lower deltaic plain of the River Mississippi. ASTER Path 21, Row 40 Centre 29.45N -89.28W 24th May 2001. Image courtesy of the Image Science and Analysis Laboratory, NASA Johnson Space Centre (http://eol.jsc.nasa.gov/) (width of image circa 50 km).

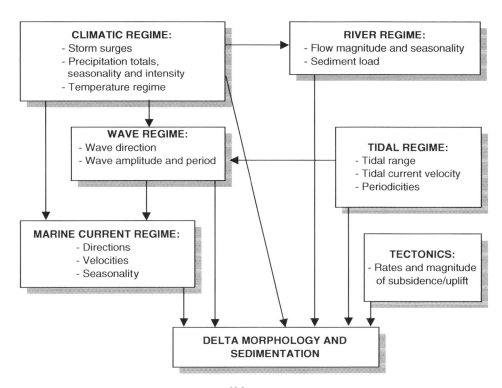

Figure 32.3   Controls on delta form (after Wright 1985)[32.2].

sediment supply to the Nile delta has been greatly reduced; the delta has entered an abandonment phase and is effectively now a wave-dominated coastal plain[32.1].

Delta sedimentation is largely controlled by the sediment load and the relative densities of the river flows and the waters of the basin into which they discharge[32.2] (Table 32.2, Fig. 32.4).

Table 32.2   River mouth flows and sedimentation.

Flow type	Relative densities	Dominant forces	Deposition
Homopycnal (inertia)	Inflow and basin waters of similar density (e.g. a lake)	Inertia of the water flows into the delta	Coarse sediment is deposited radially near the mouth as bars (e.g. Lake Geneva, Switzerland)
Hyperpycnal (friction)	Inflow denser than basin water i.e. very high sediment load	Friction between inflow and sea bed	Sediment laden freshwater flows beneath the saline water and disperse down the delta front as turbidity flows or as high-density channelised lows extending beyond the distributary mouths (e.g. the Huanghe River, China).
Hypopycnal (buoyancy)	Inflow less dense than basin water	Buoyant flows	Turbulent mixing results in coarse sediment deposited near the river mouth and finer sediments transported further offshore (e.g. Balize complex, Mississippi delta)

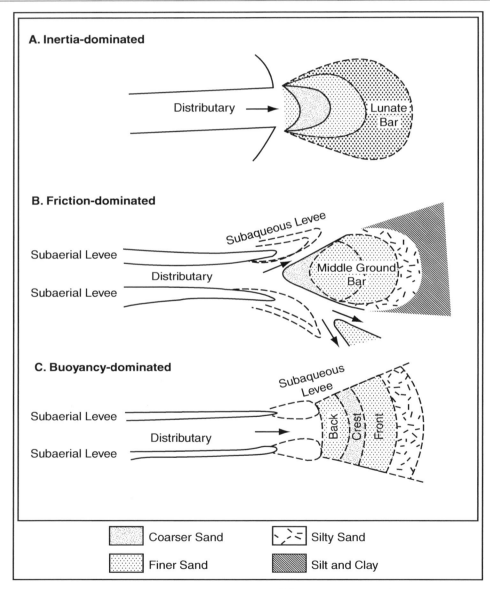

Figure 32.4   Delta mouth processes and forms (after Wright 1985)[32.2].

Behaviour

Long-term delta evolution is controlled by the basin topography (i.e. *accommodation space*), relative sea-level change and sediment availability. Over the course of glacial-interglacial cycles, delta evolution in the Gulf of Mexico has involved[32.3]:

- Sea-level fall: rapid progradation across the previously submerged continental shelf, creating thin, stacked sediments (*shelf phase deltas*). Fluvial systems incise into the delta sediments.

- Sea-level lowstands: the delta front is at or near the shelf margin i.e. increased accommodation space and seafloor gradients. *Shelf-margin deltas* are up to 150 m thick and often have hydrocarbons. Submarine mass movement is common.
- Sea-level rise (*transgressive systems*): the delta retrogresses into the incised valleys which infill with fluvial or estuarine deposits depending on the dominant sediment source. Once the valleys have infilled, a series of shelf phase deltas develop during stillstands; the each successive delta is thinner and further landward (reflecting a diminishing sediment supply as well as the relative sea-level change).

Deltas can be very sensitive to changes in sediment supply over engineering timescales:

- *Increased supply* can lead to rapid progradation e.g. growth of the Jaba delta, Bougainville in Papua New Guinea since 1972 which has been associated with the input of copper mine tailings into the river (26 Mtonnes yr^{-1}).
- *Decreased supply* can result in rapid erosion e.g. construction of the Aswan High Dam on the Nile in 1964 has been followed by rapid delta erosion – up to 120 m yr^{-1} near the Rosetta mouth[32.1].

Deltas can be prone to sudden channel switching (*avulsion*). As the delta front moves seaward, the efficiency of the channels is reduced (the length increases and gradient decreases). In time, the river will switch to a shorter, steeper distributary channel, and the original channel will be abandoned. A new *delta lobe* forms at the point where the new main channel reaches the sea. Over the past 7000 years the Mississippi has switched seven times. The last switch occurred around 600–800 years ago and

led to the formation of the modern Balize delta[32.4–5] (Fig. 32.5). The abandoned delta lobe can be prone to rapid erosion e.g. the Huanghe River, China.

Materials

Delta infill deposits can be extremely variable, reflecting the Holocene infill and the dynamic nature of contemporary processes.

Key geotechnical issues include:

- Delta deposits consist of highly porous and very highly permeable and variable materials which are extremely prone to shrinkage upon draining or drying. They have complex soil behaviour patterns that require detailed investigation prior to engineering design and construction.
- Delta sedimentary structures are partially water-supported, being in the process of active compaction; this results in continuous surface subsidence over engineering timescales. Former delta lobes tend to become submerged.
- Under- and normally-consolidated clays may be buried following the development of a drying crust. The crust effectively forms a thin over-consolidated layer (see Chapter 7).

Hazards

Delta hazards include:

- *Landslides*: the subaqueous delta slope deposits are often rich in interstitial methane gas and can have extremely high pore fluid pressures. As a result, slopes as low as 0.5° are susceptible to sub-marine landsliding, often occurring as debris flows, mudflows, mudslides and rotational failures[32.6–7] (see Chapter 19). In the Mississippi delta, landslides have damaged oil and gas pipelines, wellheads and drilling platforms e.g. mudslides generated during

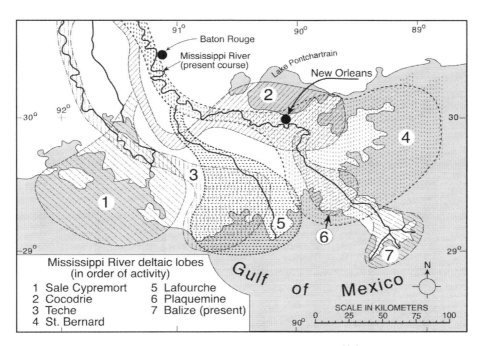

Figure 32.5 Mississippi delta channel changes (after Kolb and Van Lopik 1966)[32.5].

the 1969 Hurricane Camille resulted in the loss of a platform in South Pass Block 70[32.8].

- *Floods*: the low-lying delta plains are often densely populated and are susceptible to both river and tidal flooding. In April 1991, the combination of a massive cyclone and high tides in the Bay of Bengal resulted in a 7 m high tidal surge that flooded large parts of Bangladesh (see http://www.imd.ernet.in/section/nhac/static/). There were an estimated 140 000 deaths; many millions lost their homes and all of their possessions. Flooding can also occur as a result of tsunamis (Chapter 28).

- *Channel switching*: major river control structures above Baton Rouge are needed to prevent avulsion between the Lower Mississippi and the hydraulically more efficient Atchafalaya River, 200 km to the west. If avulsion occurred the consequences would be devastating for the region e.g. New Orleans and the major industrial complexes of the delta area would lose their freshwater supply[32.9].

- *Diapers*: local *diapirism* around the channel mouths can produce extensive and relatively rapid ground motions, including both uplift and horizontal displacements which can rapidly destroy buildings, wharfs, and pipelines e.g. the *'mudlumps'* of the Mississippi delta and the Huanghe River, China[32.10-11]. Diapirism results from the deformation and extrusion of soft, fine-grained prodeltaic sediments when the overlying sediments are removed by rapid erosion.

Management

Deltas are very sensitive to changes occurring within the river catchment and along the adjacent open coast (coastal cell). Management requires an understanding of the interactions between the river and the coastal processes; this can be very site specific.

Key issues include:

- The combination of sea-level rise and active subsidence may lead to net erosion of the delta. However, the impact may be offset by increased sediment yield from the river catchment (e.g. rivers in Java are predicted to increase their sediment yield by 43% over the next century, due to deforestation and climate change).

- Flood defence structures (e.g. the levees on the Mississippi) can prevent sediment supply to the delta plain, leading to accelerated subsidence; abandoned lobes on the Mississippi plain have been subsiding at around 2 cm yr^{-1} [32.12].

- Deltas are often internationally important wetland habitats and fishery nurseries (e.g. Ramsar sites and Special Areas of Conservation; see http://www.ramsar.org/ or http://www.jncc.gov.uk/). Nature conservation will form a key component of delta management plans.

References

32.1 Stanley, D. J. and Warne, A. G. (1993) Nile delta: recent geological evolution and human impact. *Science*, **260**, 628–634.

32.2 Wright, L. D. (1985) River deltas. In R. A. Davies (ed.) *Coastal Sedimentary Environments*. Springer Verlag, New York.

32.3 Suter, J. R. and Berryhill, H. L. (1985) Late Quaternary shelf-margin deltas, northwest Gulf of Mexico. *American Association of Petroleum Geologists Bulletin*, **69**, 77–91.

32.4 Coleman, J. M. (1988) Dynamic changes and processes in the Mississippi River delta. *Geological Society of America Bulletin*, **100**, 999–1015.

32.5 Kolb, C. R. and Van Lopik, J. R. (1966) Depositional environments of the Mississippi River deltaic plain, south eastern Louisiana. In M. L. Shirley and J. A. Ragsdale (eds.) *Deltas*, Houston Geological Society, Houston, Texas, I, 7–62.

32.6 Maslin, M. A., Mikkelsen, N., Vilela, C. and Haq, B. (1998) Sea-level- and gas-hydrate-controlled catastrophic sediment failures of the Amazon Fan: *Geology*, **26**, 1107–1110.

32.7 Reeder, M. S., Rothwell, R. G. and Stow, D. A. V. (2000) Influence of sea level and basin physiography on emplacement of the late Pleistocene Herodotus Basin megaturbidite, southeast Mediterranean Sea. *Marine and Petroleum Geology*, **17**, 199–218.

32.8 Sterling, G. H. and Strohbeck, G. E. (1975) The failure of the South Pass 70B platform in Hurricane Camille. *Journal of Petroleum Technology*, 263–268.

32.9 Penland, S. and Suter, J. R. (1988) The geomorphology of the Mississippi River Chenier Plain. *Marine Geology*, **90**, 231–258.

32.10 Morgan; J. P., James, M., Coleman, J. M. and Gagliano, S. M. (1963) *Mudlumps at the mouth of South Pass, Mississippi River; sedimentology, paleontology, structure, origin, and relation to deltaic processes*. Baton Rouge, Louisiana State University Press, 1963.

32.11 Li, G., Zhuang, K. and Wei, H. (2000) Sedimentation in the Yellow River delta. Part III. Seabed erosion and diapirism in the abandoned subaqueous delta lobe. *Marine Geology*, **168**, 129–144.

32.12 Penland, S., Boyd, R. L. and Suter, J. R. (1988) Transgressive depositional systems of the Mississippi delta plain: a model for barrier shoreline and shelf sand development. *Journal of Sedimentary Petrology*, **58**, 932–949.

33

The Coast: Fringing Beaches

Settings

Fringing beaches are spreads of sand or gravel (*shingle*) attached to the shoreline (*barrier beaches* are detached from the shoreline; see Chapter 34). They are *lag deposits* (i.e. too coarse to be removed by wave action) associated with:

- high storm-wave energy environments
- relatively steep offshore slopes.

Fringing beaches can occupy most shoreline energy settings, with the exceptions of plunging cliffs and rocky shores (see Chapter 36) or estuaries (unless there is a local source of coarse sediment). Their long-term survival in high-energy environments is due to the ability of the materials to be continuously re-shaped into forms that either reflect or dissipate wave energy.

Many fringing beaches rolled onshore in response to sea-level rise during the Holocene (i.e. post-glacial). Thus, some are almost relict features, receiving little or no sediment inputs from contemporary sources (e.g. Pensarn Beach, north Wales; see Fig. 29.7).

Forms

A typical beach profile is shown in Fig. 33.1; plan-forms can range from drift to swash-aligned (see Chapter 29). Profiles comprise four zones (not all four may be present on a single beach): the *shoaling wave zone*, the *breaker zone*, the *surf zone* and a *swash zone*. The zones migrate up and down the profile through the tide cycle and with changes in wave height.

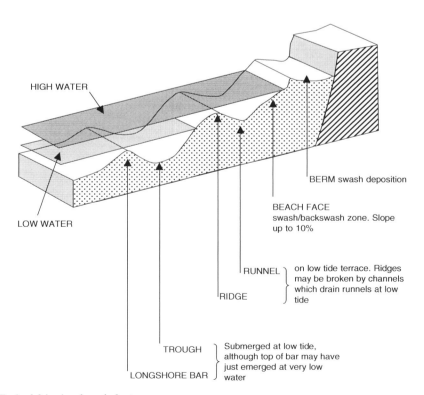

Figure 33.1 Typical fringing beach features.

213

Beach form is a reflection of wave energy and sediment type[33.1–2]; energy can either be reflected by a steep shingle beach with no bars (*reflective beaches*) or dissipated by a low-angled sand beach with a wide surf zone and multiple bars (*dissipative beaches*). Intermediate forms combine elements of both reflective and dissipative beaches and include (Fig. 33.2): *longshore bar-troughs*, *rhythmic bars*, *transverse* (*welded*) *bars* and *low-tide terraces*. This diversity of forms is related to the variety of ways in which sediment is recycled from the beach to the nearshore in storms and then returned in calmer periods.

Sediment size controls the rate of percolation of the up-rushing swash into the beach materials:

- On a shingle beach, swash water is rapidly lost into the beach from the surface flow. As uprush power > backwash power, sediment is pushed up the beach and tends not to be brought back, leading to the build up of a steep ridge.
- On a sandy beach, percolation rates are lower and the uprush and backwash volumes are similar. However, in storm conditions the backwash tends to be more powerful because it is running downslope i.e. there can be upper beach erosion and sediment accumulation on offshore bars.

Crest height of the beach is controlled by the vertical extent of swash run-up during extreme storm events.

Fringing beaches often show distinctive separation (*partition*) of the different sediment sizes, in response to backwash sorting. The more mobile sand fraction tends to occur as a broad foreshore, whereas the shingle forms a beach ridge at high tide.

Tidal range influences the intertidal bedforms. On macro-tidal beaches, many forms are 'washed out' resulting in a smooth profile. As tidal range decreases, so more distinct *roughs* and *bar troughs* develop. Micro-tidal beaches are characterised by wide (40–100 m), low amplitude (<0.5 m), parallel and sub-parallel bars, referred to as *ridge and runnel topography*.

Behaviour

There can be considerable variation in the sand beach profile (Fig. 33.3):

- *Storm waves* (steeper spilling waves and high returning velocities) tend to erode sandy sediment from the beach face and transport it offshore, where it is deposited on longshore bars (destructive waves). This results in a flat, wide profile.
- Calmer weather *swell waves* (i.e. less steep, surging swash) rebuild the beach (constructive waves), creating step-like berms and leading to a steeper profile.

The duration of these cyclic *cut and fill* changes reflects the frequency of extreme waves and the generation of *infragravity* and *edge waves* (see Chapter 28). Often the cycle can be regarded as a variation between winter and summer beach profiles. The cross sectional difference between the storm wave profile and the swell wave profile is the *sweep zone*. Over time, a persistent average profile can develop; this is known as the *equilibrium*

profile[33.3] and is a useful concept for beach management.

Nearshore bars act as breakwaters and can dissipate a large proportion of the incident wave energy. As a result the bulk of longshore transport occurs along bars. Some bars may also move onshore, which occurs as part of the calm weather profile recovery process.

During storms, shingle beaches can be subject to:

- *Overtopping* by run-up: this pushes sediment up the beach and results in an increase in crest height.
- *Overwashing* of the crest: this can result in the lowering of the crest height as sediment is 'washed over' from the crest to the back beach area. This leads to the landward 'roll-over' of the beach ridge.

The frequency of storms of different intensity and their sequence control the beach condition. Low points created by one event can then become the focus of future erosion, unless a sufficient period of repair has occurred during calmer weather.

Fringing beaches are sensitive to changes in the sediment supply. In *sediment deficit* conditions (e.g. due to a decline in sediment supply or a disruption of the transport pathway), beach material tends to accumulate at the down drift boundary and the updrift sections decline (i.e. cannibalisation; see Chapter 29). There can be rapid progradation under *sediment surplus*. For example, on the Fife coast, Scotland, excess coal mine waste was deposited on the shoreline in the early 20th century[33.4]. By the 1950s the waste had been redistributed along the shoreline, forming aprons of material 100–150 m wide and causing Buckhaven harbour to silt up.

Hazards

Changing beach levels can result in variations in erosion and flood hazards in the backshore area:

- Where beaches front coastal cliffs, their decline can lead to accelerated recession (see Chapter 36).
- Beaches can form important components of flood defence schemes, either alone or where they front the defence line. Beach decline can lead to an increased potential for overtopping and may lead to undermining of the engineering structures.

Management

Beaches need to be managed within the context of coastal cells and based on an appreciation of the sediment budget[33.5] (see Chapter 29). Most intervention is in response to sediment deficit. Key points include:

- Beach erosion can be part of the cyclic response to sequences of storms events – short-term profile lowering is not necessarily an indication of long-term decline.
- The removal of beach material for construction can have a significant impact on the beach profile, especially if there is already a sediment deficit (e.g. Hallsands, UK, see Chapter 14).
- Engineering intervention to offset the effects of sediment scarcity may only reduce the rate of beach decline and can accelerate erosion elsewhere

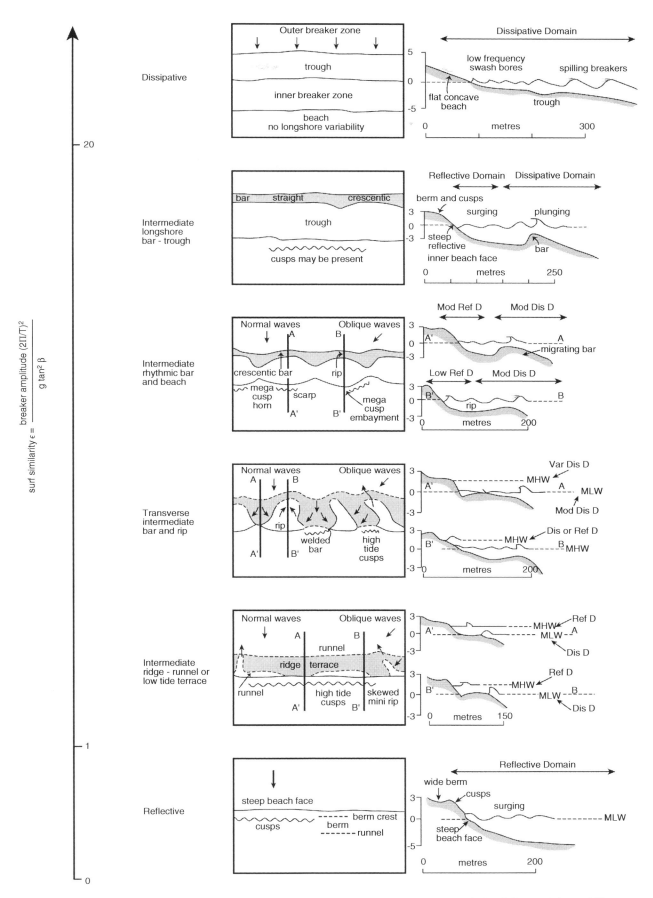

surf similarity $\epsilon = \dfrac{\text{breaker amplitude } (2\Pi/T)^2}{g \tan^2 \beta}$

Figure 33.2   Variation of beach morphology in plan and cross-section, based on surf similarity (after Wright and Short 1984)[33.1] where Ref D is Reflective Domain, Dis D is Dissipative Domain, Var variable and Mod moderate.

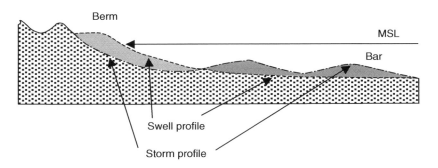

Figure 33.3 Beach profile changes under swell and storm conditions.

(Fig. 33.4). For example, groynes restrict longshore sediment transport and cause *downdrift starvation* (breakwaters can prevent all longshore transport; Fig. 33.5), but do not prevent offshore losses.

- Beach recharge (*renourishment*) by dumping suitably sized material on the foreshore must be seen as part of a regular management strategy, rather than a one-off exercise (e.g. 13 Mm^3 was placed on Miami Beach between 1976 and 1981 at a cost of $60 M). The cost of maintaining nourished US beaches during the decade to 2010 is estimated to

be $3.3 M to $17.5 M per mile (1.6 km)[33.6]. The most effective schemes are those that recharge an entire coastal cell.

- Raising the beach crest by bulldozing the profile can reduce the likelihood of storm overwashing. However, it often results in a decline in the beach face percolation rate, as the natural structure is disrupted. Sediment is drawn down the beach even under constructive swell waves and channels can be cut in the face that act as routes for overwashing and overtopping flows.

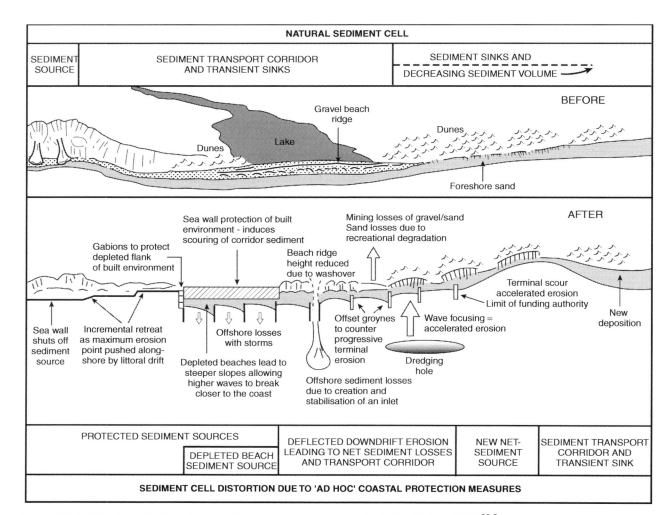

Figure 33.4 The impact of coast protection works on sediment cells (after Orford 2005)[33.5].

Figure 33.5   The effect of a breakwater on the movement of sediment towards the top of the picture along an area of the south coast of England.

References

33.1   Wright, L. D. and Short, A. D. (1984) Morphological variability of surf zones and beaches: a synthesis. *Marine Geology*, **56**, 1–18.

33.2   Short, A. D. and Wright, L. D. (1984) Morphodynamics of high energy beaches: an Australian perspective. In B. G. Thom (ed.) *Coastal Geomorphology in Australia*, Academic Press, Sydney, 43–68.

33.3   Dean, R. G. (1991) Equilibrium beach profiles: characteristics and applications. *Journal of Coastal Research*, **7**, 53–84.

33.4   Saiu, E. M. and McManus, J. (1998) Impacts of coal mining on coastal stability in Fife. In J. M. Hooke (ed.) *Coastal Defence and Earth Science Conservation*. Geological Society Publishing, 58–66.

33.5   Orford, J. (2005) Coastal environments. In P. G. Fookes, E. M. Lee and G. Milligan (eds.) *Geomorphology for Engineers*. Whittles Publishing, 576–602.

33.6   The H. John Heinz III Center for Science, Economics and the Environment (2002) *Human Links to Coastal Disasters*. Washington DC.

34

The Coast: Barrier Beaches

Settings

Barrier beaches are freestanding, linear sand or gravel (*shingle*) structures that tend to run parallel to the shoreline; they are often topped by back beach dunes (see Chapter 35). They may extend from several hundred metres to over 100 km in length. The barrier encloses the *backbarrier*, comprising tidal flats, lagoons and saltmarsh systems (Fig. 34.1). They are associated with:

- generally moderate to low wave energy environments, but exposed to episodic very high energy events (e.g. hurricanes and tropical storms)
- micro- to meso-tidal conditions (i.e. tidal range <4 m)
- high sediment availability during the development phase

- low-angled offshore slopes; many barriers coincide with offshore glacial deposits (e.g. the Wadden Sea barriers).

In general, wave energy controls the rate of sediment supply by longshore drift (i.e. the rate of barrier extension). Tidal energy controls the persistence of the tidal inlets which separate individual segments of the barrier.

The main barrier settings are:

1. Sand barriers with foredune ridges, formed on mid-latitude coasts such as the *barrier chains* of the eastern (3100 km long) and Gulf (1600 km long) coasts of the USA (e.g. the South Carolina coast, Galveston Bay).
2. Shingle barriers formed on high-latitude *paraglacial* coasts, such as the British Isles (e.g. Chesil Beach, Slapton Sands) and eastern Canada (e.g. Story Head barrier, Nova Scotia).

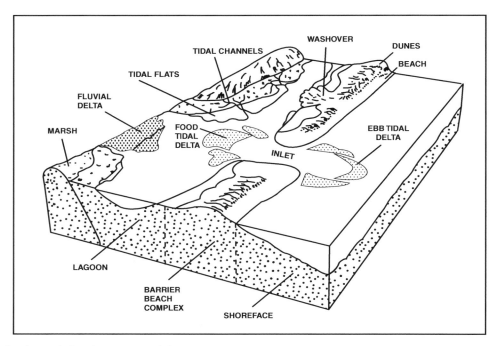

Figure 34.1 A schematic barrier coast model.

Forms

Although there is great diversity of barrier forms, 3 main types can be distinguished:

1. *Spits* attached to the mainland at one end and terminating in open water: these coincide with abrupt changes in shoreline orientation. Spits form because of the energy gradient between the open coast and a section of shoreline where wave refraction and divergence results in lower wave energy (e.g. an embayment). There is a marked reduction in the longshore wave power and sediment transport rates at the start of the bay mouth (Fig. 34.2; see Chapter 29). Sediment is deposited at this point (i.e. the rate of supply is greater than the rate of removal), initiating spit growth. Spit extension continues across the deeper water until a point is reached where sediment supply is less than the rate of erosion at the distal end: a function of wave and tidal energy (e.g. Farewell Spit, New Zealand extends 30 km from a bedrock headland).

Common spit forms include:

- *Recurved spits*, with a curved ridge at the distal spit end (e.g. Spurn Head at the mouth of the Humber, UK). The recurves are formed as nearshore bars migrate onshore and are wrapped around the spit end by refracted waves. Each ridge may represent a separate period of accelerated sediment delivery.
- *Flying spits* extend into deep water at an acute angle to the beach and often comprise a series of curved ridges (e.g. Presque Island, Lake Erie, Pennsylvania, USA, which extends around 2 km offshore).
- *Cuspate forelands* projecting out from the shoreline usually comprise a series of sand or shingle ridges, enclosing lagoons or backswamps (e.g. Dungeness, UK). They are associated with the convergence of sediment transport pathways (i.e. cell boundaries, see Chapter 29), estuary mouth tidal deltas or wave refraction zones behind offshore rocky outcrops.

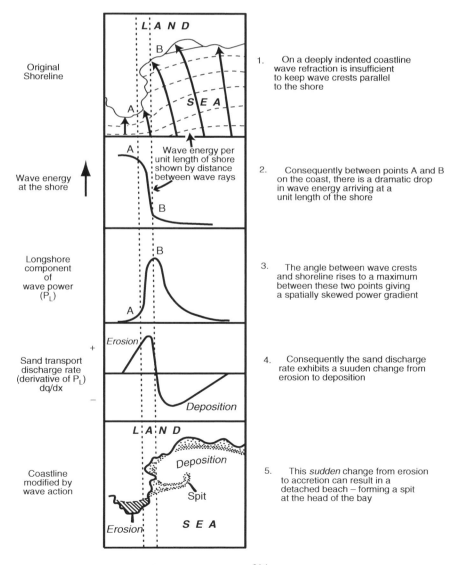

Figure 34.2   Simple spit development model (after Pethick 1984)[34.1].

- *Tombolos* provide a sediment bridge from the mainland to an offshore island (e.g. the Orbetello tombolo, on the Tyrrhenian Sea, Italy). Spits extend from the shoreline because of the wave shadow behind the island. Tombolos form where the ratio of the island's offshore distance *J* to its longshore length *L*≤1.5 ([34.2]).

2. *Welded barriers* are attached to the mainland at both ends enclosing a lagoon (e.g. Martha's Vineyard, USA) or wetlands (e.g. Slapton Sands, UK). They are common on micro-tidal coasts, where there is insufficient tidal energy to maintain a tidal inlet. Many have rolled onshore in response to rising sea-levels.

3. *Barrier islands* often occur in chains parallel to the shoreline, extending over 100 km with individual islands separated by tidal inlets (e.g. Padre Island, Texas which is over 200 km long). Inlets are the openings in barriers through which water and sediments are exchanged between the open sea and the protected embayments (lagoons) behind the barriers. The channel generally contains both ebb and flood tide delta structures (see Chapter 31); the inlets can migrate along the shore or remain fixed, dependent on the interactions between tidal prism, wave energy and sediment supply[34.3].

Barrier forms vary with tidal range (Fig. 34.3):

- Micro-tidal barriers (i.e. *wave dominated barriers*) tend to be long, narrow and continuous or with a limited number of tidal inlets (e.g. the Outer Banks of North Carolina, Maryland and New

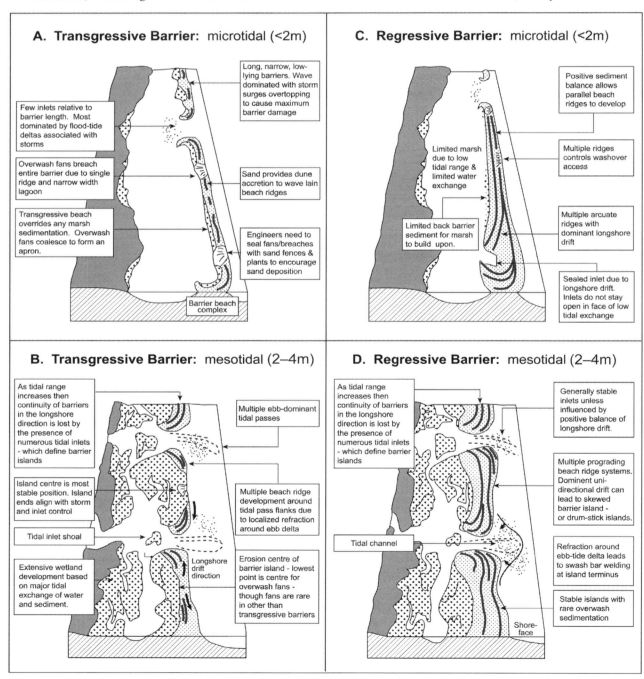

Figure 34.3   Barrier island models (after Leatherman 1982)[34.4].

Jersey, USA). The inlets often develop as washover sites during severe storms and then enlarge by the outflow of storm waters trapped in the lagoon. Longshore drift can block the inlets where there is limited tidal flow to keep the inlet open, or force the opening to migrate along the barrier.

- Meso-tidal barriers (i.e. *mixed energy barriers*) tend to be segmented by frequent tidal inlets (e.g. the East Friesian islands, Germany). Tidal flows generate ebb and flood tide deltas around the inlets (see Chapter 31). The ebb delta acts as a breakwater and provides protection to parts of the island close to the inlet; changes in ebb-delta size may result in changes in the pattern of erosion and accretion along the island frontage. Multiple beach ridges occur at island longshore ends; often the sediment is derived by erosion of the island front (i.e. cannibalisation; see Chapter 29).

Backbarrier deposition is dependent on the presence of tidal inlets:

- Impeded drainage of freshwater occurs behind *continuous barriers*, forming lagoons infilled with terrestrial deposits such as peats (e.g. Slapton Sands and Slapton Ley, UK).
- *Discontinuous barriers* allow free inflow of saltwater into the back barrier area. Sediments are mixed marine and terrestrial, including overwash deposits, marine muds and sands, and peats (e.g. Chesil Beach and The Fleet, UK).

Behaviour

The beach face profile adjusts to variations in wave energy by flattening during storms and steepening during the subsequent calm periods (i.e. the response is similar to the cyclic behaviour of fringing beaches; see Chapter 33).

Sediment availability is an important control on barrier behaviour (see Fig. 29.7):

- *Positive sediment budget* results in the barrier extending seaward (*progradation*) through the development of a series of beach ridges and dune systems (i.e. *regressive behaviour*; see Chapter 10). Each ridge builds up during a period of sediment surplus and is a former shoreline.
- *Negative sediment budget* results in erosion and cannibalisation of the beach face, increased potential for overwashing (i.e. transfer of sediment from the face to the back beach) and landward migration of the barrier (i.e. roll-over-*transgressive behaviour*; see Chapter 10).

Barrier island migration is dependent on inlet characteristics; instability of inlets results in inlet migration or closure. A simple measure of inlet stability is[34.5] (Table 34.1):

$$\text{Stability Rating} = \frac{P}{M}$$

where P is the tidal prism and M is the total annual littoral drift.

Table 34.1 Inlet stability ratings[34.5].

P/M ratio	Comment
>150	Conditions are relatively good, minor bar present, good channel scour (flushing)
100–150	Conditions become less satisfactory and offshore bar formation becomes more pronounced
50–100	Entrance bar may be rather large, but there is usually a channel through the bar
20–50	Longshore sediment typically bypasses the inlet due to wave action/tidal flows <20 Entrances become unstable; overflow channels rather than permanent inlets

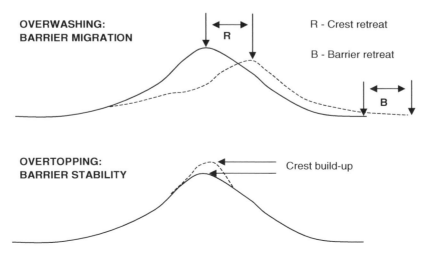

Figure 34.4 Simple shingle beach overwashing and overtopping models.

Gravel barrier retreat is controlled by the relative significance of *overtopping* and *overwashing* events (Fig. 34.4):

- Relatively low magnitude, overtopping surge transports gravel up the beach, leading to crest height increase;
- High magnitude overwashing surge carries gravel over the crest and down the backslope where it is deposited as a series of fans. Overwashing leads to barrier retreat (roll-over).

The balance of these two processes controls the frequency and magnitude of roll-over events. If the crest height becomes low relative to wave run-up, then the rate of overtopping will increase and the beach crest will rise, reducing the potential for further overwashing.

Beach behaviour involves brief episodes of overwashing and erosion, followed by longer periods of recovery during which the damage caused by storm events is 'healed'. The extent to which the beach can recover from a particular storm is conditional on the precise timing and sequence of subsequent storm events. Areas of damage (i.e. crest lowering) can be the focus of future overwashing events and, ultimately, become vulnerable to breach events.

Shingle barriers tend to respond to sea-level rise by roll-over and by increasing the crest height[34.6–7]. The barrier *retreat efficiency* (the change in retreat rate per unit increase in the rate of sea-level rise) is related to the barrier size or *barrier inertia* (i.e. cross sectional area × crest height; Fig. 34.5).

Sediment scarcity can lead to the re-orientation of the beach, from drift-aligned to swash-aligned (see Chapter 29). For example, the Porlock shingle barrier in Somerset, UK, responded to a reduction in longshore sediment supply around 4000 BP by changing from a continuous, drift-aligned structure in front of a freshwater lagoon to an unstable swash-aligned barrier prone to overwashing and breaching[34.8].

Hazards

Mid-latitude sandy barriers are popular tourist locations and often support high value property along the shorefront. The exposure to hurricanes and severe tropical storms presents a major hazard from the combined effects of storm surges and shoreline erosion. For example, in 1900 over 6000 people living on a barrier beach were killed when a hurricane crossed Galveston Island, Texas, USA (see http://www.1900storm.com/).

Many high-latitude shingle barriers provide flood defence to low-lying areas and wetlands behind. For example, Chesil Beach, UK, protects the village of Chesilton and the former Portland naval base; flooding can cause extensive damage and sever the causeway link between the mainland and the Isle of Portland[34.9] (see http://www.swgfl.org.uk/jurassic/).

In many sites, the wetlands (both fresh and brackish) behind the barrier are of international conservation importance (e.g. Slapton Sands, UK; see http://www.slnnr.org.uk/). As the barriers migrate, the wetlands area is reduced and the habitats are increasingly affected by saline water.

Management

An awareness of coastal cell behaviour, particularly the sediment budget, is essential to barrier beach management (see Chapter 29) together with a reliable system of storm warning:

- The *removal of beach material* for construction can have a significant long-term impact on the beach profile, especially if there is already a sediment deficit. For example, gravel was extracted from Chesil Beach, UK, for at least 700 years, with an estimated 1.1 M tonnes removed between the mid-1930's to 1977. These losses could represent between 1.5–5% of the estimated beach volume. The beach is a relict feature and receives no additional shingle inputs[34.10–11].
- *Tourist development* of barrier islands can lead to increasing modification of the shoreline morphology, including disruption of sediment transport and land reclamation (Fig. 34.6). Many of these changes will increase the potential for erosion and flooding during storm events.

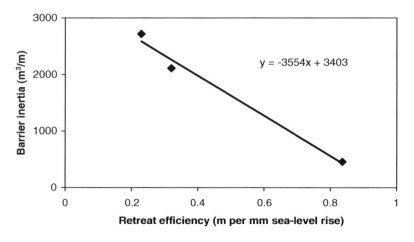

Figure 34.5   Shingle barrier retreat efficiency (after Orford *et al.* 1995)[34.7].

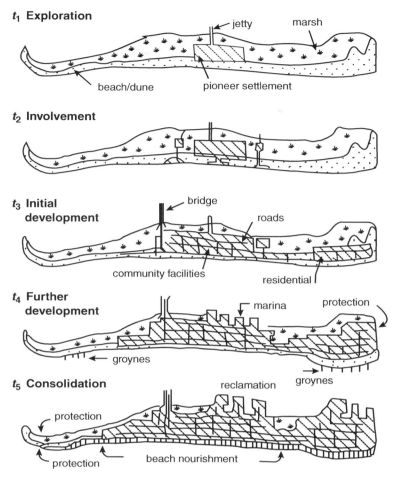

t_1 **Exploration**

jetty   marsh

beach/dune   pioneer settlement

t_2 **Involvement**

t_3 **Initial development**

bridge

roads

community facilities   residential

t_4 **Further development**

marina   protection

groynes

t_5 **Consolidation**

reclamation   groynes

protection

protection   beach nourishment

Figure 34.6   Impacts of barrier island resort development (after Woodroffe 2002)[34.12].

- *Shoreline protection measures* (e.g. rock revetments, seawalls, groynes) can be effective, but may accelerate erosion problems elsewhere along the coast. Localised protection work may result in the division of the barrier into separate sub-cells.

- *Beach recharge* is often only a temporary solution because of the high sediment transport rates along barrier frontages. Experience in the USA suggests that few recharge projects are effective for longer than 10 years, with nearly 50% ineffective within 5 years. It is not a one-off solution, as regular re-supply will be needed.

- *Storm surge protection* can be provided by increasing the barrier crest height by promoting dune growth (e.g. the use of dune fences, vegetation seeding and fixing), reducing recreation pressures on the dunes (e.g. restricting access, the use of boardwalks to prevent trampling) or re-profiling the beach face by bulldozing. However, a single high ridge is more vulnerable to wave attack than a series of ridges (see Chapter 33).

- *Damage and losses* can be reduced by the use of hurricane proofing measures (e.g. building on stilts, smooth aerodynamic designs with no eaves or balconies, brace or strengthen joists and beams with 'hurricane' clips).

- *Impact of severe storms* can be reduced by the development of contingency and evacuation plans, and raising public awareness of the dangers associated with barrier beaches.

References

34.1   Pethick, J. S. (1984) *An Introduction to Coastal Geomorphology*. Edward Arnold, London.

34.2   Sunamura, T. and Mizuno, O. (1987) A study on depositional shoreline forms behind an island. Annual Report, Institute of Geoscience, University of Tsukuba, **13**, 71–73.

34.3   US Army Corps of Engineers (2002) *Coastal Engineering Manual*. Manual No. 1110–2–1100. Washington DC.

34.4   Leatherman, S. P. (1982) *Barrier Island Handbook*. National Park Service, Boston.

34.5   Bruun, P., Mehta, A. J. and Jonsson, I. G. (1978) *Stability of Tidal Inlets: Theory and Engineering*. Elsevier Scientific Publishing Co., Amsterdam, Netherlands.

34.6   Orford, J. D., Carter, R. W. G., Jennings, S. C. and Hinton, A. C. (1995) Processes and timescales by which a coastal gravel-dominated barrier responds geomorphologically to sea-level rise: Story Head barrier, Nova Scotia. *Earth Surface Processes and Landforms*, **20**, 21–37.

34.7   Orford, J. D., Carter, R. W. G., McKenna, J. and Jennings, S. C. (1995) The relationship between the rate of mesoscale sea-level rise and the retreat rate of swash-aligned gravel-dominated coastal barriers. *Marine Geology*, **124**, 177–186.

34.8   Jennings, S., Orford, J. D., Canti, M., Devoy, R. J. N. and Straker, V. (1998) The role of relative sea-level rise and changing sediment supply on Holocene gravel barrier development: the example of Porlock, Somerset, UK. *The Holocene*, **8**(2), 165–181.

34.9   Lee, E. M. (1994) *The Occurrence and Significance of Erosion, Deposition and Flooding in Great Britain*. HMSO, London.

34.10   Carr, A. P. and Blackley, M. W. L. (1974) Ideas on the origin and development of Chesil Beach, Dorset. *Proceedings Dorset Natural History and Archaeological Society*, **95**, 9–17.

34.11   Lee, E. M. (2001) Estuaries and Coasts: Morphological adjustment and process domains. In D. Higgett and E. M. Lee (eds.) *Geomorphological processes and landscape change: Britain in the last 1000 years*. Blackwell, 147–189.

34.12   Woodroffe, C. D. (2002) *Coasts: Form, Process and Evolution*. Cambridge University Press.

35

The Coast: Dunes

Settings

Coastal dunes are associated with the combination of strong onshore winds, relatively low rainfall, an abundant source of sand-sized material and vegetation to trap the blown sand (Fig. 35.1). Tidal range is important—a high range exposes a large intertidal area which may dry out between tides and increase the potential for *saltation* (see Chapter 18). Dunes tend to be associated with dissipative beaches, although reflective beach systems do occur (see Chapter 34). Common settings include:

- windward margins of oceanic basins (e.g. California, Western Scotland)
- areas adjacent to extensive glacial deposits (e.g. the Netherlands, Arctic Canada)
- areas adjacent to rivers delivering large volumes of sandy material to the coastal zone (e.g. the Oregon and Washington rivers, USA)
- mouths of estuaries forming parts of the tidal delta complex (e.g. Dawlish Warren at the mouth of the River Exe, UK).

Most dune systems originated in the late-glacial to mid Holocene, when sea-level was lower and/or when sediment supply from the sea floor was at its peak (see Chapter 29). In NW Europe, dunes began forming in the Holocene around 5000–6000 BP[35.1]; in northern Australia dune building began around 24 000–17 000 BP and continued in phases until the mid-Holocene[35.2]. Many dune systems have experienced episodes of accretion/erosion separated by phases of stability marked by soil formation.

Forms

Foredunes occur around the MHWS line (see Table 28.4), where grasses (e.g. marram grass, sand spinifex, lyme grass) have begun to colonise incipient (*embryo*) dunes. On prograding (advancing) coasts a series of foredune ridges (the most recent is termed the *first dune*) develop with interdune hollows, the precursors of slacks. The time interval between ridges may be 70–200+ years.

Slacks are damp/wet hollows between dune ridges, formed by deflation (wind velocities are higher in the interdune valleys) down to the capillary fringe/water table.

Blowouts are unvegetated or partly vegetated hollows excavated by onshore winds. They can transform shore parallel ridges into a sequence of *parabolic dunes* (U-shaped dunes with their arms pointed into the dominant wind; Fig. 35.2). Blowouts are an example of positive feedback; after an initial foredune breach the wind is able to exploit the absence of vegetation and cause

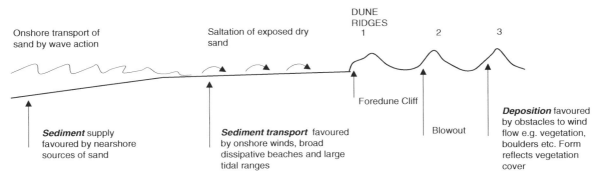

Figure 35.1 Coastal dune settings.

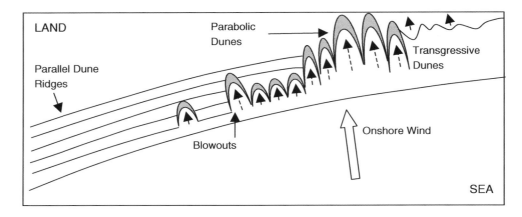

Figure 35.2 Development of blowouts into parabolic and transgressive dunes.

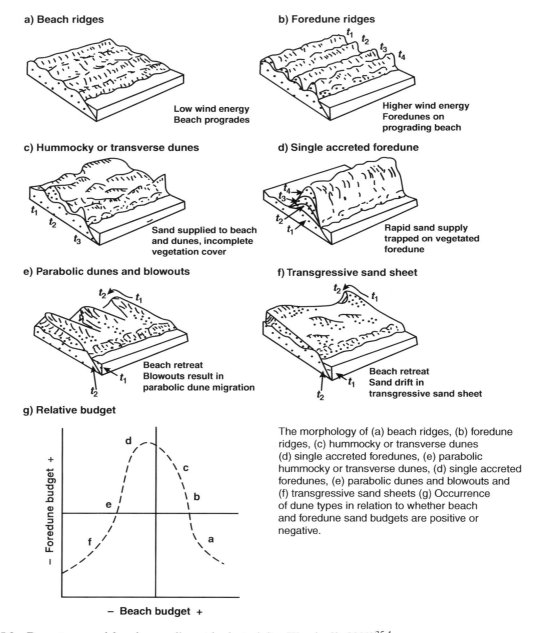

The morphology of (a) beach ridges, (b) foredune ridges, (c) hummocky or transverse dunes (d) single accreted foredunes, (e) parabolic hummocky or transverse dunes, (d) single accreted foredunes, (e) parabolic dunes and blowouts and (f) transgressive sand sheets (g) Occurrence of dune types in relation to whether beach and foredune sand budgets are positive or negative.

Figure 35.3 Dune types and foredune sediment budgets (after Woodroffe 2002)[35.4].

rapid expansion of the eroded area. Blowouts are associated with:

- change to more arid climate (i.e. vegetation decline)
- increase in wind intensity and frequency (i.e. increased storminess)
- vegetation destruction (e.g. grazing, rabbit burrowing, intense recreation use).

Climbing dunes can form where sandy beaches are backed by cliffs; these may extend upwards to form *cliff-top dunes* (relict cliff top dunes are disconnected from the original sand source).

The sediment budget and vegetation are the key controls on dune form[35.3] (Fig. 35.3):

- *Positive sediment budgets*: as wind energy increases, the dune forms change from beach ridges (insufficient wind to form dunes) to foredune ridges (a single vertically accreting foredune is associated with very dense vegetation cover) and hummocky dunes (ineffective vegetation cover).
- *Negative sediment budgets* are associated with parabolic and transgressive (i.e. moving inland) dunes.

Behaviour

Dunes are *stores* of beach sand, with material moved onto the beach or offshore during major storms. They undergo repeated phases of foredune erosion (*cliffing*) during major storms, followed by renewed accretion as sand is blown back from the beach (cyclic behaviour; see Chapter 5). Foredune erosion does not necessarily imply a retreating coastline – recovery is dependent on the sediment budget:

- positive sediment budget ⇒ foredune ridges build up between major storms (prograding coastline)
- negative sediment budget ⇒ coastal retreat during storms not compensated by subsequent accretion.

Dunes can be very sensitive to changes in storm frequency:

- increased storm frequency ⇒ dunes may not recover from the previous erosion event (net retreat)
- decreased storm frequency ⇒ foredunes build up (net progradation).

Hazards

- Severe erosion and washover (flooding) of beach front and back-shore communities can occur during storms (Fig. 35.4). For example, The Great New England Hurricane of 1938 caused over 700 deaths including 312 in the Rhode Island and the Providence Area, US; the surge tide level at Providence was 5 m above mean high water (see http://www2.sunysuffolk.edu/mandias/38hurricane/).
- Landward migration of dunes (*transgressive dunes*) can also cause problems. For example, the 1694 Culbin Sands disaster when 16 farms on the south shore of the Moray Firth, UK, were overwhelmed in a single violent storm[35.5]. The whole area was buried by up to 30m of loose sand.

Dune Management

Dunes can form part of the natural defences against coastal flooding, by forming a barrier of high ground in front of low-lying areas.

A dynamic beach-dune system is better than a static one. The following practises should be adopted:

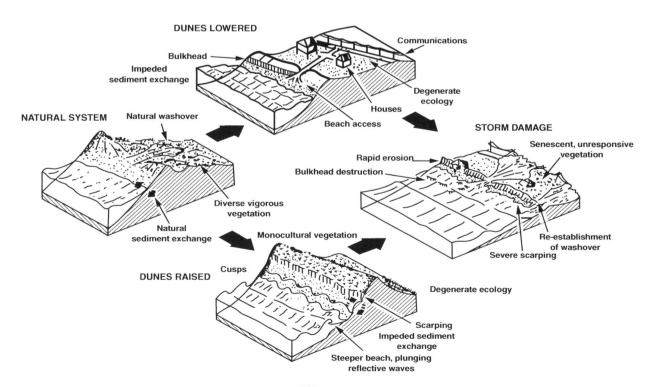

Figure 35.4 Dune barrier washover (after Carter 1988)[35.1].

- Avoid severing the beach-dune link by engineering works (e.g. seawalls) to prevent storm damage.
- Avoid flattening dunes for development; this increases the potential for washover and breaching.
- Avoid re-shaping dunes to form a single artificial high ridge; this is more vulnerable to wave attack than a series of foredune ridges.
- Avoid blocking washover channels (e.g. with artificial dunes).
- Design buildings to cope with storms (e.g. building on stilts, smooth aerodynamic designs with no eaves or balconies, brace or strengthen joists and beams with "hurricane" clips).
- Use vegetation to stabilise actively mobile dunes (e.g. grasses, trees).
- Control beach front development by the use of e.g. 'set back' lines or zoning.
- Control recreational use e.g. restricting access to walkways and limiting car park sizes.

Sand dunes are of major conservation value, for wildlife (e.g. natterjack toads) and plant species (e.g. dune gentian). Dune habitats require repeated phases of wind erosion and stability, together with marine erosion of the foredunes, to maintain a full sequence of successional stages.

Stabilisation has the potential to reduce the conservation value. Dune management should be directed towards increasing stability on unstable sand dunes. This involves a variety of well-proven methods of restoration[35.6–7].

References

35.1 Carter, R. W. G. (1988) *Coastal Environments*. Academic Press.

35.2 Lees, B. G., Yanchou, L. and Head, J. (1990) Reconnaissance thermoluminescence dating of northern Australian coastal dune systems. *Quaternary Research*, **34**, 63–75.

35.3 Psuty, N. P. (1992) Spatial variation in coastal foredune development. In R. W. G. Carter, T. G. F. Curtis and M. J. Sheeby-Skeffington (eds.) *Coastal Dunes: Geomorphology, Ecology and Management for Conservation*. Balkema, Rotterdam, 3–13.

35.4 Woodroffe, C. D. (2002) *Coasts: Form, Process and Evolution*. Cambridge University Press.

35.5 Lee, E. M. (1994) *The Occurrence and Significance of Erosion, Deposition and Flooding in Great Britain*. HMSO.

35.6 Doody, J. P. (1985) *Sand Dunes and their Management*. Focus on nature conservation, 13 NCC, Peterborough.

35.7 Ranwell, D. S. and Boar, R. (1986) *Coast Dune Management Guide*. Institute of Terrestrial Ecology, NERC.

36

The Coast: Cliffs

Settings

Coastal cliffs occur where marine erosion has cut into the seaward margins of high ground (Fig. 36.1). Cliffs occur on many coastlines, including parts of the UK, Portugal, France, Italy, the west coast of the USA, Japan, Australia and New Zealand[36.1].

Most clifflines developed in response to the post-glacial rise in sea-level. However, the inheritance from previous phases of cliff development during the last interglacial (the Ipswichian) is important. The world-wide fall in sea-level during the last glaciation stranded the Ipswichian cliffline, which was then subject to slope degradation processes. In mid-high latitudes, extensive spreads of landslide debris, periglacial deposits and scree accumulated at the foot of the abandoned clif-fline. As the sea-level rose after the glaciation, it would have gradually reached this debris which then would have been trimmed back, resulting in the creation of a new cliffline.

A number of settings occur:
- Cliffs where the pre-Holocene debris has been completely removed. Recession involves retreat of the *in-situ* cliff materials (e.g. Beachy Head, UK).
- Cliffs where pre-Holocene debris has only been partially removed. Recession involves retreat through transported debris, often with a high sand and gravel/cobble content (e.g. glacial till and head such as the drumlin shorelines of Nova Scotia).
- Cliffs where there has been no removal of the pre-Holocene debris e.g. where the coastal slopes are protected by a large shingle beach, as at Abbotsbury, behind Chesil Beach, UK.

Cliff recession is the landward retreat of the *cliff profile* (from cliff foot to cliff top) in response to cliff erosion processes. *Cliff erosion* is a four-stage process involving (Fig. 36.2):
- *detachment* of particles or blocks of material (involving landslides, seepage erosion, piping,

Figure 36.1   Coastal cliffs developed left: in volcanic rocks in the Canary Islands and right: rotational landsliding by topples and falls (photograph courtesy of Tony Waltham, Geophotos) (see Fig. 36.3).

surface erosion and wave abrasion and hydraulic action)
- *transport* of this material through the cliff system
- *deposition* on the foreshore and
- *removal* by marine action.

Recession on many cliffs can be restricted by the rate at which debris is removed from the foreshore, as the debris provides protection against further detachment (Fig. 36.2). Removal of the debris is necessary for the continuation of erosion at the base of the cliff, although cessation of basal erosion does not necessarily prevent coastal landsliding.

Forms

Coastal cliff systems comprise (see Fig. 4.4):
- The *cliff sub-system* i.e. the cliff face and cliff top. Cliff systems can range from extensive clifflines in relatively uniform materials to separate small units reflecting more complex geological settings.
- The *shoreline sub-system* i.e. the beach, foreshore and nearshore zone.

Over engineering time, the morphology of the sub-systems represents a balance between the energy inputs (e.g. wave energy) and the materials (e.g. the seabed and cliff materials). However, morphology also influences the available energy (e.g. beach height, shoreline orientation).

There is a wide range of cliff types, reflecting the complex interactions between lithology, geological structure and inland relief, and the applied forces of both marine and non-marine (i.e. slope) processes. However, a number of generic forms occur[36.2] (Fig. 36.3) including

- *simple cliff systems* (comprising a single sequence of sediment inputs from falls or slides and outputs)
- *composite systems* (comprising a partly coupled sequence of contrasting simple sub-systems)
- *complex systems* (comprising strongly linked sequences of sub-systems, each with their own inputs and outputs of sediment)
- *relict systems* (comprising sequences of pre-existing landslide units which are being gradually reactivated and exhumed by the progressive retreat of the current seacliff)
- *plunging cliffs* (where a near-vertical rock cliff passes beneath low tide level without a beach or shore platform).

Cliffs are often fronted by a *shore platform* developed across *in situ* materials. Shore platforms may be well developed, exposed features (as on the New South Wales coast, Australia) or mantled by mobile beach sediments, only becoming exposed during storm events or as a result of starvation of sediment supply.

Behaviour

The recession rate R reflects a balance between the *mass strength* (allowing for both rock strength and weaknesses

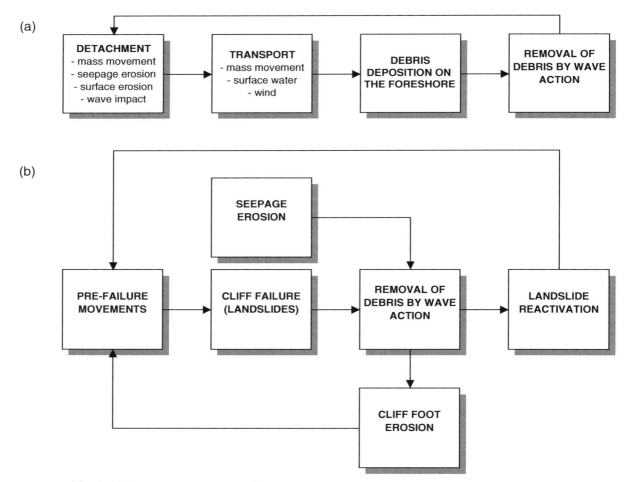

Figure 36.2 (a) The cliff recession process and (b) a simple recession model showing feedback mechanism (after Lee and Clark 2002)[36.2].

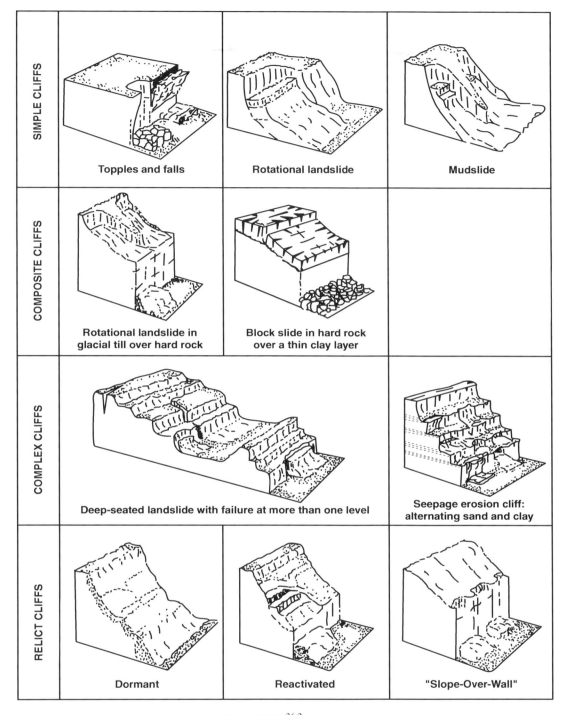

Figure 36.3   Types of coastal cliff (after Lee and Clark 2002)[36.2].

created by discontinuities) of the cliff materials *CS* and the *stresses* imposed on the cliff by gravity and the kinetic energy of waves at the cliff foot F_w [36.1]:

$$R = x \ln\left(\frac{F_w}{CS}\right)$$

where x is a constant. Variations in recession rates, either annually or over the longer term, reflect variations in the factors that control *CS* and F_w.

Cliff recession results in the gradual widening and flattening (i.e. reduced gradient) of the shore platform. Under stationary sea-level conditions this results in enhanced energy dissipation (i.e. friction losses), as waves travel through shallower water for longer distances, and a reduction in wave force F_w at the cliff foot. Over time, recession rates could be expected to decline, reaching zero when the wave force is less than the mass strength of the materials. Cliff recession involves minor events (small scale losses associated with water and wind erosion, weathering and spalling off a cliff face) and episodic landslide events, associated with the periodic failure of cliffs in response to *preparatory factors*, such as slope profile steepening

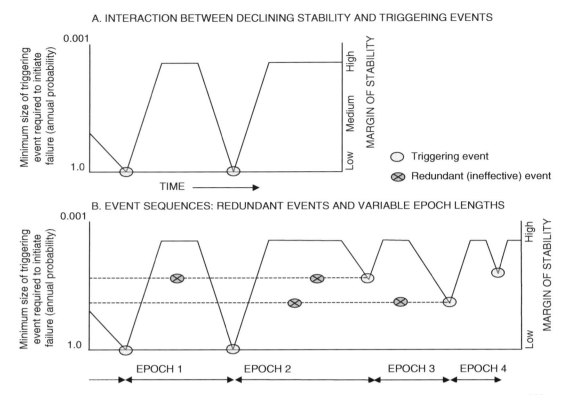

A. INTERACTION BETWEEN DECLINING STABILITY AND TRIGGERING EVENTS

B. EVENT SEQUENCES: REDUNDANT EVENTS AND VARIABLE EPOCH LENGTHS

Figure 36.4   Variable interaction between triggering events and coastal landslides (after Lee and Clark 2002)[36.2].

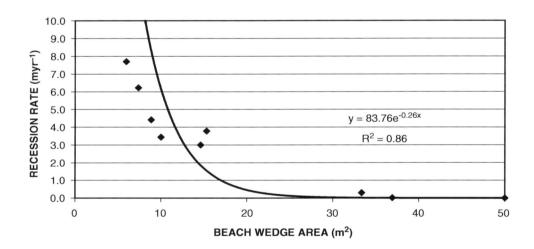

$$y = 83.76e^{-0.26x}$$

$$R^2 = 0.86$$

Figure 36.5   Beach volume and cliff recession rate: Suffolk, UK (after Lee 2005)[36.3].

and *triggering factors*, such as large storms or periods of heavy rainfall.

Recession rates should be viewed as *inherently uncertain*. As the margin of stability is progressively reduced by the operation of preparatory factors, so the minimum size of event required to trigger recession becomes smaller (Fig. 36.4). Thus, triggering events of a particular magnitude may be *redundant* or *ineffective* (i.e. do not initiate cliff recession) until preparatory factors lower the margin of stability to a critical value. This can mean variable time periods between recession events, depending on the sequences of storm or rainfall events. In addition, the same sized triggering events may not necessarily lead to recession events.

As a result of this complexity, in the *short-term* cliff recession often appears to be a highly variable process, with marked fluctuations in the annual recession rate around an average value. However, over the *medium-term* these fluctuations smooth themselves out (assuming environmental controls are stationary). Over this timescale the process can be characterised by a relatively constant mean recession rate and maintenance of cliff form, with parallel retreat of the cliff profile and a balance over time in the sediment budget (i.e. *steady state* behaviour; see Chapter 5).

Beaches control wave energy dissipation on the foreshore and, in some situations, can provide complete

Table 36.1 Examples of cliff recession rates.

Class range (m yr^{-1})	Category	Example	Rock type	Recession rate (m yr^{-1})
0–0.1	Negligible	Point Peron, W. Australia	Limestone	0.001
		Sunset Cliffs, San Diego	Pre-Miocene shales, sandstone	0.01
		Haranomachi, Japan	Pliocene sandstone, mudstone	0.02
0.1–0.5	Moderate	La Jolla, California	Alluvium	0.1–0.5
		Black Rock Point, Victoria	Tertiary sediments	0.04–0.8
		Rockaway Beach, California	Volcanic and metavolcanic rocks	
		Santa Cruz, California	Mio-Pliocene mudstone with shale	0.4
0.5–1.0	Intense	Scarborough Bluffs, Lake Ontario	Glacial deposits	0.5
		Colonial Beach, Virginia	Pleistocene deposits	0.5
		Fukushima, Japan	Pliocene siltstone	0.8
1.0–1.5	Severe	Hengistbury Head, Dorset	Tertiary sandstones and clays	1.12
		Dunwich, Suffolk	Pleistocene deposits	1.2
> 1.5	Very severe	Martha's Vineyard, Massachusetts	Glacial deposits	1.7
		Holderness, East Yorkshire	Glacial deposits	1.8
		Port Bruce, Ontario	Glacial deposits	2.2
		Covehithe, Suffolk	Pleistocene deposits	7.75
		Krakatoa, Indonesia	Pyroclastic deposits	33
		Surtsey	Lava	25–37
		Nishinoshima Island, Japan	Volcanic ejecta	80

protection from marine erosion. The relationship between the average beach volume (measured as the beach profile area above High Water Mark, the 'beach wedge') and annual cliff recession rate for 7 profile monitoring sites on the Suffolk cliffs, UK (Pleistocene Norwich Crag sediments) is shown in Fig. 36.5[36.3].

Hazards

Typical hazards associated with cliff recession include:

- Loss of life or injury to beach users as a result of rockfalls or rapid landslide events e.g. in 1996 nine parents, teachers and children attending a school surfing competition at Cowaramup Bay, Western Australia died when the limestone cliff under which they were sheltering from rain collapsed.
- Loss of cliff top land and property (see Table 36.1) e.g. the Holderness coast, UK has retreated by around 2 km over the last 1000 years, including at least 26 villages listed in the Domesday survey of

AD 1086[36.4] (see http://www.hull.ac.uk/coastalobs/ easington/erosionandflooding/).

- Reactivation of ancient, pre-existing landslides e.g. the coastal town Grottammare, Italy[36.5].

Management

Cliff recession can be influenced by both site-specific slope processes and shoreline processes operating across a coastal cell (see Chapter 29). Effective management needs to be based on an understanding of *cliff behaviour units* (CBU). These units span the nearshore to the cliff top and are coupled to adjacent CBUs within the framework provided by coastal cells[36.2, 6]. Key points include:

- Every cliff problem will be unique because of the great range of CBU forms and the variability of the cliff materials. Cliff management strategies need to be designed to reflect site conditions and cannot be provided 'off-the-shelf'.

- Cliff management needs to be supported by reliable predictions of future rates of cliff recession. The simplest approach involves the extrapolation of past recession rates. However, probabilistic methods provide a means of taking account of the uncertainty and potential variability in the process[36.7].

- The use of average annual recession rates can be misleading because of the episodic and uncertain nature of the process.

- Cliff protection will require some form of *toe protection* to prevent or reduce wave attack (e.g. seawalls, rock revetments, breakwaters). *Slope stabilisation* will often be needed to prevent the deterioration of the protected cliffs (see Chapter 20).

- Prevention of marine erosion at the cliff foot does not eliminate the potential for coastal landsliding. Internal slope processes of weathering, strain-softening, creep and the recovery of depressed pore water pressures can cause *delayed failures* many years later (e.g. the 1993 Holbeck Hall landslide at Scarborough, UK[36.8]). Large-scale failure of protected slopes can damage toe protection structures and may lead to renewal of cliff foot erosion.

- Eroding coastal cliffs are often the main source of sediment for neighbouring beaches. In England, sediment inputs from cliff recession could have declined by as much as 50% over the last 100 years because of widespread cliff protection. This has been a factor in the degradation of many beaches around the coastline[36.2, 36.6].

- It is rarely economic or desirable to protect all eroding cliffs. Alternative management strategies include *avoidance* of areas at risk (land use planning), *minimising the impact of human activity* on unstable slopes (e.g. preventing water leakage) and *minimising the risks* to public safety by the use of early warning systems[36.2].

References

36.1 Sunamura, T. (1992) *Geomorphology of Rocky Coasts*. John Wiley and Sons, Chichester.

36.2 Lee, E. M. and Clark, A. R. (2002) *The Investigation and Management of Soft Rock Cliffs*. Thomas Telford.

36.3 Lee, E. M. (2005) Coastal cliff recession risk: a simple judgement based model. *Quarterly Journal of Engineering Geology and Hydrogeology*, **38**, 89–104.

36.4 Valentin, H. (1954) Der landverlust in Holderness, Ostengland von 1852 bis 1952. *Die Erde*, **6**, 296–315.

36.5 Angeli, M.-G., Pontoni, F. (2002) Instability processes as a result of coastal and climate change at Grottammare (Central Italy). In R. G. McInnes and J. Jakeways (eds.) *Instability, Planning and Management*, 571–580, Thomas Telford, London.

36.6 Brunsden, D. and Lee, E. M. (2004) Behaviour of coastal landslide systems: an inter-disciplinary view. *Zeitschrift fur Geomorphologie*, **134**, 1–112.

36.7 Lee, E. M., Hall, J. W. and Meadowcroft, I. C. (2001) Coastal cliff recession: the use of probabilistic prediction methods. *Geomorphology*, **40**, 253–269.

36.8 Lee, E. M. (1999) Coastal Planning And Management: The impact of the 1993 Holbeck Hall Landslide, Scarborough. *East Midlands Geographer*, **21**, 78–91.

37

Methods of Investigation

Introduction

Engineering geomorphology can contribute to the investigation, design and construction of civil engineering projects (see Chapter 1):

- predicting near surface ground conditions (e.g. to support the design of ground investigations and the identification of possible sources of construction materials)
- identifying and evaluating the geological hazards and risks to the project
- identifying the possible impact of the project on the physical environment.

There are many ways in which these objectives can be addressed by engineering geomorphologists, either working alone or as part of a multi-disciplinary team (i.e. a *geoteam*). Available approaches include *mapping landforms*, *understanding and describing materials* and *modelling processes*. Most projects will be based on a combination of these generic approaches and will involve:

- defining and characterising earth surface systems, including the landforms and materials (state variables) and the energy regime (loadings; see Chapter 4)
- evaluating the contemporary system behaviour, including the processes that are forcing change (see Chapter 5)
- predicting the future system behaviour (see Chapter 12).

Mapping Landforms

Mapping of surface form (i.e. morphological mapping) and the recognition of specific landforms (i.e. geomorphological mapping; see Chapter 42) or terrain units (i.e. terrain evaluation; see Chapter 41) can provide a framework for:

- *Extrapolating geological observations across a broader area*. For example, a programme of geological and geomorphological mapping was undertaken during the investigations into ground conditions and geohazards (e.g. rockfalls, debris flows) along major iron ore railways in the Pilbara, Western Australia[37.1]. A *total geology model* (i.e. an initial terrain model; see Chapter 38) was developed to describe the expected conditions; *geomorphological units* (e.g. alluvial fans and talus slopes) provided the spatial framework for collating and extrapolating the geological observations. This model supported the design of the ground investigation, the search for construction resources and the assessment of hazards (i.e. landslides and floods). The units were defined with a predicted range of engineering characteristics; these became the *reference conditions* for the project design and construction (see Chapter 1).

  At the early stages of a project, the combination of geological and geomorphological mapping is a more effective initial approach to investigating a route corridor, such as the Pilbara railways, than a series of boreholes along an assumed route (often at fixed intervals). It can provide information relative to whatever final alignment is chosen.

- *Understanding the associations between the landform and materials in a particular area*. For example, alluvial fans are cones of poorly sorted coarse sediment (boulders, cobbles, gravels and some sands) laid down where a channel emerges from an upland area onto a plain. Recognition of fans and subsequent mapping of fan morphology can help identify potential sites for construction aggregates. This approach was used to develop desert construction materials maps in Ras al Khaimah and Bahrain[37.2].

  Landform-material associations can be important in interpreting fossil periglacial slopes in temperate regions (see Fig. 2.6). For example, recognition of solifluction lobes and bluffs in east Devon, UK, alerted engineers to potential shallow landslide problems along the route of the Axminster bypass[37.3] (see Fig. 42.1). Failure to identify these features can prove costly. Construction of the A21 Sevenoaks Bypass, UK, had to be halted in 1966 because solifluction lobes were reactivated during excavation operations; the affected portion of the route had to be realigned at a (then) huge cost of £2 M[37.4].

- *Predicting system behaviour*. Detailed mapping of the landform assemblages in an area support the development of a *dynamic system model* that identifies the inter-relationships between the landforms. Figure 37.1 (see colour section) presents a schematic system model of a large mudslide complex in the Lesser Caucasus, Georgia and highlights how the individual components interact to influence landslide behaviour[37.5]. *Loading* at the rear of the accumulation zone (Unit Id) is caused by the run-out of material from the open hillside landslides (Units IIIc, IVc, Vc). *Unloading* at the toe of the accumulation zone is caused by stream bank erosion; it results in the progressive deterioration of stability conditions at the distal end of the accumulation zone and the development of secondary mudslides (Units VIIa, VIIb). Unloading also maintains the marginally stable conditions across the accumulation zone.

- *Identifying the processes and mechanisms that have created or shaped the landforms.* For example, stream channel-form (i.e. cross section area, bed roughness and estimated water depths) was mapped to provide the basis for estimating past flood events along the 1000 m wide Wadi Dhamad, Saudi Arabia[37.6] (see Chapter 27). Surface mapping of landslides is an effective approach to identifying the failure mechanism. Geomorphological mapping of the landslide features at the Channel Tunnel portal, UK, indicated a deep-seated rotational failure with a basal shear surface around 25 m below ground level[37.7]. This failure mechanism and predicted depth was confirmed by sub-surface investigations.

Understanding and Describing Materials

Description and classification of near surface materials (i.e. bedrock, residual soils, transported soils and mobile sediments; see Chapters 6–8) form an important component of most engineering geomorphological studies. Materials need to be accurately described with a geomorphological understanding of the local landscape (Fig. 37.2; Fig. 37.3). Examples include:

- *Verification of initial terrain models.* As projects proceed through reconnaissance studies to basic engineering, design and construction it will be essential to check, refine or modify the initial terrain models developed from surface mapping studies. During the front-end engineering design (FEED) for the In Salah Gas Project, Algeria, terrain evaluation methods were used to produce terrain models of the route corridor (100 km wide, centred on the preferred route, some 500 km long; see Chapter 41). The predicted ground conditions and trench excavatability rates were made during a rapid drive-over survey[37.9] and later verified by borehole and trial pitting exercises carried out along the selected route during the detailed design and construction stages[37.10–11].
- *Estimating the nature and scale of past events.* Careful description and interpretation of transported soils can provide an indication of the history of landslide or flood events in an area. For example, debris flow deposits consist of poorly sorted, large clasts embedded in a matrix of fine material. Boulders may be concentrated at the top of the deposit (i.e. reverse grading) because of the buoyant forces and dispersive pressures within a debris flow. By contrast, flash

flood deposits tend to be cross-stratified and show fining-upwards grading. Detailed logging within an alluvial fan (48 000 m^3) at the mouth of a very steep catchment in Hong Kong revealed that the fan was probably the product of 3 separate channelised debris flow events[37.12].

- *Estimating the surface erosion potential.* The nature of the surface materials can provide an indication of their potential erodibility by water or wind (see Chapters 17 and 18). For example, the desert surface conditions around a town in Saudi Arabia were mapped to support the assessment of wind blown sand and dust hazards[37.13]. The most erodible surfaces proved to be those with a sparse cover of fine gravels over silty sandy soils. In other areas a continuous stone pavement cover provided protection against wind erosion.
- *Predicting system behaviour.* The spatial distribution of mobile sediments can indicate the ways in which a system has developed in response to changing conditions. For example, a detailed sieve analysis of shingle samples taken along the length of Chesil Beach, UK, revealed a series of distinct grading zones (Fig. 37.4). The pattern suggests that the beach may be splitting into a series of sub-units, probably because of the relatively recent decline in sediment supply from the west (i.e. updrift) following construction of the original harbour at West Bay in the mid 18th century (see below). The beach grading model was used to evaluate the potential impacts of shingle re-nourishment schemes at the western end of the beach[37.14].

Modelling Processes

Models are an attempt to understand and simplify the real world, by examining selected critical factors and sometimes ignoring others. By providing an explanation about the behaviour of system components, a model can be used to make predictions of future morphological changes.

- *Application of existing models.* A wide range of existing models are available for predicting processes such as dune mobility (see Chapter 18), landsliding (e.g. stability models, see Chapter 19), channel bed scour (see Chapter 25) and sediment transport rates (e.g. the CERC formula, see Chapter 29). Each model will have its own data requirements. The Universal Soil Loss Equation, for example, can be used to estimate soil loss by surface water erosion (Chapter 17). Field data requirements include slope angle and length, soil type (i.e. erodibility), vegetation cover and land management practice. Rainfall intensity data can be obtained from meteorological records and collected at site using a rain gauge; airfields are often a useful source of local data.
- *Development of landscape models.* Evidence of past change (e.g. historical records and monitoring results; see Chapter 40) can be organised into a model to explain the historical evolution of a system. Such models tend to be qualitative and subjective, but can provide a framework for understanding and predicting future change.

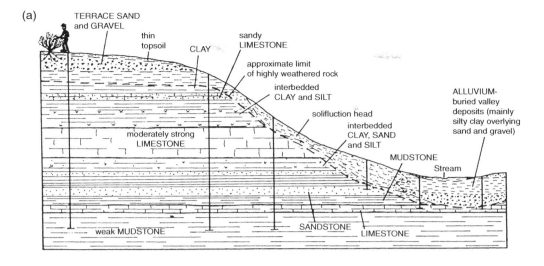

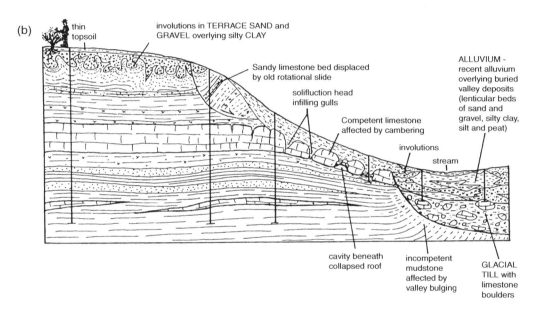

Figure 37.2 The value of a geomorphological model in interpreting SI data in relict periglacial terrain (after Fookes 1997)[37.8] (a) Some potential misinterpretations of geology of (b) based on borehole evidence; and (b) Actual geology (broadly modelled on Lower-Middle Jurassic of East Midlands, UK).

Figure 37.5, for example, is a simple evolutionary model of the East Devon coast, UK, developed to support the planning of coast protection works at Lyme Regis[37.15,18]. Historical chart evidence suggests a breakdown of established sediment cells into smaller units over the last 250 years or so. The likely sequence of events is:

1. Pre-1750: free longshore transport of shingle from Charton Bay to Lyme Regis and to the east;
2. c.1750: the connection of the Cobb to the shoreline in 1754 led to disruption of the transport pathway. This resulted in the build-up of a large shingle beach on the western side of the Cobb (Monmouth Beach).
3. 1765–1840: a large landslide in 1765 at Humble Point prevented the longshore transport of shingle

from Charton Bay into Pinhay Bay, which, at that time, contained an extensive shingle beach. A further landslide in 1840 reinforced this sediment barrier, promoting the build-up of the present day shingle berm behind the landslide debris.
4. 1900–1990s: by the turn of the century the shingle beach in Pinhay Bay appears to have been lost, replaced by the present-day boulder beach. Monmouth Beach has continued to build up behind the Cobb.

• *Development of predictive models*. The combination of records or measurements of past change and an understanding of system behaviour can form the basis for developing a model to predict future rates of change. For example, simple *factor-based models* can be used to model the impact on the historical

Figure 37.3   Trial pit section in Saudi Arabia. At the surface (A) is thin lag gravel (single particle thickness) created by deflation overlying a dark silty sandy gravel deposit (B) containing few cobbles. Below this is a lighter layer (C) that represents a nodular calcrete horizon. The calcrete nodules reach cobble size and lie within a matrix of weakly cemented silty gravelly sand.

EAST   **WEST**

Chiswell			Wyke Regis								Langton Herring				Abbotsbury				West Bexington			Cogden Beach		Burton Cliff		West Bay		

0	1	2	3	4	5	6	7	8	9	10	11	12	13	14	15	16	17	18	19	20	21	22	23	24	25	26	27	Transect Number
Zone I		Zone II									Zone III											✱		Zone IV				Crest
Zone I		Zone II										Zone III									✱			Zone IV				Mid-upper beach
Zone I		Zone II											Zone III							✱				Zone IV				MHWS
Zone I		Zone II										Zone III							✱					Zone IV				MLW
Zone I		Zone II												Zone III							✱			Zone IV				MLWS

✱ Localised reversal in grading trend

    Zone I: mean pebble size rapidly increases from west to east, with D_{50} almost doubling in size over 1–3km. D_{50} is generally in the range 35–50mm
    Zone II: mean pebble size gradually increases from west to east, with D_{50} doubling over 8–12km. D_{50} is generally in the range 10–25mm.
    Zone III: fairly constant pebble size, with D_{50} ranging from c8–15mm.
    Zone IV: pebble size gradually increases from west to east, with D_{50} rising by over 500% over 6–10km. Around transects 23–25 there is
        evidence to suggest localised reversals in the overall trend of west to east increasing pebble size. D_{50} ranges from c2–10mm.

Figure 37.4   Chesil Beach grading zones (after Lee and Brunsden 1997)[37.14].

coastal cliff recession rate of a limited number of factors that represent combinations of loading and state variables[37.16]:

Predicted Recession Rate = Historical Recession
    Rate × (S–L) × W × B × S × E

where (S–L) is the *sea level rise factor* that represents the change in average annual cliff recession rate related to change in the rate of sea level rise; W is the *winter rainfall factor* that represents the change in average annual cliff recession rate related to change in effective winter rainfall; B is the *beach level factor* that

represents the change in average annual cliff recession rate related to change in the degree of cliff protection provided by the beach; S is the *storminess factor* that represents the change in annual cliff recession related to changes in the wave energy arriving on the shoreline as a result of changes in storminess and E is the *cliff toe protection factor* that represents the change in average annual cliff recession rate related to changes in toe protection. This takes account of the influence of future shoreline management practice at the site.

This model can be used either *deterministically* (i.e. effects logically follow particular causes in a non-random manner) to generate a single recession

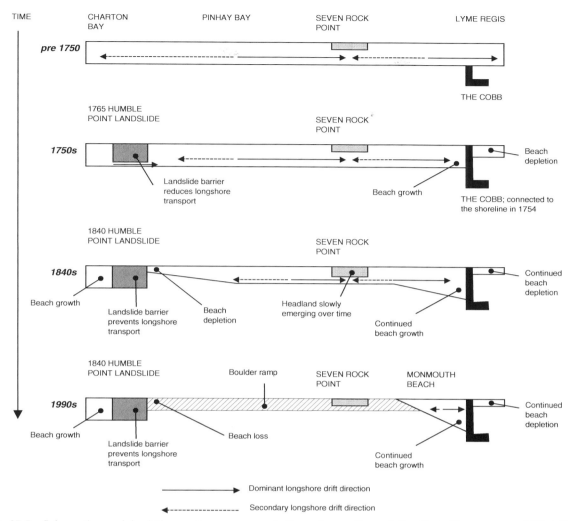

Figure 37.5 Schematic model of the contemporary evolution of a section of the East Devon coast, UK (after Lee 2001)[37.15,18].

rate or in a *probabilistic* manner to generate a probability distribution for the rate.

Schematic predictive models were developed to assess the habitat loss and gain associated with future shoreline management policies in England and Wales[37.17] (Fig. 37.6; Table 37.1). The predictions were based on an extrapolation of past trends, but taking account of the potential impact of relative sea-level rise on the rate of coastal change. Habitat loss/gain accounts were developed for internationally important conservation sites (SAC/SPA and Ramsar sites), based on a best-guess coastal defence scenario for the next 50 years (i.e. do nothing, hold the line, advance the line or managed retreat).

Investigation Techniques

The following Chapters provide a brief introduction to the range of data collection techniques that are available to the engineering geomorphologist. These range from desk-based methods (e.g. development of initial terrain models, aerial photograph and satellite image interpretation, and analysis of historical records) to field observations and measurements (e.g. geomorphological

mapping, terrain evaluation and monitoring change) and laboratory methods (e.g. dating). Chapter 43 highlights the importance of the use of judgement to develop answers to the questions posed by civil engineers.

References

37.1 Baynes, F., Fookes, P. G. and Kennedy, J. F. (2005) Total engineering geology approach to applied railways in the Pilbara, Western Australia. *Bulletin of Engineering Geology and Environment*, **64**, 67–94.

37.2 Cooke, R. U., Brunsden, D., Doornkamp, J. C. and Jones, D. K. C. (1982) *Urban Geomorphology in Drylands.* Oxford University Press.

37.3 Croot, D. and Griffiths, J. S. (2001) Engineering geological significance of relict periglacial activity in South and East Devon. *Quarterly Journal of Engineering Geology and Hydrogeology*, **34**, 269–281.

37.4 Skempton, A. W. and Weeks, A. G. (1976) The Quaternary history of the Lower Greensand escarpment and Weald Clay vale near Sevenoaks, Kent. *Philosophical Transactions of the Royal Society, London.* Vol. **A283**, 493–526.

37.5 Lee, E. M., Wise, D. J. and Champelovier, I. J. (In Press) Evaluation of the Landslide Problems at Dgvari Village, Western Georgia. *Quarterly Journal of Engineering Geology and Hydrogeology*.

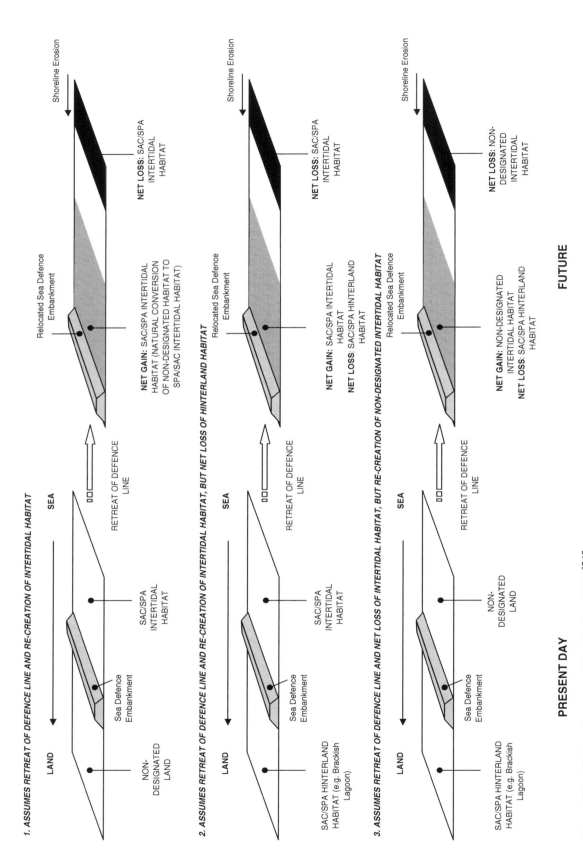

Figure 37.6 Habitat change models (after Lee 2001)[37.17].

Table 37.1 Estimated 50-year habitat loss/gain account for SAC/SPA and Ramsar sites in England and Wales[37.17].

Habitat	Estimated Loss (ha)	Estimated Gain (ha)	Balance (ha)
Mudflat/sandflat	11459	12991	1532
Saltmarsh	6996	7685	689
Shingle bank	238	110	−128
Sand dune	504	381	−123
Cliff top	133	0	−133
Soft cliff	0	0	0
Hard cliff	0	0	0
Wet grassland	3214	0	−3214
Reed bed	172	0	−172
Coastal lagoon	530	30	−500
TOTAL	23246	21197	−2048

37.6 Richards, K. S., Brunsden, D., Jones, D. K. C. and McCaig, M. (1987) Applied fluvial geomorphology: river engineering project appraisal in its geomorphological context. In K. S. Richards (ed.) *River Channels: Environment and Process*, 348–382. Blackwell, Oxford.

37.7 Griffiths, J. S., Brunsden, D., Lee, E. M. and Jones, D. K. C. (1995) Geomorphological investigations for the Channel Tunnel Terminal and Portal. *The Geographical Journal*, **161**(3), 275–284.

37.8 Fookes, P. G. (1997) First Glossop Lecture: Geology for engineers: the geological model, prediction and performance. *Quarterly Journal of Engineering Geology*, **30**, 290–424.

37.9 Fookes, P. G., Lee, E. M. and Sweeney, M. (2005) In Salah Gas project, Algeria – part 1: terrain evaluation for desert pipeline routing. In M. Sweeney (ed.) *Terrain and Geohazard Challenges Facing Onshore Oil and Gas Pipelines*, Thomas Telford, 144–161.

37.10 Shilston, D. T., Lee, E. M., Fookes, P. G. and Standish, I. (2005) In Salah Gas project, Algeria – part 2: detailed terrain evaluation for site design. In M. Sweeney (ed.) *Terrain and Geohazard Challenges Facing Onshore Oil and Gas Pipelines*, Thomas Telford, 162–174.

37.11 Sweeney, M., Pettifer, G. S., Shilston, D. T., Bel-Ford, P. and Sockbridge, M. (2005) In Salah Gas project Algeria – part 3: Prediction and performance of large chain trenchers on a desert pipeline project. In M. Sweeney

(ed.) *Terrain and Geohazard Challenges Facing Onshore Oil and Gas Pipelines*, Thomas Telford, 529–548.

37.12 Moore, R., Lee, E. M. and Palmer, J. S. (2002) A sediment budget approach for estimating debris flow hazard and risk: Lantau, Hong Kong. In R. G. McInnes and J. Jakeways (eds.) *Instability, Planning and Management*. Thomas Telford, 347–354.

37.13 Jones, D. K. C., Cooke, R. U. and Warren, A. (1986) Geomorphological investigation, for engineering purposes, of blowing sand and dust hazard. *Quarterly Journal of Engineering Geology*, **19**, 251–270.

37.14 Lee, E. M. and Brunsden, D. (1997) *West Bay geomorphological study*. Report to West Dorset District Council.

37.15 Lee, E. M. (2001) Estuaries and Coasts: Morphological adjustment and process domains. In D. Higgett and E. M. Lee (eds.) *Geomorphological Processes and Landscape Change: Britain in the Last 1000 years*. Blackwell, 147–189.

37.16 Lee, E. M. (2005) Coastal cliff recession risk: a simple judgement based model. *Quarterly Journal of Engineering Geology and Hydrogeology*, **38**, 89–104.

37.17 Lee, E. M. (2001) Coastal defence and the Habitats Directive: predictions of habitat change in England and Wales. *Geographical Journal*, **167**, 39–56.

37.18 Lee, E. M. and Brunsden, D. (1999) Coastal processes and geomorphology: Monmouth Beach. Report to West Dorset District Council.

38

Desk Study and Initial Terrain Models

Introduction

Many difficulties that arise on engineering projects are due to a lack of appreciation of ground conditions, surface processes and the potential for landform change. Potential problems need to be recognised as early as possible: *forewarned is forearmed*. However, resources may be extremely limited during the initial stages of many engineering projects, often restricting fieldwork and site appraisal to a rapid reconnaissance. It is important, therefore, that maximum benefit is derived from readily available sources of information (i.e. *desk study*). The risk of unexpected conditions (e.g. previously unrecognised hazards) or impacts (e.g. unanticipated changes) can be reduced by the development of *initial terrain models* and *maps*.

Some useful UK-based information sources are listed in Table 38.1.

Anticipating Conditions from Environmental Controls

It is possible to gain a reasonable appreciation of what to expect in an area from a basic understanding of the climate, plate tectonics and bedrock geology; this forms the *total geology/geomorphology* starting point for the eventual site model[38.1–3]. Global variations in climate, for example, are largely responsible for differences in the rate and intensity of surface processes between *morphoclimatic zones*[38.4] (see Table 2.1, Fig. 2.2, Fig. 38.1). In broad terms, the differences between these zones include[38.5]:

1. Variations in the amount and frequency of precipitation, e.g.
   - *Hot dryland environments*: rainfall can be subject to extreme spatial variability. In the Sahara, the 24-hour rainfall can exceed the mean annual rain-

Table 38.1 Desk study information sources in the UK.

Source	Materials
British Geological Survey	For the UK, 1:50,000 geological map sheets (hard copy and in digital form) with substantial coverage at 1:10,000; a range of specialist maps such as hydrogeology and earth sciences; records of 2 million boreholes; geotechnical property data from nearly 200,000 samples
Royal Geographical Society	Unique publicly accessible collection of topographic, specialist and historic maps from all over the world at a range of scales; major collection of books and journals
Geological Society	Geological maps for the UK and overseas; major collection of books and journals
Ordnance Survey	Digital maps at any reasonable map scale for any location in the UK; aerial photographs and digital orthophotos
British Museum	Publicly accessible collection of historic maps for the UK and overseas; world renowned collection of books and journals
Hydrographic Office	Coastal charts for all areas of the UK and overseas
British Library	The main UK library for all world books and journals
Local libraries	Historic, large-scale and specialist maps; local publications
National Soil Resources Institute	(Formerly the Soil Survey of England and Wales) Soil survey and land-use maps
Institution of Civil Engineers	Library is an excellent source of journals, books, and historic construction drawings

All major information repositories have web sites that should be viewed for products and services; there are also a range of web search engines that should be used to identify relevant data.

242

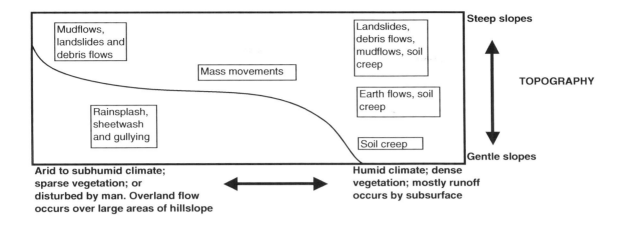

Figure 38.1 Environmental controls and characteristic geomorphological processes (after Dunne and Leopold 1978)[38.4].

fall by 2–3 times once every 20–30 years. As a result, dryland stream flows are ephemeral and flood events are rare and often unexpected[38.6].

- *Temperate environments*: decadal-scale variations in annual rainfall occur because of changes in the atmospheric circulation (e.g. the North Atlantic Oscillation), influencing the frequency and intensity of lowland flooding. Landslide activity is often related to the occurrence of wet year sequences i.e. consecutive years with an above average moisture balance[38.7].

2. Variations in the extent and character of vegetation cover, e.g.

- *Hot dryland environments*: aridity leads to severely restricted vegetation growth. Decreasing precipitation and vegetation cover result in a reduction in surface stability (vegetation cover protects the soil surface, increases surface roughness and, hence, reduced wind velocity). Bare, dry soil surfaces are susceptible to wind erosion (deflation), leading to blown sand and dust problems (see Chapter 18). Bare desert surfaces also tend to be sealed by a soil crust, resulting in low infiltration rates during infrequent rain storms, high runoff and the potential for flash floods. As the vegetation cover increases, so infiltration rates tend to increase[38.6].

- *Hot wetlands*: high temperatures and humidity result in dense tropical rainforest cover. This cover protects the surface and allows the development of deep weathering profiles (i.e. rate of weathering > rate of removal by erosion). The rate of surface erosion and landslide activity tends to be restricted, although the impact of tropical cyclones can be dramatic[38.8].

3. Variations in the soil moisture and temperature regime, e.g.

- *Hot dryland environments*: the combination of high temperatures and low precipitation causes net *evaporating* conditions. Downward leaching of

salts within the ground is limited; even highly soluble salts (e.g. gypsum and sodium chloride) can remain in the soil profile. *Capillary rise* is often very pronounced leading to the concentration of salts in the upper soil profile. This produces a highly aggressive environment in which *salt weathering* is an important factor in rock disintegration, ground heave and road and concrete attack[38.6].

- *Savannah environments*: higher precipitation and cooler temperatures results in net downward percolation and leaching of minerals and organic matter (*lessivation*)[38.9]. Soils are generally leached, highly erodible and nutrient poor (oxisols and ultisols), frequently with duricrust layers (e.g. ferricrete and silcrete; see Chapter 7).

- *Tundra environments*: *permafrost* develops where the ground temperatures remain below 0°C (dry permafrost when water is absent; wet permafrost when water is present). Above the upper surface of permafrost (the permafrost table), an *active layer* experiences annual freeze/thaw cycles. Changes in the thermal regime (e.g. due to construction of poorly insulated buildings) can cause severe engineering problems, including subsidence, frost heave, foundation flexing (especially roads) and shallow landsliding (*gelifluction*)[38.10].

Initial Terrain Models

Initial models can be based on a desk study of available sources. The objective should be to identify the relevant systems and sub-systems, along with the main environmental controls. Global scale differences in climate, tectonic activity and relative relief provide the framework for developing initial terrain models i.e. models of the landscape that incorporate: *surface processes*, *ground materials* and *landforms* (i.e. ground shape or morphology). It is important to realise that the landscape rarely reflects any one climate or period of geomorphological change; it is usually an assemblage of landforms that have superimposed histories (i.e. a *palimpsest*; see Chapter 2).

A selection of generic terrain models, largely based on climate zones, have been included to provide an indication of what should be achievable after a desk study or preliminary assessment of an area (Figs 38.2 to 38.8) (see colour section). These models are broad generalisations and need to be modified or replaced by versions that reflect particular regional or local conditions[38.5].

These generic models can be used to alert engineers to the nature and extent of the hazards and resources that could be expected in different working environments (Table 38.2). For example, an engineer working in a hot desert (e.g. the Sahara) should be aware of the aggressive ground chemistry, wind-blown sand and rare flash floods that are characteristic of this type of environment. If the hot desert is in a tectonic area (e.g. the Atacama Desert, Chile) or a mountain region (e.g. the Zagros Mountains, Iran), then additional potential hazards such as earthquakes, active faulting, liquefaction and large landsliding should be anticipated, investigated and planned for.

Initial terrain models should be the starting point in the process of ensuring that potential problems are identified, evaluated and managed by an engineering project. For example, Fig. 38.9 (see colour section) is a project specific initial model developed to assist the routing of a major highway through Himalayan mountain terrain. Typical mountain terrain issues include[38.11]: major access and routing problems, large landslides, debris flows and torrents, snow melt and flash floods, landslide dams, glacial lake outburst floods, snow and ice avalanches, availability of suitable sources for construction materials.

The terrain model should be developed progressively to become site specific as the understanding of the local conditions improves during the project. These *site specific terrain models* should be used to develop a checklist of questions to be answered during subsequent stages of the investigations.

Initial Terrain Mapping

Satellite imagery and aerial photography (see Chapter 39) can be used to develop and characterise an initial terrain map of the site/area and surrounding landscape:

- Aerial photograph interpretation (API) should be undertaken using a stereoscope and working on acetate overlays to the photographic images.
- Interpretation of digital imagery may be undertaken as an on-screen overlay during the computerised image analysis, forming a component layer of a GIS system.

This initial map will need to be modified and improved as the project progresses through subsequent phases of investigation and design, focussing in more detail on key issues and/or sites.

Preparation of initial terrain maps should involve defining (at a scale appropriate to the project):

1. *The extent and components of earth surface systems*: typical questions that should be considered include:
   - Sediment source areas: what are the dominant processes involved in sediment detachment? (e.g. surface water or wind erosion, landsliding) Is there evidence of recent activity or are the erosional features vegetated and degraded?
   - Sediment transport pathways: is there evidence of temporary storage of sediment within these pathways? (e.g. channel bars, debris lobes) Are the margins of the pathways contributing sediment to the system? (e.g. active channel bank erosion)
   - Sediment accumulation zones: where and how is the transported sediment stored? Is there any evidence of recent deposition? Are the sediment stores connected to other systems? (e.g. toe erosion of landslide debris aprons)

2. *The system environmental controls and energy regime factors* (see Chapter 4): what are likely to be the main drivers of change? (e.g. intense rainstorms, earthquakes, storm surges) What might be the impact of climate change or relative sea-level change?

3. *The system state* (see Chapter 4): what landforms are present? What are their dimensions? Can inferences be made about the materials and groundwater conditions? Are the landforms dynamic (i.e. active or unstable forms) or can they be expected to be unchanging over the project life? (e.g. stable relict or fossil forms) Which landforms are most likely to be sensitive to change?

4. *Preliminary hazard models*: what processes operating in each zone could result in hazards? What triggering factors could be involved in generating hazards? Is there evidence of past hazard events?

There will usually be a need for ground inspection to confirm the desk study results. Surface mapping techniques such as terrain evaluation (Chapter 41) and geomorphological mapping (Chapter 42) can be used to confirm the nature and extent of the systems and landforms and to determine the degree of threat that hazards may pose to the project[38.12].

Geographical Information Systems

Data management is essential to ensure that terrain models and maps can be easily updated and revised as more information becomes available. Geographical Information Systems (GIS; the term Geoscience Information Systems, GSIS is also used for geological applications) are databases that can be used to organise, analyse and display all the available spatial (*geo-referenced*) information that has been collected for particular terrain units or landform (e.g. maps and plans, aerial photographs, borehole logs, geophysical sections)[38.13, 14].

On large construction projects, GIS enables new data on ground conditions to be rapidly assimilated as excavations proceed (e.g. the UK's Channel Tunnel high speed rail link[38.15]). Variations from the anticipated ground conditions or landform changes can be rapidly assimilated and interpreted; terrain models of a site can be continually updated. This approach can support the effective use of the Observational Method and Reference Conditions (Chapter 1).

Table 38.2 Characteristic hazards associated with different climatic zones (see colour section).

Environment	Characteristic hazards
Glacial (Fig. 38.2)	High latitude 'cold ice' glaciers form relatively static ice caps. Lower latitude 'warm ice' creates faster moving active glaciers often constrained within valleys; seasonal meltwater floods with risks of outbursts where meltwater is trapped behind temporary ice or debris barriers; snow avalanches, ice and rockfalls on slopes above the ice. Glaciated terrain may consist of varves (thinly laminated sequences of clays and silts); metastable high slopes; quick clays (weakly cemented open textured marine/glacial sediments with the potential to collapse). The original landscape is often obscured by extensive glacial and meltwater outwash deposits (i.e. transported soils). Weathering tends to be mainly mechanical and actual weathering profiles are shallow.
Periglacial (Fig. 38.3)	Active ground has seasonal freeze/thaw features in upper layers, meltwater floods and summer slope instability (solifluction). Former active ground may contain relict features, e.g. pingos, cryoturbation, thermokarst, valley bulging and cambering, ice fractured parent rocks (Fig. 2.6). Sparse vegetation cover results in active wind erosion and loess deposition. Weathering tends to be mainly mechanical and actual weathering profiles are relatively thin.
Temperate (Humid mid-latitude) (Fig. 38.4)	Pre-existing landslides prone to reactivation; first-time landslides tend to be less frequent. Relict solifluted slopes with pre-existing shear surfaces prone to reactivation, especially by human disturbance. Extensive areas of fine grained (with peat) wetland; alluvial terraces with large range of variable very fine to coarse materials; buried valleys under alluvial or glacial cover; extensive areas of periglacial and glacial deposits from former glaciers. Metastable windblown soil cover (loess, some sand) locally; karst on limestone terrain. Seasonal floods in lowlands.
Temperate (Dry-continental) (Fig. 38.4)	Mudflows and similar old and new slope instability locally common; seasonal floods. Soils commonly finely granular (typically with high loess content) and highly erodible, especially when disturbed by engineering. Sabkha (salt) conditions may be present on low coasts, occasionally inland. Vegetated dunes; pseudokarst locally present, especially on degraded steppe.
Mediterranean (Hot semi dry; semi arid tropical) (Fig. 38.5)	Flash floods along ephemeral streams; seasonal floods on larger rivers; high sediments loads. Widespread landsliding on mudrock and clay slopes; mudslides and mudflows. Little rock weathering but some duricrusts possible on flat-lying surfaces. Some swelling soils locally extensive and terra rossa soils on karstic chalky limestone. Summer salts on coastal lowlands possible.
Hot drylands (Arid tropical) (Fig. 38.6)	Flash floods along ephemeral streams. Landsliding confined to rocky escarpments and mountains (rock slope failures, debris flows). Very little vegetation; sand dune mobility and wind blown sand/dust problems. Sheet and dune sands, poor bearing and borrow potential. Salty-rich groundwater high rates of evapotranspiration result in capillary rise and evaporitic salts on the ground surface (detrimental to aggregates and concrete reinforcement). Falling water tables may occur in urban areas due to over-extraction. Metastable fine soils deposited by wind and flash floods; water and wind erosion particularly where surface cover disturbed. Surface and buried duricrusts with possible evaporite salts redistributed by wind.
Savannah (Hot wet dry; humid and tropical) (Fig. 38.7)	Seasonal flooding of large rivers, flash flooding of ephemeral streams. Landsliding generally confined to steep slopes and rocky escarpments (rock slope failures, debris flows). Granular deposits poorly graded, fine to coarse; often quickly eroded with gullying, piping and badlands. Tropical iron rich weathering profile to most bedrock but highly variable in geometry and state, laterites locally common; sometimes strong catenary relationships exist on slopes.
Hot Wetlands (Humid tropical) (Fig. 38.8)	Slopes protected by dense vegetation but may be prone to frequent landsliding, usually only of the saprolith and/or regolith (e.g. shallow debris slides, channelised debris flows). Landslides occur typically every few decades and are related to high rainfall and progressive development of critical soil depths (slope ripening). Severe surface erosion common in intense rainfall events, particularly where the natural surface cover has been disturbed. Residual soil deep with irregular profile and weathering front, wide joint spacing rocks may develop corestones and residual tors on hillslopes. Presence of corestone boulders and relict discontinuities may be of critical engineering importance for investigation, excavation and stability. Very high rainfall and saturated soils gives groundwater tables at the surface and regular flooding of all lowland areas. Groundwater often rich in organic acids leading to weathering of all unprotected materials (e.g. concrete and reinforcement, steel and masonry culverts).

There are two contrasting GIS structures, both suitable for different applications. The most suitable needs to be defined before time and money are spent collecting and inputting the data:

- *Vector systems*. Vector data comprises lines, arcs or polygons. Terrain units are defined by their boundaries, creating a series of polygons. Thematic information can be stored that describes each polygon. Vector GIS can be used in coastal change, river network analysis and flood flow routing studies, but they are of most value as an automated cartography technique providing landscape models and maps that can easily be updated.
- *Raster systems*. Raster data is an abstraction of the real world where spatial data is expressed as a matrix of cells or pixels, with spatial position implicit in the ordering of the pixels (Fig. 38.10). The spatial data is not continuous but divided into discrete units. This makes raster data particularly suitable for certain types of spatial operation, for example overlays or area calculations.

Each pixel corresponds to a square parcel of land on the ground which has defined attributes, creating a series of overlays (Fig. 38.11). Areas containing the same attribute value (e.g. a terrain unit) are recognized as such; however, raster structures cannot identify the unit boundaries as polygons. Raster methods are appropriate for hazard modelling where the potential is a function of a variety of different factors (e.g. geology, slope angle, soil type).

There are techniques that allow data to be transferred between the two systems; this results in some loss of definition (*resolution*) or accuracy.

Awash Valley Soil Erosion Model, Ethiopia[38.16]

Terrain evaluation methods (Chapter 41) were used to provide the framework for estimating the relative soil erosion hazard (using the USLE, see Chapter 17) within the 120 000 km^2 River Awash Catchment. A simple *raster* GIS was used to store and analyse the data. Once the GIS was set up, the effects on soil erosion rates of both land degradation and soil conservation measures were modelled. This supported the identification of areas that would benefit most from the limited resources available to undertake a soil preservation programme.

Landslide Hazard Mapping in Barbados[38.17]

This was a study of landslide and erosion problems within the 60 km^2 Scotland district of Barbados. By combining a landslide inventory with maps of the controlling factors on landslide occurrence (slope, soil, geology, land use, wet areas and drainage), correlations were established that allowed the production of a *raster* landslide susceptibility map. A landslide hazard map with five ordinal classes was developed to provide a land capability and planning guidance map.

Coastal change, Cumbria, UK[38.18]

Vector-based GIS was used to integrate historical and baseline surveys of the Cumbrian coast, UK, in order to

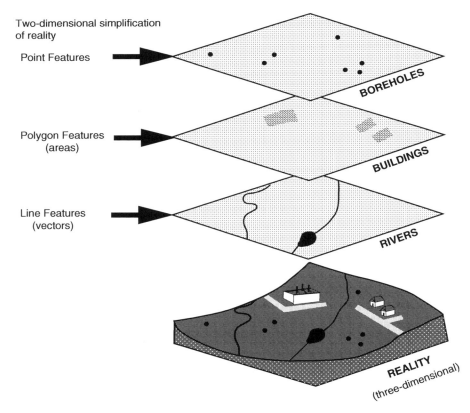

Figure 38.10 GIS – the overlay concept.

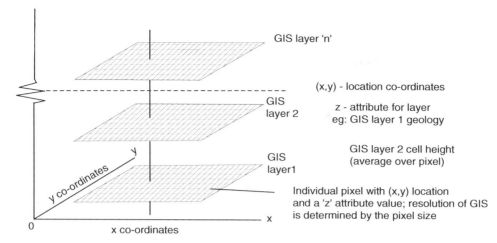

Figure 38.11 Three dimensional arrays in a raster GIS.

analyse coastal change at a nuclear power site. At selected profiles, the position of shoreline features (e.g. cliff edge, dune front) was derived from digitised maps and orthorectified aerial photographs of different dates, to derive a series of vector lines. Changes between dates were measured precisely using a defined datum at the landward edge of each profile. As the accuracy of the maps and aerial photographs could be calculated by error analysis, any changes could be quantified in terms of root mean square (RMS) errors; this provided an indication of the confidence that could be placed in the results.

References

38.1 Fookes, P. G. (1997) Geology for engineers: the geological model, prediction and performance. *Quarterly Journal of Engineering Geology*, **30**, 290–424.

38.2 Fookes, P. G., Baynes, F. J. and Hutchinson, J. N. (2000) Total geological history: a model approach to the anticipation, observation and understanding of site conditions. *GeoEng 2000*, an International Conference on Geotechnical and Geological Engineering, **1**, 370–460.

38.3 Fookes, P. G., Baynes, F. J. and Hutchinson, J. N. (2001) Total geological history: a model approach to understanding site conditions. *Ground Engineering*, March, 42–47.

38.4 Dunne, T. and Leopold, L. B. (1978) *Water in Environmental Planning*. Freeman & Co, San Francisco.

38.5 Fookes, P. G., Lee, E. M. and Milligan, G. (eds.) (2005) *Geomorphology for Engineers*. Whittles Publishing, Caithness.

38.6 Lee, E. M. and Fookes, P. G. (2005) Hot drylands. In P. G. Fookes, E. M. Lee and G. Milligan (eds.) *Geomorphology for Engineers*. Whittles Publishing, 419–454.

38.7 Gregory, K. J. (2005) Temperate environments. In P. G. Fookes, E. M. Lee and G. Milligan (eds.) *Geomorphology for Engineers*. Whittles Publishing, 400–418.

38.8 Douglas, I. (2005) Hot wetlands. In P. G. Fookes, E. M. Lee and G. Milligan (eds.) *Geomorphology for Engineers*. Whittles Publishing, 473–500.

38.9 Thomas, M. F. (2005) Savannah. In P. G. Fookes, E. M. Lee and G. Milligan (eds.) *Geomorphology for Engineers*. Whittles Publishing, 455–472.

38.10 Walker, H. J. (2005) Periglacial forms and processes. In P. G. Fookes, E. M. Lee and G. Milligan (eds.) *Geomorphology for Engineers*. Whittles Publishing, 376–399.

38.11 Charman, J. and Lee, E. M. (2005) Mountain environments. In P. G. Fookes, E. M. Lee and G. Milligan (eds.) *Geomorphology for Engineers*. Whittles Publishing, 501–534.

38.12 Brunsden, D. (2002) Geomorphological roulette for engineers and planners: some insights into an old game. *Quarterly Journal of Engineering Geology and Hydrogeology*, **35**, 101–142.

38.13 Bonham-Carter, G. F. (1997) *Geographical Information Systems for Gescientists: Modelling with GIS*. Pergamon/Elsevier, Rotterdam.

38.14 Culshaw, M. G. (2005) The seventh Glossop lecture – from concept towards reality: developing the attributed 3-D geological model of the shallow subsurface. *Quarterly Journal of Engineering Geology and Hydrology*, **38**, 231–284.

38.15 Beckwith, R. C., Rankin, W. J., Blight, I. C. and Harrison, I. (1996) Overview of initial ground investigations for the Channel Tunnel high speed rail link. In C. Craig (ed.) *Advances in Site Investigation Practice*, Thomas Telford, London, 119–144.

38.16 Griffiths, J. S. and Richards, K. S. (1989) Application of a low-cost database to soil erosion and soil conservation studies in the Awash Basin, Ethiopia. *Land Degradation and Rehabilitation*, **1**, 241–262.

38.17 Hearn, G. J., Hodgson, I. and Woody, S. (2001) GIS-based landslide hazard mapping in the Scotland District, Barbados. In: J. S. Griffiths (ed.) *Land Surface Evaluation for Engineering Practice*. Geological Society Engineering Geology Special Publication No. **18**, 151–158.

38.18 Moore, R., Fish, P., Koh, A., Trivedi, D. and Lee, A. (2003) Coastal change analysis: a quantitative approach using digital maps, aerial photographs and LiDAR. *Coastal Management 2003*. Thomas Telford, 197–211.

39

Satellite Imagery and Aerial Photographs

Introduction

Aerial photographs and digital imagery obtained from satellite and aircraft scanners are basic tools for engineering geomorphology e.g. to interpret ground conditions and identify the presence of hazards (Table 39.1). The quality of the information obtained depends largely on the skill of the interpreter[39.1]. Ground truthing is required to verify the results.

Recent developments have provided opportunities for using increasingly detailed data sets:
- The availability of high resolution digital data (e.g. airborne spectral imagery, IKONOS and QuickBird).
- Developments in ship-borne seabed scanners[39.2], particularly by the hydrocarbons industry.

The Electro-magnetic Spectrum

All remote sensing relies on sensors recording information carried by electromagnetic radiation (EMR, measured in watts, W). The part of the electromagnetic spectrum used in remote sensing is shown in Fig. 39.1:
- The *radiant flux* is the power emanating from a body which, if measured per unit area, is known as the radiant flux density (W m^{-2}).
- The radiant flux density falling on a given surface is known as the *irradiance*, whereas that leaving is the *emittance*.

- In most analytical work the main interest is in the quantity of EMR measured at a specific wavelength or wavelength range, i.e. *spectral radiant flux* or *spectral radiance*.

Remote sensing information may be derived from:
- *Passive approach*: the measurement of *solar* radiation reflected from the ground surface. The spectral characteristics (i.e. wavelengths) of *reflectance* vary with the nature of the ground.
- *Active approach*: the measurement, by a sensor, of radiation reflected by the ground surface *which that sensor has generated*. This work tends to occur in the longer electromagnetic wavebands, such as microwave and radio (e.g. radar).

EMR data are collected in a number of ways:
- *Photographic images* recorded on light sensitive paper of visible light or the near infrared, collected from aerial surveys.
- *Digital data* compiled in a grid (pixel) format, generally associated with satellite platforms.

Aerial Survey

The most useful aerial photographs are taken vertically downward with adjacent photographs overlapping by 60% to enable them to be viewed *stereoscopically* (i.e. in 3D). Where ground control information is available

Table 39.1 Aerial photograph and remote sensing interpretation: indication of the information that should be identifiable.

Topography	Main areas of hills and mountains; rolling or undulating ground; scarps and cliffs; typical hill footslopes; cols, ridges, floodplains and low-lying areas; coastline
Land cover	Forest; woodland; scrubland; grassland; mangrove; bare ground (e.g. rocks, dunes, soil); arable land; human settlement, including roads; derelict land; quarries etc
Drainage	Watersheds; drainage network (rivers, streams, gullies, canals, drainage ditches etc); direction of water flow; lakes; reservoirs; canals; poorly drained areas; swamps; dry rivers; dry valleys
Geomorphology	Landforms and processes
Geology	Structural features (e.g. faults, folds, unconformities, bedding); evidence of rock type, neotectonics

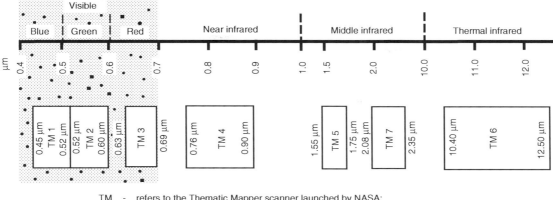

TM - refers to the Thematic Mapper scanner launched by NASA;
TM1, TM2 etc refers to the wavebands
The ground resolution of the TM scanner is c. 30 metres

Figure 39.1 The region of the electro-magnetic spectrum used in remote sensing studies.

(i.e. target sites of known location and height), these photographs can be used to create accurate maps of the ground.

The scale of an aerial photograph can be established in two ways:

$$\text{Scale} = \frac{\text{distance on photograph}}{\text{distance on the ground}}$$

$$\text{or} \quad \frac{\text{focal length of camera lens}}{\text{flying height of aeroplane}}$$

The scale is only *completely accurate* in the centre of the photograph (*focal point*); distortion increases away from this point.

A photo mosaic of controlled aerial photographs can be compiled into a spatially correct image know as an *orthophoto*; these provide an excellent base map for geomorphological mapping (Chapter 42).

Oblique images may also be viewed in stereo and have particular value in studies of steep cliffs and road cuttings.

Airborne Digital Data

High resolution airborne digital imagery covering a wider part of the electromagnetic spectrum is becoming increasingly available. Typically, airborne digital image surveys will acquire data at 20 wavelength ranges (wavebands) with a ground resolution of one metre. Two widely used airborne systems are ATM (airborne thematic mapper) and CASI (compact airborne spectrographic imager).

The data from the scanners are provided as pixels, where each pixel is associated with a digital number (DN). Each pixel represents a grid cell on the ground, the *size* of which defines the resolution and effectively determines the scale of the useable image that can be produced. The value of each pixel, or DN, represents the *spectral attribute* of the ground surface i.e. the intensity of response or average reflectance in a particular waveband for that location.

Airborne data can be subject to a range of digital enhancement techniques (see below). Data can be used

for detailed measurement of the terrain, and monitoring of ground movements if areas are regularly surveyed[39.3–4]. Data can be collected from very narrow wavebands.

Satellite Platforms

Originally developed for military purposes, satellite imagery collected in digital format has provided vast amounts of earth science information. Initially, this was via the ongoing LandSat series of US satellites, but data are now available from French (SPOT), European (ERS), Russian, Indian and Chinese satellites (Table 39.2). Data are available from the mid-70s, and much of the US material can be easily downloaded from the internet (e.g. http://worldwind.arc.nasa.gov/ or http://www.resmap.com/).

Images appropriate for *detailed* geomorphological surveys include:

1. IKONOS data has 1–4 m resolution and can be viewed stereoscopically. IKONOS images are equivalent to a 1:5000 scale aerial photographs, in terms of the ground features that can be identified.
2. QuickBird was launched in 2001 and provides the largest swath width, largest on-board storage, and highest resolution of any currently available commercial satellite. It collects both multi-spectral (2.44 m resolution) and panchromatic (0.6 m resolution) imagery.

Google Earth is a free download (http://earth.google.com) that provides world wide satellite image cover. The package also contains digital terrain model data collected by NASA's Shuttle Radar Topography Mission, allowing areas to be viewed in 3D. It has proved a useful tool for reconnaissance assessments of remote regions, although the image resolution is generally too low for reliable geohazard evaluation.

Digital Image Analysis

Digital data are numeric in form, comprising a multidimensional array of pixels with defined spatial coordinates. DNs range from 0 to 255, representing the spectral attributes for each waveband. As numeric data, digital images are subject to a range of statistical manipulations for the correction of distortions and enhancement of specific details:

- *Pre-processing*: the imagery is initially corrected for radiometric effects, the earth's rotation and platform viewing angle. The data may also be converted to a particular map projection e.g. Universal Transverse Mercator (UTM) and the individual bands georeferenced to each other using *ground control points*.
- *Contrast stretching*: also referred to as *histogram equalisation*, this involves stretching the values of the DNs over the full 0–255 range. This has the benefit of improving the overall contrast within the image, and can be applied to each individual wave band.
- *Spatial filtering*: different types of filtering techniques exist with the aim of enhancing image features e.g. smoothing areas or extracting edges or lines.
- *Multispectral image analysis*: a combination of three bands of data, each assigned a primary colour, are displayed simultaneously to improve the visual interpretability of the image. A True Colour Composite (TCC) is formed if the three bands of the multispectral image consist of the three visual primary colour

Table 39.2 Satellite imagery resolution and spectral spread.

Satellite	Launched	Resolution	Wave bands
Landsat multispectral scanner (MSS)	1972	80 m	4: visible, near infrared, infrared
Landsat thematic mapper (TM) • thermal infrared	1982	30 m 120 m	6: visible–infrared thermal infrared
Landsat enhanced thematic mapper (ETM+) • panchronmatic • thermal infrared	1999	30 m 15 m 60 m	6: visible–infrared visible thermal infrared
SPOT • multispectral • panchromatic	1986	20 m 10 m	3: visible and v. near infrared visible
IKONOS • multispectral • panchronmatic	1999	4 m 1 m	4: blue, green, red, near infrared visible light
QuickBird • multispectral • panchronmatic	2001	2.44 m 0.6 m	4: blue, green, red, near infrared

Table 39.3 Digital image analysis (Thematic Mapper* wavebands).

Wave band	Application
Visible blue-green (TM 1)	Vegetation and forest; sediment laden water (coastal and inland); surface properties of snow and ice; bathymetry
Visible green (TM 2)	Suspended sediment; pedology
Visible red (TM 3)	Healthy vegetation; soils and agriculture; soil and rock boundaries; rock types; human interference in the landscape
Near infrared (TM 4)	Vegetation mapping; drainage network mapping; landform delineation
Middle infrared (TM 5)	Plant moisture and plant stress; differentiating solid and drift geology; ratio of TM bands 5/7 allows clay mineral differentiation; TM 4/5 ratio separates hydrous and iron-rich rocks
Middle infrared (TM 7)	Discriminating metamorphic rocks
TM 3, 2, 1	Simulated natural colour image
TM 4, 5, 3 and 4, 5, 6	False colour composite for soils and geology
TM 2, 4, 5	False colour composite for vegetation, soils and wetlands
TM 4, 5, 7 and 7, 5, 4	False colour composite for geology
TM 1, 3, 5	False colour composite for soils and sediment studies

* Thematic Mapper(TM) scanner mounted on LANDSAT.

bands e.g. bands 1 (blue), 2 (green) and 3 (red) of an IKONOS image. A False Colour Composite (FCC) is formed when the primary colours are assigned image bands arbitrarily e.g. Landsat ETM+ wave bands 4, 5 and 7, sharpened using the higher resolution panchromatic data, were used to create a FCC for mapping landslides in Nepal and Bhutan[39.5].

- *Principal Components Analysis* (PCA): a mathematical transformation that compresses the multi-band imagery, which may be highly correlated, into fewer (new) uncorrelated bands. Since the original bands may be displaying the same features, PCA minimises data redundancy and produces a new data set which may better classify the land surface.

Details of the techniques available are contained in a range of specialist texts[39.6-8]. Table 39.3 provides a list of waveband combinations that have proved of value in geomorphological and geological analysis.

References

39.1 Fookes, P. G., Dale, S. and Land, J. M. (1991) Some observations on a comparative aerial photography interpretation of a landslipped area. *Quarterly Journal of Engineering Geology*, **25**, 313–330.

39.2 Badman, T. D., Gravelle, M. A. and Davis, G. M. (2000) Seabed imaging using a computer mapping package: an example from Dorset. *Quarterly Journal of Engineering Geology*, **33**, 171–175.

39.3 Chandler, J. H. (2001) Terrain measurement using automated digital photogrammetry. In J. S. Griffiths (ed.) *Land Surface Evaluation for Engineering Practice*. Engineering Geology Special Publication, No. 18, 13–18.

39.4 Lane, S. N., Chandler, J. H. and Richards, K. S. (1998) Landform monitoring, modelling and analysis: landform in geomorphological research. In: S. N. Lane, K. S. Richards and J. H. Chandler (eds.) *Landform Monitoring, Modelling and Analysis*. Wiley, Chichester.

39.5 Hart, A. B., Hearn, G. J., Petley, D. N., Tiwari, S. C. and Giri, N. K. (2003) *Using remote sensing and GIS for rapid landslide hazard assessment: potential public sector uptake in Nepal and Bhutan*. Report to the UK Department for International Development (DFID).

39.6 Graham, R. and Koh, A. (2002) *Digital Aerial Survey: Theory and Practice*. Whittles Publishing, Caithness.

39.7 Mather, P. M. (2004) *Computer processing of remotely-sensed images: an introduction*. 3rd Edition, Wiley, London.

39.8 Rencz, A. N. and Ryerson, R. A. (1999) *Remote Sensing for the Earth Sciences. Manual of Remote Sensing, Volume 3*. Wiley, London.

40

Analysis of Change

Historical Records

There can be a wide range of sources that provide information on the past occurrence of events, including: aerial photographs, topographic maps, satellite imagery, public records, local newspapers, consultants' reports, scientific papers, journals, diaries, oral histories and interviews with local people. It is essential to consider the reliability of the data source[40.1]:

- What type of event is being recorded, and with what detail?
- Who is making the report, in particular how qualified are they to comment on the event? Is their report based on personal observation and experience, or on an editing of reports from other people, who themselves may have modified the facts i.e. a plausible rumour, or even a complete invention or falsification?
- In the light of knowledge of this type of event, is the report credible as a whole, only in part, or not at all?

The problems with historical data can be minimised[40.2]:

- Never assume that the whole event population is represented.
- Regard surveys and diaries as a time-specific sample of the data and base judgements on the quality of the observations.
- Use as many data sources as possible and compare trends or extremes.
- Compare the data with other independently collected data series (e.g. climatic records).
- Always assume changes in reporting quantity and quality over time.
- Never assume that the present standard of recording is better than in the past.

Historical Maps

Historical topographical maps and charts provide a record of the former positions of various features. In many cases, historical maps and charts may provide the only evidence of landform change over the previous 100 years or more. When compared with recent surveys or photographs, they can provide the basis for estimating cumulative land loss and the average annual erosion rate

between map/survey dates. Great care is needed in their use because of the potential problems of accuracy and reliability[40.3].

In the UK, for example, the Ordnance Survey map-based record of historical landscape change consists of a number of separate measurements made, typically, five times or less over the last 200 years or so (Table 40.1). The record is often insufficient to identify the pattern and size of individual events that led to the cumulative land loss between the measurement dates; it reveals only a partial picture of the past processes.

Two main methods are often used to measure changes of positions of features from historical sources (Fig. 40.1):

- *Measurement of distance changes* along evenly spaced transect lines drawn normal to the feature (e.g. landslide backscar or cliffline). An average rate of change is obtained by dividing the distance change by the time interval between surveys. A frequent problem using this method is the need to identify fixed points (e.g. the corner of a building, property boundary) that are common to all map editions.
- *Measurement of areal changes* between selected positions at different survey dates, along segments of uniform length. The area of land loss or gain between each successive position can be measured using a planimeter or by counting squares, and is converted to an average annual rate by dividing by the segment length and the time interval between surveys.

Historical Map Accuracy

Among the most important problems when using historical maps are:

1. *Plotting errors*: although the positional accuracy of many defined objects on Ordnance Survey maps in Great Britain is around ± 0.8 m, inaccessible features of marginal importance situated away from settlements may not be mapped with comparable accuracy. The error x at any one check point on a map is defined as the vectorial distance between the

Table 40.1 Historical topographic maps available in Great Britain.

1. One-inch maps (1:63 360 scale) commenced by the Ordnance Survey in 1805 and available for most of southern England and Wales by 1840. Superceded by the 1:50 000 scale series.
2. Six-inch maps (1:10 560 scale) commenced by the Ordnance Survey in 1840. First revision took place 1891–1914. Superceded by the 1:10 000 scale series.
3. 25-inch maps (1:2500 scale) commenced by the Ordnance Survey in 1853; the first detailed survey was completed in 1895. First revision took place 1891–1914.

'County' maps date from Elizabethan times to the 19th century and are of variable scale and quality.

Large-scale tithe, enclosure and estate maps are available in manuscript for many areas.

From 1836 to 1860 a series of *Tithe Survey* map was prepared in connection with the Tithe Commutation Acts. These maps are very detailed topographical surveys, usually at a scale of 13.3 or 26.7 inches/mile, and they exist for thousands of parishes. The *Enclosure Maps*, often at a similar scale to the Tithe Maps, are generally earlier, often dating from the first decades of the nineteenth century.

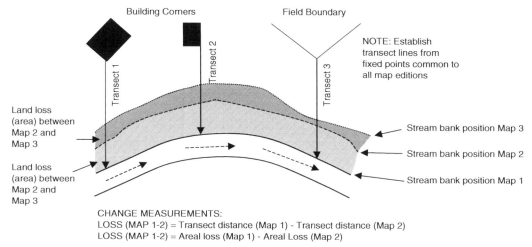

Figure 40.1 Examples of measurement of change from historical maps.

point on the ground (E, N) and the point on the map (E', N'):

$$\delta E = E - E'$$
$$\delta N = N - N'$$
$$x = \sqrt{\delta E^2 + \delta N^2}$$

The *root mean square* (RMS) error is used to evaluate the overall accuracy of a survey in plan and elevation. RMS error is defined as:

$$r = \sqrt{\frac{\Sigma x^2}{n}}$$

where $x_1, x_2, \dots x_n$ are the errors at n check points.

If the errors are random then $x = 0$ and the *consistency* of the survey is the RMS error. If $x \neq 0$ there is a systematic error s in the data given by

$$s = \bar{x} = \frac{\Sigma x}{n}$$

The consistency of the survey is found by taking the systematic error from the individual errors at each check point to give the *standard error* σ.

$$\sigma = \sqrt{\frac{\Sigma (x - \bar{x})^2}{n}} = \sqrt{r^2 - s^2}$$

The standard error removes the systematic component of the inaccuracy to provide an estimate of the range either side of the accepted true observation within which true results may be expected to lie. Measures of the accuracy of Ordnance Survey maps in Great Britain are presented in Table 40.2.

2. *Interpretative errors*: when the feature to be mapped is not clearly defined in the field, its position may be based on a surveyor's perception of its form; plotting on different editions or different sheets of the same edition may be sensitive to operator variance.
3. *Revisions*: not all features are revised for each new map edition, so it is sometimes uncertain exactly when a particular feature was last revised.
4. *Accuracy of comparisons*: the validity of the measurements can be defined by the plotting errors associated with the different map editions. Error estimates can be produced for each map period as follows:

$$E = \frac{eT_1 + eT_2}{T}$$

where E = error estimate associated with the map period (m yr^{-1}); eT_1 = plotting error of the feature on map edition 1; eT_2 = plotting error of the feature on map edition 2 and T = time period between map editions.

Table 40.2   Typical errors on Ordnance Survey maps in Great Britain.

Type of survey	RMS error on ground	RMS error on map
1:1250 resurvey and continuous revision	0.4 m	0.32 mm
1:2500 resurvey and continuous revision	0.8 m	0.32 mm
1:2500 overhaul and continuous revision	2.5 m	1.00 mm
1:10 000 resurvey and continuous revision	3.5 m	0.35 mm

Table 40.3   A selection of error estimates on UK Ordnance Survey Maps at different locations.

Location	Map period	Map scale	eT_1 (m)	eT_2 (m)	T (years)	E (m)
A	1850–1890	1:10 560	3.5	3.5	40	0.18
B	1880–1946	1:10 560	3.5	3.5	66	0.11
C	1901–1960	1:2500	0.8	0.8	59	0.03
D	1909–1975	1:2500	0.8	0.8	66	0.02
E	1893–1977	1:2500	0.8	0.8	84	0.02
F	1893–1966	1:2500	0.8	0.4	73	0.01
G	1872–1976	1:1250	0.4	0.4	104	0.008

The error estimate E indicates the minimum change rate that can be resolved; this will vary with map scale and the period between editions (Table 40.3).

When E is greater than or equal to the average annual rate measured between the map editions, no reliable estimate is possible. Greater precision is possible for longer map periods because plotting errors become proportionally less as the cumulative change increases; accurate data are needed to resolve recession over short time periods, especially when rates are slow.

5. *Distortions*: in many cases it will not be possible to work directly from the original historical maps, as they are rare, fragile and valuable documents. There will be distortions in the photocopying or reprinting process of around $\pm$ 1% across a map sheet.

Visual Inspection

In many situations *visual inspection* can help in identifying future problem sites and keeping them under review until a more formal measurement and recording strategy becomes necessary. As the written word and the unaided memory are inadequate for detecting anything other than gross changes, such inspections should be carried out systematically with data recorded using proformas and accompanied by a timed and dated photographic record. The comparison of sequential photography taken during successive inspections can be helpful in defining the nature and extent of change in active systems[40.4].

Ground Survey

Conventional *ground survey* methods can give detailed measurements of change; this may involve simple taping from fixed points, traditional levelling or a total station or Global Positional System (GPS). A number of strategies can be adopted, including:

1. *Occasional surveys* of the relevant features after specific events (e.g. a major landslide) or periods of active river bank erosion.
2. *Formal systematic surveys* of the position of features (e.g. channel cross sections, beach profiles or cliff tops) at fixed points, undertaken at regular intervals (e.g. on an annual or 6 monthly basis). Establishment of a network of reference points can provide a baseline against which future changes can be compared (Fig. 40.2).

For example, on the Holderness coast, UK, the local authority installed a series of 71 marker posts at 500 m intervals along 40 km of coastline, each post located at a distance of between 50 m and 100 m normal to the coast. These posts are replaced further inland from time to time if they become too close to the cliff top. Annual measurements from each post to the cliff top, defined as the lip of the most recent failure scar, commenced in 1953; average annual recession rates of 1.8 m yr^{-1} have been recorded[40.5].

Analytical Photogrammetry

Photogrammetry is the process of obtaining accurate measurements, maps and digital elevation models from photographs. The technique can be used for quantifying the nature and extent of landform changes between aerial photographs of different dates by comparing the 3D

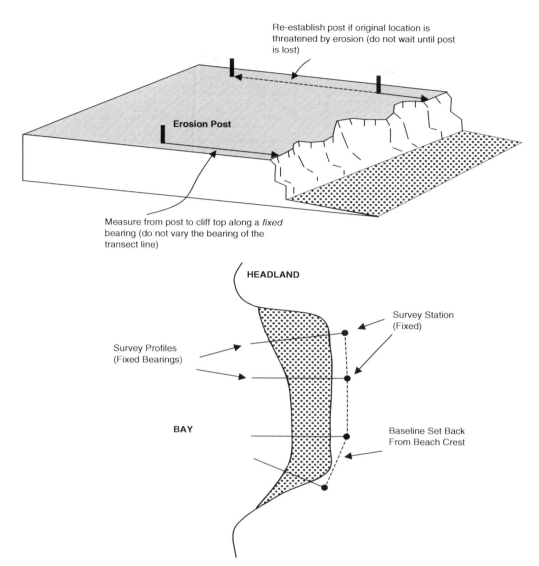

Figure 40.2 Erosion posts (top) and beach profile surveys (bottom).

co-ordinates of the same points. The method was used to evaluate the nature and scale of slope instability and cliff recession over a 40-year period in a highly active coastal cliff on the west Dorset coast, UK[40.6]. A series of *digital surface models* were compiled. Surface models of the differences between photograph dates were generated by subtracting one surface model from another. These were contoured and used to directly identify areas experiencing erosion and deposition over the intervening years.

Analytical photogrammetry can be combined with data held in GIS to create 3D ground models. Regular surveys can then be used to create a 4D model that incorporates changes through time, and can be used to extrapolate future situations e.g. recession of the cliffs at Barton-on-Sea, UK[40.7].

LiDAR

Light Detection and Radar survey (LiDAR) is a relatively cheap, laser system that records distance measurements from the scanning laser on an aircraft to surface features. The LiDAR system effectively 'paints' the ground surface of the earth with laser pulses that record the elapsed time until the return of the pulse:

Range = (Speed of Light × Time of Pulse) × 0.5

When these data are combined with information about the position and attitude of the sensor the result is a dense sampling of elevation points. LiDAR can measure object-heights (producing *digital surface models*) and ground heights (producing *digital terrain models*) in a single scan. Repeated scans allow the development of digital surface change models.

LiDAR data can be collected day or night. Horizontal accuracy ranges (depending on flying height) from 0.4–1.5 m. Vertical accuracy ranges from 0.1–0.6 m. Annual LiDAR surveys have been used to monitor patterns of erosion and accretion on the Cumbrian coast, UK[40.8].

Movement Monitoring

A wide range of instrumentation and techniques may be applied to measure surface movement of unstable slopes[40.9, 10]. These range from low-cost routine measurement of *pins* and *peglines* to more sophisticated automatic monitoring and early warning systems such as extensometers, tiltmeters and settlement cells. Measurement of sub-surface movement can be determined using inclinometer systems and slip indicators.

Dating

When historical records are not available, dating methods often provide the only means of establishing the pattern of events through time and defining links with potential causes. For example, a clustering of events around a particular date might be associated with deteriorating climatic conditions or a specific triggering event such as a large earthquake[40.11].

Relative Dating

Relative dating can be established through morphological characteristics, degree of surface weathering, re-vegetation rate and stratigraphic relationships revealed in pits and exposures. Shallow debris slides, for example, can be dated with a reasonable degree of precision through their presence or absence on aerial photographs taken at different times. In Hong Kong, the Natural Terrain Landslide Inventory used vegetation cover and re-vegetation rates on the scars of old landslides to provide an indirect measure of the date of shallow hillside failures[40.12]. Landslide source areas in Hong Kong start to re-vegetate within 2–3 years and reach 70% vegetation after 15–30 years. A 4-fold classification was developed:

- Class A: totally bare of vegetation, assumed to be less than 3 years old
- Class B: partially bare of vegetation, assumed to be between 2–3 and 30 years old
- Class C: completely covered in grasses, assumed to be more than 15 years old
- Class D: covered in shrubs and/or trees, assumed to be more than 25 years old.

Absolute Dating

Absolute dating (i.e. obtaining a reasonably precise date for the event) is potentially more valuable and a number of methods are available. They fall into two main categories: incremental and radiometric.

Incremental dating methods include:

- *Dendrochronology*, whereby the age of death of a tree (e.g. buried beneath a landslide) may be determined by cross reference to a master tree ring chronology for that region (in the USA, a continuous bristlecone pine tree ring chronology extends back to 8200 BP). The oldest rings of trees that have colonised a now stabilised slide or debris fan will give a minimum age for the event.
- *Varve chronology*: varves (lake bed seasonally alternating bands or laminations of fine sands and silts/clays) arise from annual variations in the supply of sediment. Where the varve record can be tied into a calendar timescale the absolute dates can be identified. Suitable varve deposits are often associated with glaciolacustrine sediments and have been used in studies on the late-glacial and Holocene periods (<15 000 BP)[40.13].
- *Lichenometry*: Lichens are complex organisms consisting of algae and fungi in a symbiotic relationship that grow at defined rates. The dating limit is around 4500 years in extremely cold and dry continental regions; elsewhere the practical limit is 500 years or less[40.14].

Radiometric dating methods include:

- *Radioactive dating*, in which measurement of the ratio of stable (^{12}C) to radioactive (^{14}C) isotopes in an organic sample (e.g. charcoal, wood, peat, roots, macrofossils) allows the determination of the time elapsed since biological or inorganic fixation. The age range for radiocarbon dating is from a few centuries to around 40 000 years. For example, the Holocene landslide history of the Lake Zurich basin, Switzerland was established by radiocarbon dating of organic material from the lake bed sediments[40.15].
- *Luminescence dating*, based on estimating the time since a sediment was last exposed to daylight, which 'zeroes' the previously accumulated radiation damage to minerals (e.g. quartz or feldspar) in the sample. The age of a sample is derived from:

$$\text{Age} = \frac{\text{palaeodose}}{\text{dose rate}}$$

where the palaeodose is the accumulated radiation damage and the dose rate is the rate at which the sample absorbs energy from the immediate proximity.

Palaeodoses are calculated by thermoluminescence (TL) – energy is supplied by an oven – or optically stimulated luminescence (OSL). The maximum age range for these methods depends on the mineral. For quartz it is around 100 000–150 000 years; for feldspars it is around 800 000 years. The OSL signal is reset by exposure to daylight more completely than the TL signal; as a result OSL has provided more precise ages and allows younger samples to be dated (e.g. within the last 50–100 years). Loesses have been successfully dated up to 800 000 years old. Other applications include dating alluvial fans and aeolian deposits. A large colluvium lobe (1 Mm^3) on North Lantau, Hong Kong, was shown to have been deposited within the last 2000–10 000 years[40.16].

- *Uranium series dating* is applicable for the time range <1000 to over 500 000 years. It involves testing surface sediments to establish the radioactive disequilibrium in the decay chains of Uranium and related isotopes such as Thorium (e.g. the ratio between ^{230}Th and ^{234}U). The method has been used to date corals[40.17], calcretes and halite deposits, or crusts beneath landslides.
- *Cosmogenic nuclide dating*, whereby the concentration of cosmogenic nuclides (e.g. ^{3}He, ^{10}Be, ^{14}C, ^{21}Ne, ^{26}Al and ^{36}Cl) produced by *in situ* nuclear reactions (the interaction between cosmic rays and

terrestrial atoms) in the upper few metres of the ground surface are detected by accelerator mass spectrometry. The concentrations can be used to determine the length of time that material has spent at or near the ground surface. These methods have been used to date rockfalls, basalt flows and alluvial fan surfaces. The datable range is from around 1000 years to several million years.

- *Cation-ratio rock varnish dating method*, based on the observation that the ratio of certain cations (K, Ca, Titanium) in rock varnish decreases with age as a result of preferential leaching. Cation-ratios can be calculated by photon-induced x-ray emission spectrometry, electron microprobe or inductively coupled plasma spectroscopic (ICP) methods. Comparing ratios from a number of samples can generate relative ages. Calibrated ages can be determined by comparison with curves generated for samples of known age.

- *Amino acid dating methods* provide relative ages by measuring the extent to which certain amino acids within protein (e.g. bones, mollusc shells, eggshells) residues have transformed from one of two chemically identical forms to the other (e.g. the transformation from L-form to D-form amino acids) until equilibrium is reached. The methods can be used for timescales from a few years to hundreds of thousands of years.

Interpretation and Reliability

Interpretation of dating results needs to be based on:

1. An appreciation of the *context* for the dated sample i.e. the position within the feature, the type of event, its morphology and expected dynamics;
2. Awareness of the *precision* (i.e. how well the result was determined) and *accuracy* (i.e. how close the result is to the true value) – these are independent of each other. It is important to be mindful that the standard deviation is as informative as the mean value; there is a 68.3% probability that the event will have occurred in the time span from $t - 1$ standard deviation to $t + 1$ standard deviation.

References

40.1 Potter, H. R. (1978) *The use of historical records for the augmentation of hydrological data*. Institute of Hydrology Report No. 46. Wallingford.

40.2 Brunsden, D., Ibsen, M.-L., Lee, E. M. and Moore, R. (1995) The validity of temporal archive records for geomorphological purposes. *Quaestiones Geographicae Special Issue* **4**, 79–92.

40.3 Hooke, J. M. and Redmond, C. E. (1989) Use of cartographic sources for analysing river channel change with examples from Britain. In G. E. Petts, H. Moller and A. L. Roux (eds.) *Historical Change of Large Alluvial Rivers: Western Europe*, 79–94. John Wiley and Sons.

40.4 Grainger, P. and Kalaugher, P. G. (1991) Cliff management: a photogrammetric monitoring system. In R. J. Chandler (ed.) *Slope Stability Engineering*. Thomas Telford, 119–124.

40.5 Pethick, J. (1996) Coastal slope development: temporal and spatial periodicity in the Holderness Cliff Recession. In M. G. Anderson and S. M. Brooks (eds.) *Advances in Hillslope Processes*, **2**, 897–917.

40.6 Chandler, J. H. and Brunsden, D. (1995) Steady state behaviour of the Black Ven mudslide: the application of archival analytical photogrammetry to studies of landform change. *Earth Surface Processes and Landforms*, **20**, 255–275.

40.7 Moore, R., Fish, P., Glennerster, M. and Bradbury, A. (2003) Cliff behaviour assessment: a quantitative approach using digital photogrammetry and GIS. Proceedings of the *38th DEFRA Conference of River and Coastal Engineers*, 08.3.1–08.3.13.

40.8 Moore, R., Fish, P., Koh, A., Trivedi, D. and Lee, A. (2003) Coastal change analysis: a quantitative approach using digital maps, aerial photographs and LiDAR. *Coastal Management 2003*. Thomas Telford, 197–211.

40.9 Mikkelsen, P. E. (1996) Field instrumentation. In A. K. Turner and R. L. Schuster (eds.) *Landslides: Investigation and Mitigation*, Transportation Research Board, Special Report 247, National Research Council, National Academy Press, Washington DC, 278–318.

40.10 Lawler, D. M., Thorne, C. R. and Hooke, J. M. (1997) Bank erosion and instability. In C. R. Thorne, R. D. Hey and M. D. Newson (eds.) *Applied Fluvial Geomorphology for River Engineering and Management*, John Wiley and Sons Ltd., 137–172.

40.11 Lang, A., Moya, J., Corominas, J., Schrott, L. and Dikau, R. (1999) Classic and new dating methods for assessing the temporal occurrence of mass movements. *Geomorphology*, **30**, 33–52.

40.12 Evans, N. C., Huang, S. W. and King, J. P. (1997) *The Natural Terrain Landslide Study: Phase* III. Hong Kong GEO Special Project Report SPR 5/97.

40.13 O'Sullivan, P. E. (1983) Annually-laminated lake sediments and the study of Quaternary environmental changes – a review. *Quaternary Science Reviews*, **1**, 245–313.

40.14 Innes, J. L. (1983) Lichenometric dating of debris flow depsoits in the Scottish Highlands. *Earth Surface Processes and Landforms*, **8**, 579–588.

40.15 Strasser, M., Schnellmann, M. and Anselmetti, F. S. (2005) Seismic vs. climatic triggering of mass movements: the Lake Zurich event stratigraphy. *NGF Abstracts and Proceedings*, Geological Society of Norway, **2**, 83–84.

40.16 King, J. P. and Choi, A. S. W. (2003) *Luminescence Dating of Colluvium and Landslide Deposits in Hong Kong*. GEO Report No. 134.

40.17 Muhs, D. R. (2002) Evidence for timing and duration of the last interglacial period from high precision uranium-series ages of corals on tectonically stable coastlines. *Quaternary Research*, **58**, 36–40.

41

Terrain Evaluation

Introduction

Terrain Evaluation is a method for organising and communicating earth science information or intelligence to civil and military engineers, land use planners, agriculturalists and many others[41.1–2]. Terrain evaluation has proved to be a cost and time effective technique for covering relatively large areas quickly and providing the basis of an initial terrain model (Chapter 38). It has been used for:

- Rapid assessments of remote, inaccessible regions where the basic data sources are limited, especially for dams and reservoirs, bridge crossings, road and pipeline routes[41.3–4].
- Major infrastructure projects in developed regions (e.g. the Channel Tunnel high speed rail link[41.5]).

Terrain evaluation (TE) involves classifying the landscape by dividing it into landform *assemblages* with similarities in terrain, soils, vegetation and geology (i.e. *land systems mapping*; Fig. 41.1). TE generally does not involve ground exploration by excavation (except using small hand-dug pits or hand auger holes) or geophysics.

Principles

The principles of terrain evaluation are[41.6–7]:

- It is possible to define areas of terrain that have similar physical characteristics i.e. a typical range of topographic, hazard, constructability or landscape sensitivity factors.
- These areas of terrain can be considered to present a consistent level of challenges that will need to be addressed by a project.
- Areas can be ranked in terms of the degree of relative difficulty they present to a project e.g. from *showstoppers* to *easygoing*. When the areas are classified and put on a map it provides a readily accessible picture of what challenges the project needs to address in particular regions.

At the broadest scale, landscape types (*terrain models*) can be defined (e.g. mountains, desert plateaux, coastal plains); this level of sub-division may be suitable for pre-feasibility overviews of very large areas.

Within a landscape type it will be possible to identify a variety of landform assemblages (*terrain units*) such as river floodplains, escarpment faces, extensive areas of unstable hillslopes and plateaux surfaces; this level of detail may be sufficient, for example, in route corridor assessment. Within a *terrain unit* there will be numerous individual landforms (*terrain sub-units*) that will each present slightly different challenges to a project. An escarpment face, for example, may contain a variety of sub-units, including bare rock faces and discrete landslide systems separated by stable ridges and spurs. The distribution of these sub-units will be relevant to the more detailed studies.

Terrain Evaluation Process

Terrain evaluation involves a number of discrete stages[41.6]:

1. *Desk study*: development of preliminary terrain models from available published materials and anticipating the potential problems that may be expected in the area. Often, it will not be practical to collect and collate all the existing data. It can prove helpful to categorise data sources into those that are likely to be *essential*, *desirable* or *useful* to the project. Experience has shown that essential sources include topographic maps, geological maps and memoirs, and some form of remote sensing data (e.g. satellite imagery or aerial photographs).

2. *Refining the terrain models*, through field visits and further investigations during the engineering planning and design process. Field verification of the checklists and terrain maps produced during the desk study stage is essential. Effort should be directed towards identifying whether the anticipated terrain problems actually occur within the area of interest.

   A combination of fly-over (e.g. helicopter) and drive-through or walk-over should prove sufficient to obtain an overview of the issues that are likely to influence decision-making through subsequent stages of the project. Issues may come to light during a field visit that could not have been inferred from the desk-study data; other issues may prove to be less relevant than originally perceived.

3. *Evaluating the terrain problems*; further investigations should concentrate on evaluating the relative

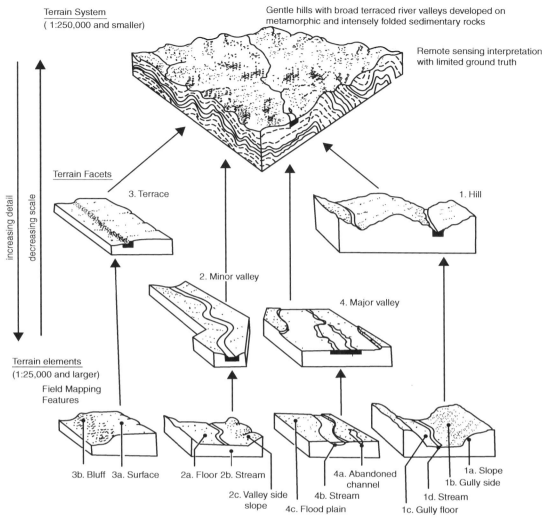

Terrain System
(1:250,000 and smaller)

Gentle hills with broad terraced river valleys developed on
metamorphic and intensely folded sedimentary rocks

Remote sensing interpretation
with limited ground truth

increasing detail
decreasing scale

Terrain Facets

3. Terrace

1. Hill

2. Minor valley

4. Major valley

Terrain elements
(1:25,000 and larger)

Field Mapping
Features

3b. Bluff 3a. Surface

2a. Floor 2b. Stream

2c. Valley side
slope

4a. Abandoned
channel

4b. Stream

4c. Flood plain

1a. Slope
1b. Gully side

1d. Stream

1c. Gully floor

Figure 41.1 Relationship between terrain systems, terrain facets and field mapping of terrain elements (after Cooke and Doornkamp 1990)[41.2].

significance of different terrain problems within the area of interest. A variety of approaches are available for evaluating the significance of geohazards and engineering cost drivers, ranging from qualitative scoring systems to semi-quantitative and quantitative risk assessments (see Chapter 12). The focus should be on providing estimates of the potential impact on the construction costs (e.g. the implications of very strong rock on the excavation rate) or operational integrity (e.g. the likelihood of pipeline rupture or exposure). Terrain units should form the spatial framework for these estimates.

4. *Communicating with the project team*: a successful terrain evaluation exercise is one that raises the awareness of important terrain issues within the engineering project team. This should help ensure that potential problems are not overlooked but given adequate attention at the appropriate stage.

The TE output is usually a suite of maps specific to the client (see Table 41.1), often stored as a series of overlays in a GIS (Chapter 38). An extended legend is normally attached to each of the maps and most terrain evaluation

studies would include an interpretative report. 3D block diagrams of the main terrain models have proved to be a useful vehicle to convey the basic message at early stages of a project.

Example 1: Pipeline Route Selection, Algeria[41.8]

The objective of this study was to produce an overview of the terrain conditions for the alignment corridor of a proposed 500 km pipeline in a hyper-arid part of central Algeria. The work involved the identification of generic terrain models, or landscape types (e.g. Fig. 41.2). Each terrain unit was described in terms of the anticipated occurrence of hazards to construction, using a 5-class scoring system (Table 41.2). Terrain maps were produced at 1:100 000 scale.

The work was mainly based on analysis of 1:100 000 scale SPOT and Landsat TM satellite imagery. This was followed by 20 man-days ground-truth mapping in a 4-wheel drive vehicle at an average of 20 km day^{-1} involving a 3-person team (engineering geologist, engineering geomorphologist, geotechnical pipeline engineer). During the drive-over, judgements were made on the ground conditions and recorded while on the move. This

Table 41.1   Terrain evaluation maps.

Terrain Evaluation Map Category	Examples of Typical Maps
Element maps: factual maps that record the actual ground conditions	• morphology • topography • bedrock geology • superficial geology • lithology • vegetation • pedology • land use • geotechnical characteristics • location of sites of special scientific interest • location of exploratory holes and wells • hydrology – surface water
Derivative maps: interpretative maps that are obtained by either combining element maps or are based on an interpretation of the element maps	• slope steepness using topography to classify maps into distinct groups based on morphology and topography • depth to bedrock utilising data from geological maps • geomorphology • hydrogeology based on the geology, hydrology and well data • depiction of various types of resources, such as sand and gravel or brick clay • foundation conditions for engineering structures • geotechnical zoning i.e. areas of homogeneous ground conditions • hazards such as subsidence, landslides, flooding or contaminated land • previous industrial usage
Summary maps: combine a range of derivative and element maps	• development potential • potential resources • planning constraints, including statutory protected land • construction constraints • summary hazards

was supplemented by field walk-over inspections every 5–10 km. The approach was more cost and time effective than a systematic borehole investigation over the whole length of the route; this was proved during the subsequent design and construction estimates.

Example 2: Landslide Hazard in the Rio Aguas Catchment, Southeast Spain[41.9–10]

Terrain evaluation methods were used to support a landslide hazard assessment for development planning in the region. A landslide inventory of the 550 km^2 catchment was compiled through field mapping and remote sensing interpretation, using 1: 13 333 colour and 1:30 000 scale black and white aerial photographs (Chapter 39). Geomorphological and geological data was compiled for each terrain unit and stored in an ArcView© GIS allowing the landslide distribution to be analysed and related to the terrain characteristics of the area. The results were summarised using a terrain model (Fig. 41.3).

Example 3: N6 Kinnegad to Athlone Dual Carriageway, Ireland[41.11]

Terrain evaluation methods were used to support a major investment programme (£3.65B; 2000 prices) to improve the Irish National Road System. The work involved the identification of landform assemblages, which would have similar constraints to the design and construction of a new highway section.

The investigation for a new dual carriageway between Kinnegad and Athlone covered some 600 km^2 in a corridor with a maximum width of 30 km. Reconnaissance mapping was carried out at 1:50 000 scale, defining a range of terrain units. Information was summarised on the geomorphology and terrain character, materials (i.e. likely ground conditions), constraints to road development and the construction resources. On the basis of this reconnaissance work the corridor was reduced to four potential route alignments that could be studied in greater detail.

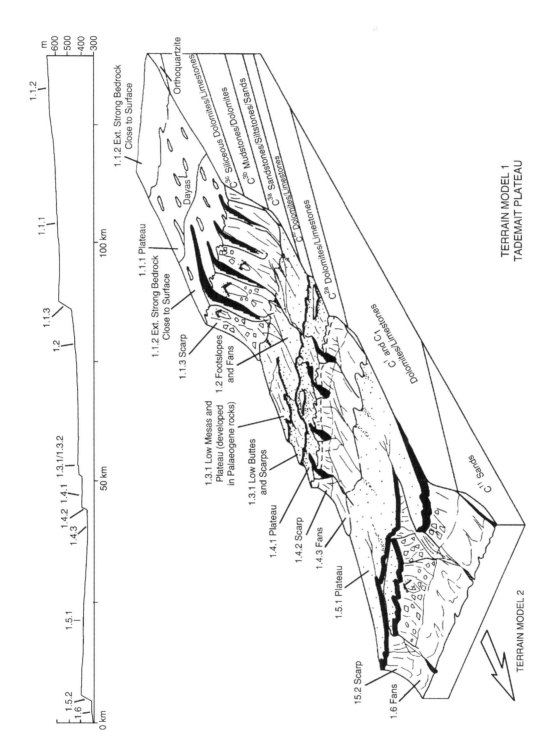

Figure 41.2 Tademait Plateau terrain model, Algeria (after Fookes *et al.* 2001)[41.8].

Table 41.2   In Salah gas project, Algeria: Terrain evaluation classes[41.8].

	Class 1	Class 2	Class 3	Class 4	Class 5
Undulations/ Relief	Sensibly planar	0–2 m over 0.1–1 km	2–10 m over 0.1–1 km	5–50 m over 0.25–2 km	20–100 m over 0.25–2 km
Fluvial systems	Sheet flow and infiltration	As 1 plus fluvial channels < 5 m wide, 1 m deep	As 2 but channels < 20 m wide, 3 m deep	As 3 but channels < 100 m wide, 5 m deep	As 4 but channels > 100 m wide, 5 m deep
Flooding	None	Minor, not significant	Likely, but minor realignment training works required	Very likely, training works/realignment required	Very likely, best avoided
Trafficability (cf. Land Rover)	Reasonable road	Tracks may exist, usually no bogging down. Some need for low ratio 4×4	No tracks. All low ratio 4 × 4, some bogging down	As 3 but lots of bogging down and/or many detours required	Not sensibly viable
Slope instability	No instability seen or expected	None or very little instability seen, although could be expected in certain circumstances	As 2 with some instability seen (avoid or take the risk)	Instability commonly seen and expected (avoid or engineer against)	Very extensive chaotic instability; frequent failure (avoid)
Gypsum heave	None seen or expected	None or very little instability seen, although could be expected in certain circumstances	Heave seen, with vertical movement up to 0.5 m	Heave common, with vertical movement up to 2 m	As 4, but movement > 2 m
Pavements and surfaces	Sand sheet surface, no pavement	1 layer of gravel, particles not touching	1 layer of gravel, particles touching	River alluvium	Fan deposits
Duricrust	None seen or expected	Crust (i.e. BS 5930 up to MW e.g. gypcrete)	Floating crete (i.e. BS 5930 > MW e.g. calcrete, siliceous calcrete)	Caprock (i.e. surface enriched bedrock)	Crete and caprock

Gypsum heave can also be considered a 'salt aggression' index (1⇒ not a problem; 5 ⇒ severe problem), but salt aggression also depends on moisture and other salts (e.g. chlorides). MW = moderately weak.

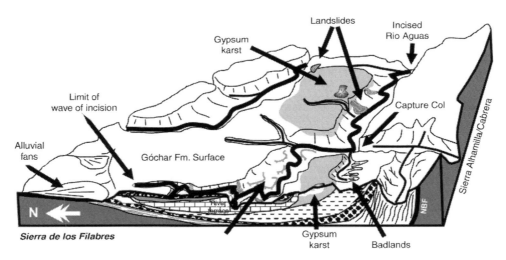

Figure 41.3   Landscape model of the Rio Aguas area, southeast Spain (after Hart 2004)[41.10].

References

41.1   Mitchell, C. W. (1973) *Terrain Evaluation*. Longman.

41.2   Cooke, R. U. and Doornkamp, J. C. (1990) *Geomorphology in Environmental Management*. 2nd Edition. Oxford University Press.

41.3   Kreig, R. A. and Metz, M. C. (1989) Recent advances in route selection and cost estimating methodology for pipelines in Arctic and Subarctic regions. In *International Symposium on Geocryological Studies in Arctic Regions*, Yamburg, USSR, August 1989.

41.4   Kreig, R. A. and Reger, R. D. (1976) Preconstruction terrain evaluation for the Trans-Alaska Pipeline Project. In Coates, D. R. (ed.), *Geomorphology and Engineering*, the 7th Annual Geomorphology Symposium, New York, Sept, 1976, 55–76.

41.5   Waller, A. M. and Phipps, P. (1996) Terrain systems mapping and geomorphological studies for the Channel Tunnel rail link. In C. Craig (ed.) *Advances in Site Investigation Practice*, 25–41.

41.6   Charman, J. H., Fookes, P. G., Hengesh, J. V., Lee, E. M., Pollos-Pirallo, S., Shilston, D. T. and Sweeney, M. (2005) Terrain ground conditions and geohazards: evaluation and implications for pipelines. In M. Sweeney (ed.) *Terrain and Geohazard Challenges Facing Onshore Oil and Gas Pipelines*, Thomas Telford, 78–94.

41.7   Griffiths, J. S. and Edwards, R. J. G. (2001) The development of land surface evaluation for engineering practice. In J. S. Griffiths (ed.) *Land Surface Evaluation for Engineering Practice*, Geological Society Special Publication 18, Geological Society Publishing House, Bath, 3–9.

41.8   Fookes, P. G., Lee, E. M. and Sweeney, M. (2001) Terrain evaluation for pipeline route selection and characterisation, Algeria. In J. S. Griffiths (ed.) *Land Surface Evaluation for Engineering Practice*, Geological Society Special Publication 18, Geological Society Publishing House, Bath, 115–121.

41.9   Griffiths, J. S., Hart, A. B., Mather, A. E. and Stokes, M. (2005) Assessment of some spatial and temporal issues in landslide initiation within the Río Aguas Catchment, Southeast Spain. *Landslides*, **2**, 183–192.

41.10 Hart, A. B. (2004) *Landslide investigations in the Rio Aguas catchment, Southeast Spain*. Unpublished PhD thesis, University of Plymouth.

41.11 Phipps, P. J. (2002) Engineering geological constraints for highway schemes in Ireland: N6 Kinnegad to Athlone dual carriageway case study. *Quarterly Journal of Engineering Geology and Hydrogeology*, **35**, 233–246.

42

Geomorphological Mapping

Introduction

Geomorphological mapping is used to record the nature and properties of the landforms and surface processes present in the landscape. The methodology is applicable to both aerial photograph interpretation (API; Chapter 39) and field mapping. Mapping based on API should be checked in the field (*ground-truthed*) where possible.

Geomorphological mapping for engineering is used for a variety of purposes, including:

- Development of an initial terrain model to act both as a framework for designing the ground investigation plan and to allow interpolation and extrapolation of the sample data derived from the results of the ground investigations.
- Identification of the general terrain characteristics of a route corridor or proposed construction site, including suggestions for alternatives alignments or sites.
- Distribution of materials that might potentially be of use in construction (i.e. potential borrow sites).
- Assessment of hazards (both within and outside the site/corridor) that might impact on the project e.g. landsliding, flooding, subsidence, active faults.
- Description of the drainage characteristics, location and pattern of surface and subsurface drainage, nature of drainage measures required, identified flood stage heights, area of unstable channels.
- Characterization of the nature and extent of weathering.

There are two main stages in geomorphological mapping:

- morphological mapping;
- geomorphological interpretation.

The process is described by referring to an initial terrain model, compiled by mapping the landscape in the Axminster area of East Devon[42.1] (Fig. 42.1) (see colour section). The maps derived from this are presented in Fig. 42.2 (see colour section).

Morphological Mapping

(See Fig. 42.2A, colour section.) This involves systematically recording the shape (*morphology*) of the ground surface on a map, or orthophoto. The objective is to sub-divide the land surface into *facets* separated by gradual changes or sharp breaks in slope. The changes and breaks in slope are identified as either *concave* or *convex* in nature and recorded using decorated lines, and arrows indicate the slope direction and angle. This is the factual data collection or 'element map' stage identified in terrain evaluation procedures (see Chapter 41).

The accuracy of the map is dependent on the quality of the base maps used, the techniques employed for identifying location in the field and the experience of the mapper. Depending on the scale of the final map that is needed, mapping onto a base plan using tape, compass and clinometer can provide data of sufficient accuracy for most purposes. The use of a hand-held Global Positioning System (GPS) has proved to be a major step forward in improving mapping accuracy without the need to employ more expensive geodetic survey techniques.

Geomorphological Interpretation

(See Figs 42.2B to 42.2F, colour section.) All the breaks and changes of slope recorded in the morphological mapping have to be interpreted. They represent the boundaries within and between landforms. These are a reflection of either the underlying geology (e.g. lithological benches), past surface processes (e.g. landslides) or human activity (e.g. quarrying). Interpretation is a subjective exercise and the accuracy of the interpretation is dependent on the skills and experience of the geomorphologists carrying out the work.

To assist in the interpretation it can be useful initially to produce a series of maps using the morphological map as a base:

1. Morphographic map: graphical representation and description of the landforms (Fig. 42.2B) (see colour section)
2. Morphochronological map: age of processes, landforms and materials (Fig. 42.2C) (see colour section)
3. Morphogenetic map: relict and active processes that have created the observed landscape (Fig. 42.2D) (see colour section)
4. Resource Map: displaying the range of available natural resources (Fig. 42.2E) (see colour section)

Castle Hill Landslide -Geomorphological Interpretation

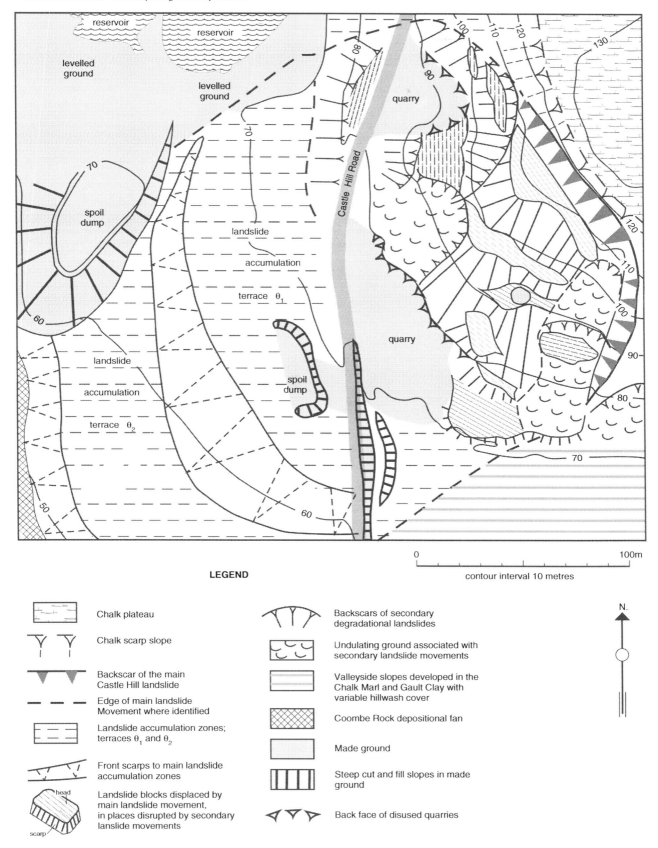

LEGEND

	Chalk plateau
	Chalk scarp slope
	Backscar of the main Castle Hill landslide
	Edge of main landslide Movement where identified
	Landslide accumulation zones; terraces θ₁ and θ₂
	Front scarps to main landslide accumulation zones
	Landslide blocks displaced by main landslide movement, in places disrupted by secondary lanslide movements
	Backscars of secondary degradational landslides
	Undulating ground associated with secondary landslide movements
	Valleyside slopes developed in the Chalk Marl and Gault Clay with variable hillwash cover
	Coombe Rock depositional fan
	Made ground
	Steep cut and fill slopes in made ground
	Back face of disused quarries

Figure 42.3 Example of a large-scale geomorphological map: Channel Tunnel portal (after Griffiths *et al.* 1995)[42.8].

5. Hazard Map: location of the natural or man-made hazards (Fig. 42.2F) (see colour section).

The geomorphological interpretation should be based on a clear understanding of the bedrock and superficial geology and the industrial archaeology (quarrying and mining has had a major impact in many countries). Standard geomorphological symbols can be used on all these maps[42.2–3]. In many situations geomorphological maps are produced with a bespoke legend for the client.

Geomorphological Map Scales

Three forms of geomorphological map can be identified, generally based on their scale:

1. *Small to medium-scale regional surveys of terrain conditions* (1:1 000 000 to 1:25 000), generally for feasibility studies, land use planning or baseline studies for environmental impact assessment. For example, a 1:100 000 scale geomorphological map of part of the Grand Erg Occidental, Algeria was produced from satellite image interpretation and a helicopter reconnaissance survey. This map was used to assist the routing of the In Salah Gas pipeline around this sand sea[42.4]. The combination of remote sensing and limited ground-truth checks is typical in the production of these regional surveys.

2. *Medium-scale assessments of resources or hazards* (1:50 000 to 1:10 000). An example of this type of work is provided by the Bahrain Surface Materials Resources Survey (1:10 000 scale)[42.5] and the Earth Science Mapping series (1:25 000 scale) produced for the UK Department of the Environment during the 1980s and 1990s[42.6]. The maps of the St Helens area, UK, provide the basic earth science data for land use planning. They also provided engineers with a comprehensive dataset for use in the initial desk study stages of a project.

3. *Specific-purpose large-scale.* These surveys delineate and characterise particular landforms and have direct engineering applications[42.7]. For example, geomorphological mapping was carried out as part of the investigations at the Channel Tunnel Portal and Terminal areas near Folkestone[42.8] (Fig. 42.3). The objective of the mapping exercise was to delimit the nature of past and contemporary landslide activity around the site. Field mapping was undertaken at a scale of 1:500, supported by the interpretation of 1:5000 scale aerial photographs. The mapping was carried out using simple equipment (a 100 metre tape, prismatic compass and clinometer). The boundaries of the landslides shown on earlier geological maps were refined; detail was provided on the form and complexity of the landslide units. This complexity could then be allowed for during the interpretation of borehole data and for the engineering design. It helped define the parts of the various landslides that needed to be drained and established where toe loading could be most effectively placed.

Mapping at a large-scale has to be predominantly field-based, although it will often be supported by remote sensing studies. A typical rate for large-scale field mapping in landslide terrain by experienced personnel will be less than 0.5 km^2 per person per day.

References

42.1 Croot, D. and Griffiths, J. S. (2001) Engineering geological significance of relict periglacial activity in South and East Devon. *Quarterly Journal of Engineering Geology & Hydrogeology*, **34**, 269–281.

42.2 Cooke, R. U. and Doornkamp, J. C. (1990) *Geomorphology in Environmental Management*. 2nd Edition, Oxford University Press, Oxford.

42.3 Demek, J. and Embleton, C. (eds.) (1978) *Guide to Medium-scale Geomorphological Mapping*. International Geographical Union, Stuttgart.

42.4 Fookes, P. G., Lee, E. M. and Sweeney, M. (2005) In Salah Gas project, Algeria – part 1: terrain evaluation for desert pipeline routing. In M. Sweeney (ed.) *Terrain and Geohazard Challenges Facing Onshore Oil and Gas Pipelines*, Thomas Telford, 144–161.

42.5 Doornkamp, J. C., Brunsden, D., Jones, D. K. C. and Cooke, R. U. (1980) *Geology, Geomorphology and Pedology of Bahrain*. GeoBooks, Norwich.

42.6 Smith, G. J. and Ellison, R. A. (1999) Applied geological maps for planning and development: a review of examples from England and Wales. *Quarterly Journal of Engineering Geology*, **32**, S1–S44.

42.7 Brunsden, D., Doornkamp, J. C., Fookes, P. G., Jones, D. K. C. and Kelly, J. M. H. (1975) Large scale geomorphological mapping and highway engineering design. *Quarterly Journal of Engineering Geology*, **8**, 227–253.

42.8 Griffiths, J. S., Brunsden, D., Lee, E. M. and Jones, D. K. C. (1995) Geomorphological investigations for the Channel Tunnel terminal and portal. *The Geographical Journal*, **161**(3), 275–284.

43

Uncertainty and Expert Judgement

Uncertainty

Some degree of *uncertainty* will exist in the knowledge of any site or area. Key types of uncertainty include:

1. *Data uncertainty*: there will inevitably be limitations to the accuracy and precision of models of complex systems. Data uncertainty can arise because of *measurement errors* (random and systematic) or *incomplete data*. In many situations the available measurements will not correspond to the processes or event types that need to be addressed. As a result, the conditions will need to be inferred (e.g. interpolated, extrapolated or analytically derived) from other information. There may be imperfect understanding regarding the processes involved or the applicability of transferring knowledge from one site to another.

2. *Environmental* (*'real world'*) *uncertainty*: some aspects may defy precise predictions of future conditions. For example:
   - It may not be possible to predict the full range of events that might occur because of the limited understanding of the nature and behaviour of complex systems.
   - Future choices by governments, business or individuals will affect the socio-economic and physical environments in which surface processes and hazards operate. There is little prospect of reliably predicting what these choices will be.

Geomorphological assessments need to be supported by a clear statement of the uncertainties in order to "inform all the parties of what is known, what is not known, and the weight of evidence for what is only partially understood"[43.1].

This is not a straightforward process as it is difficult to characterise uncertainty without making the problems appear larger or smaller than the experts believe it to be. Careful and elaborate characterisation of the uncertainties might be incomprehensible to non-specialists and unusable by decision-makers. A balance needs to be found between providing sufficient information on uncertainty to make engineers aware of the issues, and diverting attention away from the reality of the situation by dwelling on the unknown.

Judgement

Many assessments will rely on *expert judgement*, based on available knowledge plus experience gained from other projects and sites. Engineering geomorphologists need to have the ability to make rational judgements in the face of imperfect knowledge. Such judgements are inevitably *subjective*.

Reliable estimates can be developed by:
- proposing a range of possible scenarios
- systematic testing of these scenarios through additional investigation and group discussion
- elimination of non-credible scenarios
- establishing agreement between different team members.

If a project requires a *calculated risk* to be taken then the probabilities of different scenarios must be clearly identified[43.2].

Problems Associated with Expert Judgement

There are problems associated with the use of the subjective judgemental approach, especially where it is undertaken by single individuals:

- *Poor quantification of uncertainty*: where uncertainty concerning surface processes is ignored or not expressed in a consistent fashion. For example, if it is estimated that there is a 90% probability of an event in a given year, it should be expected to happen on average 9 times out of 10; the probability could be lower in reality.
- *Poor problem definition*: where the focus of the investigation is directed towards one element of a system at the expense of another, because of the assessor's experience and background (e.g. a flood specialist providing a landslide hazard assessment).
- *Cognitive bias*: where the assessor's judgement does not match the available facts, thereby introducing bias. For example, greater weight may be given to recent laboratory test results and stability analyses than to the known historical record of events and the performance history of the slope. There is also a common tendency for an assessor to underestimate the *uncertainty* associated with values of parameters

produced by laboratory testing, such as shear strength[43.3].

Managing the Expert Judgement Process

A range of techniques are available for reducing the effects of these potential problems. These techniques (arranged in order of perceived robustness) will help ensure that the judgements are *defensible* (e.g. in legal proceedings) should it become necessary to resolve controversy at a later date (Fig. 43.1).

- *Self assessment* where the rationale behind the judgement is well documented, including a description of the available information and the methods of analysis and interpretation, to enhance the defensibility of the judgement. The method may not, however, overcome many of the biases inherent in an individual's perception[43.4]. However, this is the most commonly used approach in construction claims or litigation wherever an adversarial legal processes is used (e.g. in the UK under Common Law, and in the USA).
- *Independent review* (*informal expert opinion*): obtaining a second opinion from an expert or colleague. As for self assessment, the expert's assessment should be well documented and open for review. Although an improvement over self assessment, similar problems may remain, especially if the expert is influenced by the same biases as the original assessor.
- *Calibrated assessments* (*formal expert opinion*): involving an independent review of the original assessment and an assessment of the individual's biases, for example, by a peer group review. The increased cost and difficulties in identifying and objectively quantifying the biases are the main drawbacks of this approach. This approach is now widely used in major international projects, such as those funded by the World Bank.
- *Open forum*: this relies on the open discussion between team members to identify and resolve the key issues related to the hazard. The results can be distorted by the dynamics of the group, such as domination by an individual because of status or personality.
- *Delphi panel* is a systematic and iterative approach to achieve consensus. Each individual in the group is provided with the same set of background information and is asked to conduct and document (in writing) a self-assessment. These assessments are then provided anonymously to the other assessors, who are encouraged to adjust their assessment in light of the peer assessment. Typically, the individual assessments tend to converge. Such iterations are continued until consensus is achieved. The technique maintains anonymity and independence of thought; it reduces the possibility that any one member of the panel may unduly influence any other.
- *Probability encoding*; involving the training of staff to produce reliable assessments of the probability of various events in a formal manner. This is the most systematic and defensible approach to developing subjective probability assessments, but also the most expensive.

Defensibility and Accountability

Judgements about hazards, risks and impacts may provoke considerable disagreement and controversy. In an increasingly litigious world there is a need to demonstrate that geomorphologists have acted in a professional manner appropriate to the circumstances. In Fig. 43.1 increasing defensibility can be seen to be linked with increasing sophistication of the assessment approach[43.5].

When decisions are based on judgements it is important that effort is directed towards ensuring that the judgements can be justified through adequate documentation, allowing any reviewer to be able to trace the reasoning behind particular estimates, scores or rankings. Ideally the assessment process should involve a group of experts, rather than single individuals, as this facilitates the pooling of knowledge and experience, as well as limiting bias.

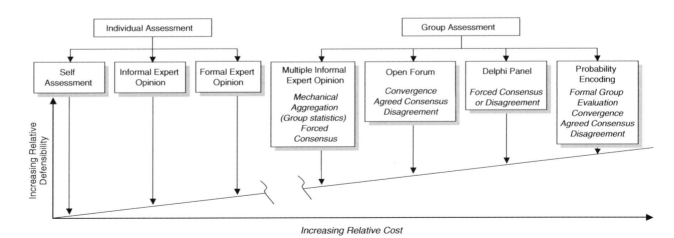

Figure 43.1   Defensibility and approach to investigation and judgement (Note: it is unlikely that there will be a linear increase in defensibility with cost) (after Fookes 1997)[43.5].

References

43.1 Stern, P. C. and Fineberg, H. V. (eds.) (1996) *Understanding Risk: Informing Decisions in a Democratic Society*. National Academy Press, Washington DC.

43.2 Kiersch, G. A. and James, L. B. (1991) Errors of geologic judgement and the impact on engineering works. In G. A. Kiersch (ed.) *The Heritage of Engineering Geology: the First Hundred Years*, Geological Society of America Centennial Special Volume 3, Boulder, Colorado, 516–558.

43.3 Whitman, R. V. (1984) Evaluating calculated risk in geotechnical engineering. *Journal of Geotechnical Engineers*, **110**, 145–168.

43.4 Fookes, P. G., Dale, S. G. and Land, J. M. (1991) Some observations on a comparative aerial photograph interpretation of a landslipped area. *Quarterly Journal of Engineering Geology*, **24**, 249–266.

43.5 Fookes, P. G. (1997) First Glossop Lecture: Geology for engineers: the geological model, prediction and performance. *Quarterly Journal of Engineering Geology*, **30**, 290–424.

Key References

Essentials

Brunsden, D. (2002) Geomorphological roulette for engineers and planners: some insights into an old game. *Quarterly Journal of Engineering Geology and Hydrogeology*, 35, 101–142.

Cooke, R. U. and Doornkamp, J. C. (1990) *Geomorphology in Environmental Management*. 2nd Edition. Oxford University Press.

Cooke, R. U., Brunsden, D., Doornkamp, J. C. and Jones, D. K. C. (1982) *Urban geomorphology in Drylands*. Oxford University Press.

Fookes, P. G. (1997) Geology for engineers: the geological model, prediction and performance. *Quarterly Journal of Engineering Geology*, 30, 290–424.

Fookes, P. G., Baynes, F. J. and Hutchinson, J. N. (2000) Total geological history: a model approach to the anticipation, observation and understanding of site conditions. GeoEng 2000, an International Conference on Geotechnical and Geological Engineering, 1, 370–460.

Fookes, P. G., Lee, E. M., and Milligan, G. (eds.) (2005) *Geomorphology for Engineers*. Whittles Publishing, Caithness.

Hutchinson J. N. (2001) The Fourth Glossop Lecture: Reading the Ground: Morphology and Geology in Site Appraisal. *Quarterly Journal of Engineering Geology and Hydrogeology*, 34: 7–50.

Lee, E. M. and Jones, D. K. C. (2004) *Landslide Risk Assessment*. Thomas Telford.

Part 1: Landform change

Ang, A. H.-S. and Tang, W. H. (1984) *Probability Concepts in Engineering Planning and Design. Volume II Decision, Risk and Reliability*. John Wiley and Sons, Chichester.

Barry, R. G., Chorley, R. and Chase, T. (2003) *Atmosphere, Weather and Climate*. Routledge, London.

Benjamin, J. R. and Cornell, C. A. (1970) *Probability, Statistics and Decision for Civil Engineers*. McGraw-Hill Book Company, New York.

Brunsden, D. (1996) Geomorphological events and landform change. *Zeitschrift fur Geomorphologie*, 40, 273–288.

Chorley, R. J. and Kennedy, B. A. (1971) *Physical Geography: a Systems Approach*. Prentice-Hall, London.

Chorley, R. J., Schumm, S. A. and Sugden, D. E. (1984) *Geomorphology*. Methuen, London.

Coates, D. R. (1976) *Geomorphology and Engineering*. George Allen and Unwin.

Coates, D. R. (ed.) (1976) *Urban Geomorphology*. Geological Society of America Special Paper 174.

Cooke, R. U. (1984) *Geomorphological Hazards in Los Angeles*. London Research Series in Geography 7. George Allen and Unwin.

Costa, J. E. and Fleisher, P. J. (eds.) (1984) *Developments and Applications of Geomorphology*. Springer, Berlin.

Craig, R. G. and Croft, J. L. (1982) *Applied Geomorphology*. George Allen and Unwin, London.

Glasson, J., Therivel, R. and Chadwick, A. (1998) *Introduction to Environmental Impact Assessment (Natural and Built Environments)*. Routledge, London.

Goudie, A. S. (1992) *Environmental Change*. Oxford University Press, Oxford.

Griffiths, J. S. and Culshaw, M. G. (2004) Seeking the research frontiers for UK engineering geology. *Quarterly Journal of Engineering Geology and Hydrogeology*, 37, 317–325.

Griffiths, J. S. and Hearn, G. J. (1990) Engineering geomorphology: a UK perspective. *Bulletin of the International Association of Engineering Geology*, 42, 39–44.

Hardy, J. (2003) *Climate Change: Causes, Effects and Solutions*. J Wiley and Sons, Chichester.

Hearn, G. J. (2002) Engineering geomorphology for road design in unstable mountainous areas: lessons learnt after 25 years in Nepal. *Quarterly Journal of Engineering Geology and Hydrogeology*, 35, 143–154.

Hurrell, J. (1995) Decadal trends in the North Atlantic Oscillation: regional temperature and precipitation. *Science*, 269, 676–679

Knill, J. L. (2003) Core values: the First Hans Cloos Lecture. *Bulletin of Engineering Geology and the Environment*, 62, 1–34.

Lamb, H. H. (1995) *Climate, History and the Modern World*. 2nd Edition. Routledge, London.

Lowe, J. J. and Walker, M. J. C. (1997) *Reconstructing Quaternary Environments*. Longmans, Harlow.

McCall, G. J. H, Laming, D. J. C. and Scott S. C. (eds.) (1992) *Geohazards: Natural and Man-made*. Chapman and Hall, London.

McGuire, B., Mason, I. and Kilburn, C. (2002) *Natural Hazards and Environmental Change*. Arnold, London.

Peck, R. B. (1969) Advantages and limitations of the observational approach in applied soil mechanics. *Géotechnique*, 19, 171–187.

Peltier, W. R. (2002) On eustatic sea level history: Last Glacial Maximum to Holocene. *Quaternary Science Reviews*, 21, 377–396.

Philander, S. G. H. (1990) *El Niño, La Niña and the Southern Oscillation*. Academic Press, San Diego, CA.

Phillips, J. D. (1999) *Earth Surface Systems: Complexity, Order and Scale*. Malden, MA: Blackwell.

Pirazzoli, P. A. (1996) *Sea-level Changes: the last 20,000 years*. John Wiley and Sons, Chichester.

Pye, K. (ed.) (1994) *Sediment Transport and Depositional Processes*. Blackwell, Oxford.

Roberts, N. (1998) *The Holocene*. 2nd edition, Blackwell, Oxford.

Robson, A. J. and Reed, D. W. (1999) *Flood Estimation Handbook Vol 3: Statistical procedures for flood frequency estimation*. Institute of Hydrology, Wallingford.

Schumm, S. A. (1977) *The Fluvial System*. Wiley, New York.

Smith, K. (2001) *Environmental Hazards: Assessing Risk and Reducing Disaster*. 3rd Edition. Routledge, London.

Summerfield, M. A. (1991) *Global Geomorphology*, Longman Group Ltd., Harlow.

Walker, M. (ed.) (2002) *Guide to the construction of reinforced concrete in the Arabian Peninsula*. Concrete Society/CIRIA. 214.

Part 2: Slopes

Bromhead, E. N. (1992) *The Stability of Slopes*. 2nd edition, Blackie & Son, Glasgow.

Brunsden, D. (1973) The application of system theory to the study of mass movement. *Geologica Applicata E Idrogeologia*, **VIII**, 185–207.

Brunsden, D and Prior, D. B. (eds.) (1984) *Slope Instability*. John Wiley and Sons, New York.

Clark, A. R., Lee, E. M. and Moore, R. (1996) Landslide Investigation and Management in Great Britain: A Guide for Planners and Developers. HMSO, London.

Crozier, M. J. (1986) *Landslides: Causes, Consequences and Environment*. Croom Helm, London.

Cruden, D. and Fell, R. (eds.) (1997) *Landslide Risk Assessment*. Balkema, Rotterdam

Jenning, J. N. (1985) *Karst Geomorphology*. Blackwell, Oxford

Lancaster, N. (1995) *Geomorphology of Desert Dunes*. Routledge, London.

McKee, E. D. (ed.) (1979) *A Study of Global Sand Seas*. US Geological Survey Professional Paper 1052.

Sidle, R. C., Pearce, A. J. and O'Loughlin, C. L. (1985) *Hillslope Stability and Land Use*. American Geophysical Union, Washington DC.

Stipho, A. (1992) Aeolian sand hazards and engineering design for desert regions. *Quarterly Journal of Engineering Geology*, **25**, 83–92.

Turner, A. K. and Schuster, R. L. (eds.) (1996) *Landslides: Investigation and Mitigation*. Transportation Research Board, Special Report 247, National Research Council, National Academy Press, Washington DC.

US Soil Conservation Service (1972) National engineering handbook, Section 4: Hydrology. US Department of Agriculture, Washington.

Part 3: Rivers

Bridge, J. S. (2003) *Rivers and Floodplains*. Blackwell, Oxford.

Chanson, H. (2004) *Environmental Hydraulics for Open Channel Flow*. Elsevier Butterworth Heineman, Amsterdam

Environment Agency (1998) River Geomorphology: a Practical Guide. R & D Publications, Environment Agency, Bristol.

Ives, J. D. (1986) *Glacial lake outburst floods and risk engineering in the Himalayas*. ICIMOD Occasional Paper No. 5, Kathmandu, Nepal: International Centre for Integrated Mountain Development.

Knighton, A. D. (1998) *Fluvial Forms and Processes: a New Perspective*. Arnold, London.

Miller, E. W. and Miller, R. M. (2000) *Natural Disasters: Floods — A Reference Handbook*. Contemporary World Issues S. ABC-CLIO.

Munich Re. (1997) *Flooding and Insurance*. Munich Reinsurance Company, Munich.

Penning-Rowsell E. C., Parker D. J. and Harding D. M. (1986) *Floods and drainage*. George Allen and Unwin.

Richards, K. S. (1982) *Rivers: form and process in alluvial channels*. Methuen Press.

Smith, K. and Tobin, G. A. (1979) *Human adjustment to the flood hazard*. Longman.

Thorne, C. R., Hey, R. D. and Newson, M. D. (eds.) (1997) *Applied Fluvial Geomorphology for River Engineering and Management*. John Wiley and Sons Ltd.

US Army Corp of Engineers (1993) *Engineering and Design: River Hydraulics*. US Army Engineering Manual EM 1110-2-1416, Washington DC.

US Army Corp of Engineers (1994) *Engineering and Design: Channel Stability Assessment for Flood Control Projects*. US Army Engineering Manual EM 1110-2-1418, Washington DC.

Part 4: Coasts

Bird, E. (2000) *Coastal Geomorphology: an introduction*. John Wiley and Sons, Chichester.

Carter, R. W. G. (1988) *Coastal Environments*. Academic Press.

Carter, R. W. G. and Woodroffe, C. D. (eds.) (1994) *Coastal Evolution: Late Quaternary Shoreline Morphodynamics*. Cambridge University Press.

Carter, R. W. G., Curtis, T. G. F. and Sheeby-Skeffington, M. J. (eds.) (1992) *Coastal Dunes: Geomorphology, Ecology and Management for Conservation*. Balkema, Rotterdam.

CIRIA (1996) *Beach Management Manual*. Report 153. Construction Industry Research and Information Association, London.

Davies, J. (1972) *Geographical Variation in Coastline Development*. Oliver and Boyd.

Dronkers, J. and van Leussen, W. (eds.) (1988) *Physical Processes in Estuaries*. Berlin, Springer-Verlag.

Dyer, K. R. (1986) *Coastal and Estuarine Sediment Dynamics*. Chichester, John Wiley and Sons.

French, P. W. (1997) *Coastal and Estuarine Management*. London, Routledge.

Hayes, M. O. (1979) Barrier island morphology as a function of tidal and wave regime. In S. P. Leatherman (ed.) *Barrier Islands*. Academic Press, New York, 1–27.

Komar, P. D. (1998) *Beach Processes and Sedimentation*. 2nd Edition. Prentice Hall, Upper Saddle River.

Leatherman, S. P. (1982) *Barrier Island Handbook*. National Park Service, Boston.

Nummedal, D. (1983) Barrier Islands. In P. D. Komar (ed.) *CRC Handbook of Coastal Processes and Erosion*. CRC Press, Inc., Boca Raton, FL, 77–121.

Orford, J. D., Carter, R. W. G. and Jennings, S. C. (1996) Control domains and morphological phases in gravel-dominated coastal barriers. *Journal of Coastal Research*, **12**, 589–605.

Pethick, J. S. (1984) *An Introduction to Coastal Geomorphology*. Edward Arnold, London.

Silvester, R. and Hsu, J. R. C. (1993) *Coastal stabilisation: innovative concepts*. Prentice Hall.

US Army Corps of Engineers (2002) *Coastal Engineering Manual*. Manual No. 1110-2-1100. Washington DC.

van der Meulen, F., Jungerius, P. D. and Visser, J. H. (eds.) (1989) *Perspectives in Coastal Dune Management*, SPB Academic Press.

Woodroffe, C. D. (2002) *Coasts: Form, Process and Evolution*. Cambridge University Press.

Wright, L. D. (1985) River deltas. In R. A. Davies (ed.) *Coastal Sedimentary Environments*. Springer Verlag, New York.

Part 5: Investigation techniques

Anon (1972) The preparation of maps and plans in terms of engineering geology. *Quarterly Journal of Engineering Geology*, **5**, 293–381.

Burrough, P. A. and McDonnel, R. A. (1998) *Principles of GIS*. University Press, Oxford.

Carr, A. P. (1980) The significance of cartographic sources in determining coastal change. In R. A. Cullingford, D. A. Davidson and J. Lewin (eds.) *Timescales in Geomorphology*, 67–78. John Wiley and Sons.

Dearman, W. L. (1991) *Engineering Geological Mapping*. Butterworth-Heinemann, Oxford.

Fookes, P. G., Sweeney, M., Manby, C. N. D. and Martin, R. P. (1985) Geological and geotechnical engineering aspects of low-cost roads in mountainous terrain. *Engineering Geology*, **21**, 1–152, Elsevier, Amsterdam.

Griffiths, J. S. (compiler) (2002) Mapping in Engineering Geology. *Key Issues in Earth Sciences*, **1**, The Geological Society, London.

Griffiths, J. S. (ed.) (2001) *Land Surface Evaluation for Engineering Practice*. Engineering Geology Special Publication, No. 18.

Hooke, J. M. and Kain, R. J. P. (1982) *Historical Change in the Physical Environment: a Guide to Sources and Techniques*. Butterworth.

James, L. B. and Kiersch, G. A. (1991) Failures of engineering works. In G. A. Kiersch (ed.) *The Heritage of Engineering Geology: the First Hundred Years*. Geological Society of America Centennial Special Volume 3, Boulder, Colorado, 481–516.

Lawrance, C. J., Byard, R. J. and Beaven, P. J. (1993) *Terrain Evaluation Manual*. State of the Art Review No.7, Transportation Research Laboratory, HMSO, London.

Paine, D. P. and Kiser, J. D. (2003) *Aerial photography and image interpretation*. 2nd Edition, Wiley, London.

Peck, R. B. (1973) Influence of non-technical factors on the quality of embankment dams. In R. C. Hirschfield and S. J. Poulos (eds.) *Embankment Dams: Casagrande Volume*. John Wiley and Sons, New York, 201–208.

Peck, R. B. (1980) Where has all the judgement gone? *Canadian Geotechnical Journal*, **17**, 584–590.

Read, R. and Graham, R. (2002) *Manual of Aerial Survey*. Whittles Publishing, Caithness.

Roberds, W. L. (1990) *Methods for developing defensible subjective probability assessments*. Transportation Research Record 1288, 183–190.

Vick, S. G. (2002) *Degrees of Belief: Subjective Probability and Engineering Judgement*. ASCE Press.

Index

275

List of Websites

Chapter 1
http://www.ice.org.uk/
http://www.asce.org/

Chapter 2
http://pubs.usgs.gov/publications/
text/dynamic.html

Chapter 5
http://www.fairlightcove.com/
http://www.ngi.no/
http://www.hull.ac.uk/coastalobs/
easington/erosionandflooding/

Chapter 6
http://www.geolsoc.org.uk/
http://earthquake.usgs.gov/learning
http://nsmp.wr.usgs.gov/GEOS/IDO/
idaho.htm

Chapter 9
http://www.ncdc.noaa.gov/paleo/abrupt/
story3.html
http://www.elnino.noaa.gov/
http://tao.atmos.washington.edu/pdo/
http://vulcan.wr.usgs.gov/Volcanoes/
Indonesia/
http://www.ipcc.ch/
http://www.ukcip.org.uk/

Chapter 10
http://www.pol.ac.uk/home/
http://www.ukcip.org.uk/

Chapter 12
http://www.hse.gov.uk/

Chapter 13
http://www.concrete.org.uk/

Chapter 14
http://www.pbs.org/wgbh/amex/
dustbowl/
http://www.dfd.dlr.de/app/land/aralsee/
http://www.hallsands.org.uk/

http://europa.eu.int/comm/
environment/eia/

Chapter 17
http://www.unesco.org/
http://www.fao.org/
http://www.ars.usda.gov/Research/
http://www.nrcs.usda.gov/

Chapter 18
http://pubs.usgs.gov/gip/deserts/eolian/
http://pubs.usgs.gov/gip/deserts/
desertification/
http://www.unccd.int/

Chapter 19
http://landslides.usgs.gov/
http://volcanoes.usgs.gov/Hazards/
What/Lahars/ RuizLahars.html

Chapter 20
http://vulcan.wr.usgs.gov/Glossary/
Tsunami/
http://neic.usgs.gov/neis/eq_depot/usa/
http://www.fema.gov/hazard/landslide/
index.shtm

Chapter 21
http://water.usgs.gov/ogw/karst/

Chapter 23
http://www.tva.gov/

Chapter 24
http//eol.jsc.nasa.gov/

Chapter 25
http://majuli-assam.tripod.com/
http://www.environmentagency.gov.uk/

Chapter 26
http://www.fema.gov/hazard/flood/
http://www.environment-agency.gov.uk
http://www.oxfam.org.uk/
http://eol.jsc.nasa.gov/

http://www.exmoor-
nationalpark.gov.uk/
http://pubs.usgs.gov/prof/p1386i/peru/
hazards.html/
http://www.jaha.org/
http://www.raunvis.hi.is/alexandr/
glaciorisk/

Chapter 27
http://www.pbs.org/wgbh/amex/flood/

Chapter 28
http://walrus.wr.usgs.gov/tsunami/
http://vulcan.wr.usgs.gov/Glossary/
Tsunami/
http://www.prh.noaa.gov/ptwc/

Chapter 30
http://en.wikipedia.org/wiki/Hurricane_
Katrina

Chapter 32
http://gulfsci.usgs.gov/missriv/
http://eol.jsc.nasa.gov/
http://www.imd.ernet.in/section/nhac/
static/
http://www.ramsar.org/
http://www.jncc.gov.uk/

Chapter 34
http://www.1900storm.com/
http://www.swgfl.org.uk/jurassic/
http://www.slnnr.org.uk/

Chapter 35
http://www2.sunysuffolk.edu/
mandias/38hurricane/

Chapter 36
http://www.hull.ac.uk/coastalobs/
easington/erosionandflooding/

Chapter 39
http://worldwind.arc.nasa.gov/
http://www.resmap.com/
http://earth.google.com/